T0258141

# Encyclopedia of Ionizing Radiation Research

# Volume III

# Encyclopedia of Ionizing Radiation Research
# Volume III

Edited by **Peggy Sparks**

New York

Published by NY Research Press,
23 West, 55th Street, Suite 816,
New York, NY 10019, USA
www.nyresearchpress.com

**Encyclopedia of Ionizing Radiation Research**
**Volume III**
Edited by Peggy Sparks

International Standard Book Number: 978-1-63238-143-9 (Hardback)

Printed in the United States of America.

# Contents

# Preface

Every book is a source of knowledge and this one is no exception. The idea that led to the conceptualization of this book was the fact that the world is advancing rapidly; which makes it crucial to document the progress in every field. I am aware that a lot of data is already available, yet, there is a lot more to learn. Hence, I accepted the responsibility of editing this book and contributing my knowledge to the community.

X-rays were discovered by Roentgen in 1895, and ever since, the ionizing radiation has been largely used in various industrial as well as medical applications. However, its harmful impact is gradually recognized through accidental uses. The experience of nuclear power plant mishaps in Fukushima and Chernobyl, has unveiled the risk of ionizing radiation and its prevalence in our contemporary society. Therefore, it is extremely necessary that more engineers, students and scientists become known to ionizing radiation research regardless of the research field they are engaged in. On the basis of this idea, this book was formulated to analyze the current achievements in this field inclusive of "Detection and measurement" and "Effects on Materials".

While editing this book, I had multiple visions for it. Then I finally narrowed down to make every chapter a sole standing text explaining a particular topic, so that they can be used independently. However, the umbrella subject sinews them into a common theme. This makes the book a unique platform of knowledge.

I would like to give the major credit of this book to the experts from every corner of the world, who took the time to share their expertise with us. Also, I owe the completion of this book to the never-ending support of my family, who supported me throughout the project.

Editor

# Part 1

## Detection and Measurement

# Nanoscale Methods to Enhance the Detection of Ionizing Radiation

Mark D. Hammig
*University of Michigan,*
*USA*

## 1. Introduction

The dominant modern methods by which ionizing radiation is sensed are largely based on materials that were developed decades ago, in the form of single crystal scintillators (such as sodium iodide (NaI(Tl))) or semiconductors (silicon (Si) or high purity germanium (HPGe) principally), and gas-filled counters (e.g. $^3$He for neutron detection). These legacy materials have survived and flourished because they have delivered adequate performance for many medical imaging, military, and plant-monitoring applications, and there was no low-cost replacement materials that delivered equivalent or superior performance. Thus, most of the research effort throughout the latter half of the 20th century was focused on the implementation of single-crystal solid and gas-filled detectors into various radiation-detection niches.

The elevation of the concern over the threat presented by secreted radiological and nuclear-weapons has, however, driven a need for superior detection media. If one wishes to have the ubiquitous deployment of radiation sensors, then existing materials are unsuitable because they have either poor efficiency (Si), demanding logistical burdens (liquid nitrogen associated with HPGe), or unduly high cost (cadmium zinc telluride (CZT)). For simple *counting* applications, emerging solutions exist in the form of boron-coated straw based neutron detectors (Lacy, et al., 2010) and liquid scintillators; however, if high resolution imaging of the radiation field is to be accomplished, then one prefers a material that makes a highly accurate conversion of the particle's energy into the information carriers in the detection media.

The dominant sensing technologies of *ionizing* radiation depend on its namesake; that is, they sense the non-equilibrium charge states induced in the interaction media. Thus, either charges, in the form of electron-ion or electron-hole pairs, can be sensed by their effects on the electric field created by the surrounding device architecture, or the light that accompanies the radiative-recombination of that charge is monitored using scintillation photon detectors. Unfortunately, in the process of creating that charge, information is lost to phonon-creation which doesn't participate in signal formation.

One of the main motivations that drives the development of nanostructured materials, whether nanoscintillators or nanosemiconductors, is that the phonon-assisted loss-processes can be suppressed to a larger degree than is possible in single-crystal materials, such that more of the incident information is converted into the information carriers (charge, light) that participate in signal formation.

As we will discuss, realizing those properties requires careful control over: (1) the structure, size, and uniformity of the nanoparticles themselves, (2) effective coupling of the nanoparticles into a colloidal solid through which the information can flow unimpeded, and (3) suitable device design such that the information can be coupled into a readout circuit. Given the challenging tasks associated with deploying nanostructured media, one might ask whether one can eschew the charge-conversion modality altogether.

If the loss of information to heat is a challenge that requires careful control over the energy-band structure of the solid, why doesn't one just measure the heat directly with a sensitive thermometer? This, in-fact, is the approach that delivers the best energy-resolution currently achievable, in the form of microcalorimeters. In fact, the resolution is several orders-of-magnitude better than that produced by HPGe or the best nanocrystalline detectors. However, the temperature variations that accompany radiation impact are sufficiently slight that small detection-volumes bounded to superconducting readout circuits are required, whether one uses transition edge sensors (Ullom et al., 2007) or magnetic microcalorimetry (Boyd et al., 2009). Useful interaction rates can be achieved by multiplexed arrays of the underlying sensor, but the microcalorimetry technologies magnify the cost and logistical concerns associated with using HPGe detectors.

The second motivation for employing nanoparticle approaches is thus that high energy-resolution performance can be achieved with a lower-cost, larger-area alternative to detectors based on single-crystal materials. Although one can deploy high-vacuum, precisely controlled chamber-based equipment (e.g. molecular beam epitaxy) to create and grow nanoparticles, the most common approach is one of wet-chemistry, in which atmospheric pressure or low-vac processes are used to create the colloidal dispersions. Furthermore, the resulting solutions can then be deposited into solids of various form by the utilization of self-assembly during the solvent drying phase. Thus, the initial capital cost is far less than that required in a fabrication methodology based on microelectronic processing equipment, such as single-crystal growth furnaces. For instance, the current cost of deploying a cubic centimeter of the leading room-temperature single crystal semiconductor, CZT, is roughly $1,000, while we estimate that an equivalent volume of a nanostructured lead-selenide detector can be deployed for less than $10, based on labor and materials costs. In Section 2, we discuss some of those processing steps required to make the nanocrystalline detectors, and we show that they can produce excellent energy response utilizing large detection volumes.

As will be shown, experimental results reveal that structuring *semiconductors* in a solid nanocrystalline composite can suppress the heat-loss mechanism relative to charge-creation processes, resulting in better estimates of the initial energy of the radiation than those produced by single-crystalline materials. In principle, the uncertainty in the energy-measurement process can be reduced the size of the band-gap, which serves as the fixed increment that governs the number of charge-carriers created. However, we currently measure ~1 keV energy uncertainty rather than typical band-gap values of 1 eV, and substantial improvement must therefore be achieved to turn the principle into practice. In fact, the precision of the nanosemiconductor is such that much of the uncertainty is governed by the electronic noise of the readout circuit, but if that is quenched, then one expects the measurement uncertainty to be governed by the non-uniformities in the underlying material.

One might ask: can one do better even than the uncertainty produced by the *electronic* band-gap? Microcalorimetry provides one pathway because information-carrier creation is governed by the phonon energy structure. One can also utilize sensors which gauge the momentum of the incident particle, and therefore avoid the energy-loss pathways inherent to energy-conversion devices, as summarized in (Hammig et al., 2005). In fact, *mechanical radiation detectors* are close cousins to microcalorimeters because small detection volumes must be cooled to quench competing thermal noise sources, and arrays of sensors must be deployed to realize large detection efficiencies. However, holigraphic methods can be brought to bear to read-out an array of vibrating elements in parallel.

When capturing the incident radiation and transforming its physical parameters into a measurable quantity, modern sensing technologies of *neutral* particles depend on an abrupt modality, in which the impinging photon or neutron is converted into a secondary charge-carrier in a point interaction, which subsequently generates heat, charge, or momentum during its slowing in the detection medium. One might prefer to avoid the point conversion altogether and take advantage of the wave-mechanics of the incident particle. In Section 3, we will discuss some of the proposed methods by which radiation can be guided and bent such that its inherent information is not processed through inefficient information-carrier creation processes.

## 2. Nanosemiconductors

### 2.1 Why not nanoscintillators?

For nanostructured media, the dominant modality currently being investigated by the radiation-detection community consists of composite arrays of nanoparticles that scintillate upon their excitation. The physics motivation is that in comparison to single crystalline materials, nanostructured scintillators exhibit: (1) enhanced light emission due to the suppression of non-radiative loss processes, and (2) more rapid emission kinematics due to a higher degree of intradot charge coupling, both of which yield higher imaging performance. Nanoscintillators consisting of lead-iodide (Withers et al., 2009), rare-earth-doped fluorides, such as $CaF_2(Eu)$ and $LaF_2(Eu)$ (Jacobsohn et al., 2011), and rare-earth oxides, such as $LuBO_3$ (Klassen et al., 2008) among many other materials are currently being studied, many exhibiting accelerated decay responses and high light conversion efficiency on a nanoparticle basis.

The main challenge presented by the composite media is the optical self-absorption in the nanoparticles, the matrix, and the innumerable interfaces as the photons meander their way to the readout surface. To date, no nanocrystalline scintillator detector has exhibited comparable characteristics to the best single-crystal media, whether composed of NaI(Tl) or brighter scintillators such as $LaBr_3(Ce)$. More to the point, one doesn't anticipate that an optimized light-emission media will ever produce a greater number of information carriers than a semiconductor equivalent because of the losses associated with converting the charge into scintillation light- and back again, at the photocathode or photodiode readout.

Thus, our initial investigations have been into nanosemiconductor composites, in which we have determined if one can overcome their expected limitations; namely, (1) charge loss during the slowing-down of the charged-particle in the matrix, (2) charge trapping during subsequent electron and hole transport, and (3) small detector volumes.

Regarding the latter point, small thin detectors typically accompany nanosemiconducting devices because of the inevitable colloidal defects that accumulate as a greater number of layers are cast onto the sample. Furthermore, biasing the device such that a high electric field is realized throughout the volume can, *a priori*, be challenging because of either poor

Schottky junctions or low-resistivity substrates. One might reasonably anticipate that depleting several centimetres of charge might be impossible without high bias voltages. Fortunately, these reasonable expectations are not met and sizable detector volumes can be realized over which rapid, clean charge transport can be achieved, as we will discuss next.

## 2.2 Cadmium telluride NC detector responses

Lead chalcogenide (PbS, PbSe, and PbTe) quantum dots have favorable properties for their use in NC applications. In contrast to a colloidal cadmium salt, such as CdTe, the large bulk Bohr radii of its excitons (e.g. 46 nm for PbSe) enables strong quantum confinement in relatively large NC structures. Since the conductivity improves sharply as the degree to which the particles are both mono-disperse and close-packed (cf. Remacle et al., 2002), larger particles result in better charge transport because the growth techniques are approaching the precision corresponding to the addition of an atomic layer. Another relative advantage of lead chalcogenides compared to CdTe is the nearly identical and small effective masses of the electrons and holes that can be attributed to the symmetric structure of the conduction and valence bands, which gives rise to relatively simple and broadly separated electronic energy levels that in turn, leads to much slower intraband relaxation (Schaller et al., 2005). Moreover, previous studies on lead chalcogenide NCs have shown that these materials have unique vibration states, weak electron-phonon coupling, and negligible exchange and Coulomb energies; thus, they have temperature-independent energy band-gap (e.g. Du et al., 2002; Olkhovets et al., 1998).

Our main focus therefore lies with the lead salts. However, the cadmium chalcogenides have, at present, more mature fabrication recipes and more importantly, their bulk band-gap is higher- 1.44 eV for CdTe compared to 0.26 eV for PbSe. This latter point is important because for low (< ~1 eV) band-gap semiconductors the fluctuations in the thermal noise compete vigorously with the small energy depositions that accompany radiation interaction, and they can swamp the measurement so that x-rays and low-energy particles are unresolvable.

### 2.2.1 Detector fabrication

CdTe is a II-VI semiconductor material that has been extensively investigated as a single-crystal semiconductor radiation detector. NCs of cadimum chalcogenides are also widely studied for the detection of optical photons. In fact, the fabrication methods are geared toward thin films consistent with the stopping of non-ionizing poorly-penetrating radiation. Although less favorable than the lead chalcogenides for the intended application, the synthesis processes are more fully characterized, and we have therefore investigated both systems.

One method to produce monodisperse NCs involves the mixing of all the reaction precursors in a vessel at a low temperature and then moderately heating the solution to grow the particles. An accelerated chemical reaction induced by increasing the temperature of the solution gives rise to supersaturation, which is relieved by the nucleation burst. The subsequent control of the solution temperature will induce further growing of NCs (Kim et al., 2009).

For instance, NCs in the ~3 nm range can be grown in 20 minutes, whereas larger particles can take 24 hours or more, the size judged during growth via the UV-photoluminescence.

For instance, a synthesized NC dispersion may exhibit green-colored photoluminescence of 540 nm average wavelength, which corresponds with a NC particle size of 3 - 5 nm. According to the chemistry of the synthetic process, the resulting dispersion contains NCs with a CdTe crystalline core surrounded by deprotonated –OH and –COOH groups of the thioglycolic acid; thus, the NCs have negative charge (Rogach, 2002).

For electronic testing, assemblies composed of ~4 nm NCs were prepared by two methods: the layer-by-layer (LBL) method and via drop-casting the NC dispersion on aluminum and gold-metalized glass substrates. The LBL method is a well-known method for efficiently depositing NC colloidal dispersions into high quality and stable thin-film layers on the substrate, while preserving the distinctive optoelectrical and magnetic properties of the size-quantized states of the NCs. For the LBL deposition, poly(diallyl dimethyl ammonium chloride) (PDDA), was induced as a polycation used for adsorbing nanoparticles. Thioglycolic-acid (TGA)-stabilized CdTe nanocrystal solution was spread on the PDDA layer and adsorbed. By an alternatively adsorbing procedure, a bilayer consisting of a polymer/nanocrystal composite was developed and the cycle was repeated from 1 to 30 times, to obtain a multilayer film of the desired thickness.

Unfortunately, in order to stop the high-energy charged particles associated with ionizing radiation one desires a thickness of at least 10's of micrometers if not several centimeters, which requires repetitive and lengthy casting procedures; for instance, roughly 80 hours of aqueous layer-by-layer deposition is required to achieve 10's of micrometers thickness. This can be enabled by robotic dipping tools, but for characterization studies, one can utilize sub-micrometer layers to evaluate the material.

The result of spin-casting the NC assembly onto a glass 4" diameter wafer is shown in Fig. 1A. The varying color reflects the non-uniform thickness deposited towards the outside of the wafer. If the uniform, inner-section of the assembly is diced and bounded by metallic electrodes, then 1 x 1 cm$^2$ detectors can be realized, as shown in Fig. 1B. In order to determine the thickness of the resulting NC assembly, we examine the cross-section of the layer and utilize energy dispersive x-ray spectroscopy to identify the constituent layers, as shown in Fig. 2. Note that the gold, indium, and cadmium signals overlap but a careful examination of the various structures reveals that the layer is only 130 nm thick, achieved after 48 drops of the LBL process was repeated (cf. Fig. 2C).

## 2.2.2 Spectroscopic ion measurements

*A priori*, one expects only slight energy depositions within a 130 nm layer, particularly for weakly ionizing electrons. Even if one exposes the assembly to 5.485 MeV alpha particles (from $^{241}$Am), the energy deposited from the transmission of the ion through the layer is less than 30 keV. This is enough energy to produce a measurable signal, as shown in Fig. 3, which is very promising because gamma-rays, for instance, can produce energy depositions that are 10 times as big if there is enough volume to fully stop the products of the interaction.

Thus, even though the cadmium salts may not be as favorable as their lead equivalents- at least in their physics properties- they represent a promising material upon which nanostructured ionizing radiation detectors can be developed. Furthermore, detectors fabricated using the LBL method are more reproducible in their behavior than those that are solidified using gross drop- or spin-casting methods.

(A)                                                    (B)

Fig. 1. (A) Gold-metalized glass wafer upon which a colloidal dispersion of CdTe NCs is deposited via the LBL method utilizing spin-casting. (B) The 1 x 1 cm² detectors after indium evaporation and dicing.

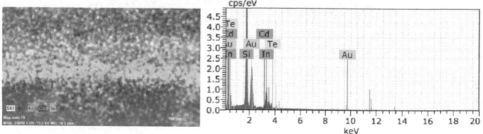

Fig. 2. (A) SEM micrograph of a cross-section of the NC CdTe detectors, showing the 329 nm gold layer and 130 nm CdTe assembly, assignments verified using energy dispersive x-ray spectroscopic mappings of (B) and (C).

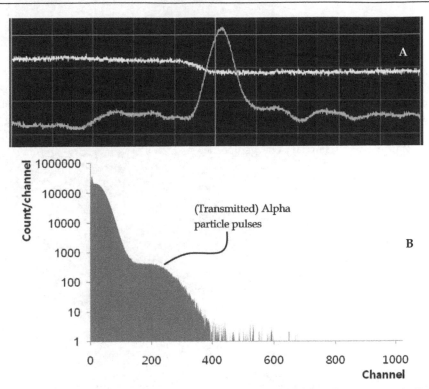

Fig. 3. (A) Preamp (yellow) and shaped (blue) signals derived from the passage of a 5.485 MeV alpha particle through a 130 nm CdTe nanocomposite detector, in which the induced charge signal is measured with a Ortec 142A preamp. The time per division is 10 µs and the volts per division are 200 mV for the preamp and 5 V for the shaped pulse. (B) An example alpha particle spectrum in which the alpha pulses are shown as a hump above the noise.

Nevertheless, instead of using the dozens of hours required to grow detectors with good intrinsic detection efficiency, one can produce exceptional detector behavior using chemistry based on fast-drying organic solvents and through deposition procedures that result in detectors that can be realized in minutes instead of hours. Furthermore, the lead salts, with their higher atomic numbers and densities, have inherntly higher stopping power to either charged or neutral ionizing radiation, as shown next.

## 2.3 Evaluation of the relative efficiencies of competing materials

Fig. 4A shows that when one compares various common single-crystal detection media and the lead chalcogenides under study, PbSe and PbTe in both single-crystal and nanocrystalline form are superior in terms of their combined mass densities and atomic numbers, Z, both of which combine to yield a higher specific detection efficiency.

The nanocrystalline forms of PbSe and PbTe are consistent with some of our current colloidal solids, in which spherical 10 nm PbSe nanoparticles or spherical 5 nm PbTe dots are spun cast into a face-centered-cubic (FCC) nanocrystalline solid with an interstitial para-MEH-PPV polymer of order 1. In realized detectors, we vary the relative blending of the polymer and NCs such that higher polymer orders are typically employed, and we grow rods and cubes, as well as spheres. Furthermore, the geometry of the solid deviates from a perfect FCC structure to a degree depending on the size dispersion of the colloidal solution; nevertheless, the reduction in densities between the single-crystal lead chalcogenides and the nanocrystalline forms reflects the values that are achieved when the ordering and spacing are optimized for spherical NCs. As shown in the figure, even in nanocrystalline form, the lead chalcogenides maximize the probability of interacting with impinging high-energy photons, and more generally, they exhibit superior characteristics for the efficient stopping of primary or secondary charged particles.

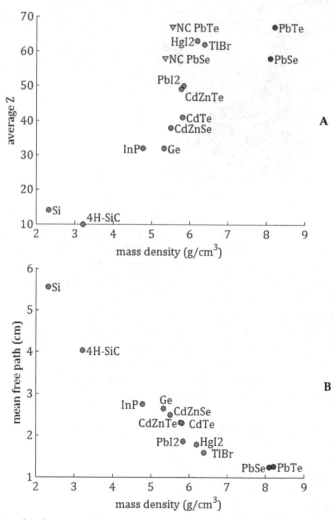

Fig. 4. (A) Relationship between the average atomic number $Z$ and the mass density for various semiconducting materials. (B) The mean free path for a 662 keV photon interacting with single-crystal forms of the semiconducting media (Hammig & Kim, 2011).

As an example, Fig. 4B shows the photonic mean free path, as derived from the density and the macroscopic *photoelectric-absorption* linear attenuation coefficient at a photon energy of 662 keV, which is consistent with the gamma-ray emission from [137]Cs. A comparison of PbSe or PbTe with CZT- the most common room-temperature gamma-ray detection medium- shows that the mean free path in CZT is 1.85 times longer than that of either lead salt. On a path-length normalized basis therefore, the lead chalcogenides can form a highly efficient gamma-ray detection medium. Nevertheless, the question is: in a field in which device thicknesses are measured in nanometers to hundreds of nanometers, can the solid be formed into electronic structures with enough active thickness and lateral extent to take advantage of the inherently high interaction probabilities and stop charged particles that may range over hundred of micrometers? As described in Section 2.41, the drop-, dip- and spun-cast solids can be formed into thicknesses which yield highly efficient detector configurations.

One not only desires high probabilities of interaction with the incident quanta, but the multiplicity of the subsequent charge-creation should be maximized and the trapping of the charge-ensemble should be minimized so that the information conversion is optimized. As summarized in Section 2.42, the charge transport can vary across the lateral area as well as in depth because of the varying field or mobilities that accompany different packing regions. Nevertheless, one can bias the detector such that the charge velocities are saturated, resulting in uniform collection across the ensemble.

As shown in Section 2.43, even with a colloidal solid that is not optimized from a charge-separation and transport perspective, the resulting energy-resolving behavior is comparable to or better than state-of-the-art single-crystal semiconductors- Si, CZT, and high-purity germanium operated at liquid nitrogen temperatures- which reflects the promise of the nanostructured approach to semiconductor device design, extending beyond the sensing application reported.

## 2.4 Lead selenide NC detector responses
### 2.4.1 Detector fabrication
The chemical recipe used to form the colloidal dispersion, the subsequent characterization of the nanoparticles, and their formation into solid assemblies is described in (Kim & Hammig, 2009). Colloidal PbSe NCs were synthesized through the high-temperature solution-phase routes (Murray et al., 2001), as illustrated in the Fig. 5. Lead oleate and selenide-dissolved trioctylphosphine (TOP-Se) were used as the organo-metallic precursor and diphenyl ether (DPE) as the high-temperature organic solvent. The lead oleate precursor was prepared by dissolving lead acetate, $Pb(CH_3COO)_2$, into a mixture of DPE, TOP and oleic acid, heated up to 85 °C for 1 hour and cooled down. Mixed with 1 M TOP-Se solution, precursors were rapidly injected into rigorously stirred DPE at various reaction conditions ($T_{inj}$) in an inert environment. PbSe NCs were grown in $T_{gr}$ for 3 - 6 minutes and cooled.

In the fabrication of the NC assembly detector, one forms a mold in an inert substrate which acts as a basin into which the colloidal dispersion is dripped. Furthermore, both the top and bottom metal contacts must be accessible in order to connect each face of the detector to the readout electronics. Although we have used silicon, printed circuit board, glass, and plastic substrates, we typically use glass substrates for thin detectors (10's of micrometers to 100's of micrometers) and employ plastic substrates for millimeter or centimeter thick devices. Fig. 6 shows the typical process if a glass substrate is used, which makes use of standard photolithography processes.

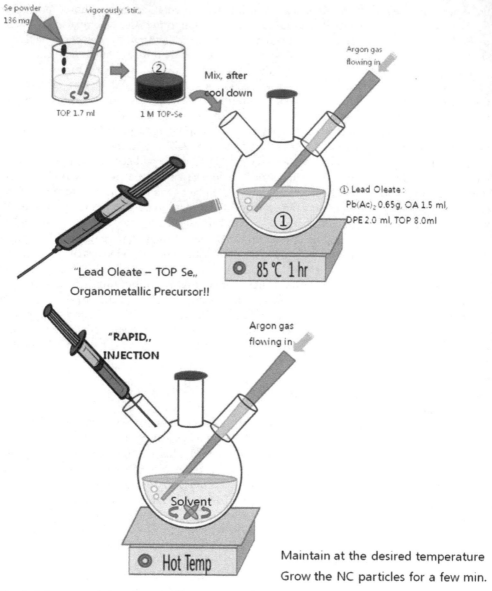

Fig. 5. Schematic of the PbSe NC dispersion synthesis.

Several fabricated substrates are shown in the Fig. 7. The lines stretching horizontally and vertically are due to the dicing of the substrate. Blue colored tape is attached to the backside, before the substrate is diced. The glass wafer-based substrates were designed in a circular shape to make the spin-casting procedure more efficient. Once the metalized molds are realized, the NC assembly can be deposited to form the active region of the sensor.

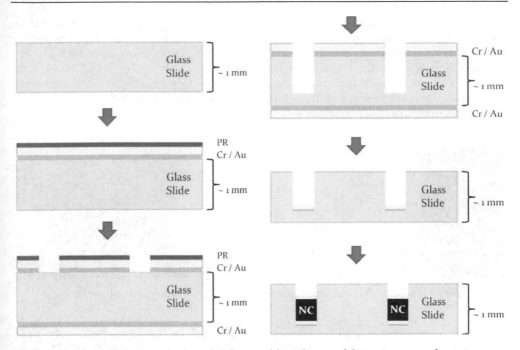

Fig. 6. Schematic of glass wafer-based NC assembly substrate fabrication procedure.

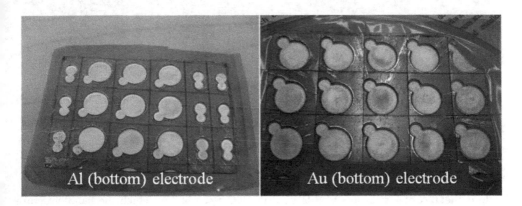

Fig. 7. Glass wafer-based NC assembly substrate with Al and Au bottom contacts.

Synthesized NC particles inherently self-assemble, if coordinated with organic ligands properly. NC particles dispersed in the chloroform solvent can be deposited on the substrate using many methods, but we have focused on drop- and spin-casting. Drop-casting is the process of depositing a droplet of the NC solution on the substrate and repeating the procedure, while allowing time for the solvent to evaporate between applications. Spin-casting is the process of dripping a droplet of the solution on a spinning substrate, which allows more uniform spreading of the NC solution as well as more rapid drying.

In order to realize suitable charge transport, the PbSe NC solution is mixed with a conductive polymer, for instance, para-MEH-PPV, which is used as a hole transporting agent. Greater polymer concentrations formed a brighter, more yellow assembly, whereas the pure PbSe NC assembly exhibits a dark assembly, as shown in Fig. 8 and Fig. 9. Note that Fig. 9 shows the assemblies deposited into ~5 mm deep holes with 7 mm diameters milled into plastic substrates, and the detectors therefore form active depths that are several millimeters thick.

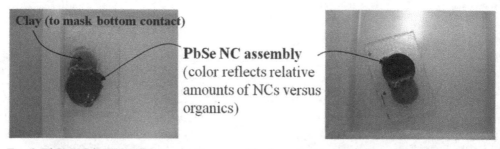

Fig. 8. PbSe NC/MEH-PPH composite assembly deposited on the glass substrate.

Fig. 9. PbSe NC/MEH-PPH composite assembly deposited on plastic substrates, before the top metal contact is formed (left) and after gold evaporation (right).

Various metal contacts can be used to abut the NC assembly, in which the band-structure can vary significantly with the particle size. In general, the MEH-PPV is p-type and we therefore use metal contacts with higher work functions, such as Pt (5.6 eV) and Au (5.2 eV), for the ohmic contact, and we form the rectifying contact with lower work function metals, such as In (3.8 eV) and Al (4.2 eV). Various combinations of samples were made in order to study the effects of the metal contacts, including Pt/NC/Pt, Pt/NC/In, Au/NC/Au, Au/NC/In, etc.. Nevertheless, the behavior of the resulting detector does not depend strongly on the bounding Schottky barrier; rather, the charge creation and transport is more dependent on the uniformity of the colloidal solid underlying the electrodes, some features of which will be described next.

## 2.4.2 Charge transport characteristics

As mentioned in Section 2.2, a close-packed solid with a high degree of size uniformity results in high conductivity. The size-dispersion in the NC solution depends on procedures used to prepare the liquid dispersion of coordinated-NCs. One can achieve monodisperse (particle size with less than 5 % uncertainty) solutions either directly from the nucleation and growth procedure- with finely controlled reaction conditions- or by employing size-selective precipitation procedures. For instance, Fig. 10 shows the size separation for CdTe NCs that can be elicited using incompatible and compatible solvents along with centrifugation, into a size-stratified solution (Fig. 10A). The separated layers (Fig. 10B) then exhibit different particle size via their size-dependent band-gap in their color or in their excitation spectra. The size can be further refined by repeated application of the procedures if needed.

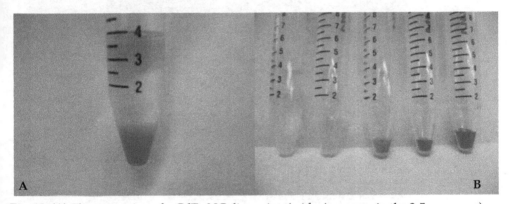

Fig. 10. (A) The separation of a CdTe NC dispersion (with size range in the 3-5 nm range) into different precipitation bands following centrifugation. (B) The separation of the bands into different vials through pipetting.

The importance of monodispersity is not only in the electrical characteristics but in the mechanical ordering properties (Coe-Sullivan et al., 2005). A large (> 10 %) distribution in size results in random packing, but narrow distributions that approach the uncertainty due to the addition of an atomic layer (< 5 %) result in high degrees of order.

Over microscopic areas therefore, monodispersity and the coordination chemistry can be employed to realize close-packed assemblies. However, when one examines the solid over larger volumes, then defects are apparent, ranging from quantum dot (QD) crystal defects, such as vacancies and dislocations, to large-scale cracks, such as those shown in Fig. 11. The large-scale cracking, which we see in all of our samples is a natural consequence of stress-relief during drying of the solvent. The formation of domains of differing degrees of NC order can be reflected in the charge transport characteristics.

For instance, if the bias if held low enough (a value depending on the diode design), then Fig. 12 shows that following the interaction of a single quanta, the holes move in fits and starts, as reflected in the induced charge pulses (in yellow) along with their amplified and shaped counterparts (in blue). This reflects an underlying variation in the drift velocity as the charge cloud traverses different depths of the active volume, with can be caused by either variations in the electric field or the mobility. Although these variations could potentially be problematic if one were to use pulses such as those shown in Fig. 12 to extract the energy deposited by the radiation, one can increase the voltage such that all of the slow-drift regions are eliminated, and the pulse rises sharply and cleanly, such as that shown in Fig. 13. One explanation for this behavior is that the charge-velocities saturate across the various domains in the NC detector and the transport variations are therefore of no consequence.

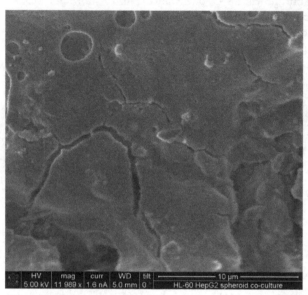

Fig. 11. The cracked and stacked dried mud-puddle look of the NC assembly at a scale of 10's of micrometers.

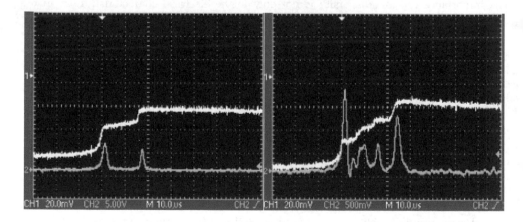

Fig. 12. Two examples of preamp (yellow) and shaped (blue) signals derived following the stopping of an [241]Am alpha particle within an under-based colloidal PbSe detector.

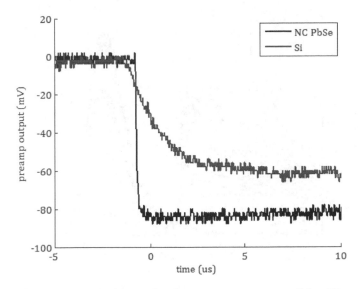

Fig. 13. The pulse shape derived from the charge-sensitive preamplifier (Ortec 142a) when an alpha particle from $^{241}$Am is impinged upon a silicon PIN or NC PbSe detector, both operated at 0 V applied bias. The PbSe nanoparticles are spherical with a mean diameter of 7.4 nm ($\pm$ 0.5 nm).

Thus, although the dispersion in not monodisperse and the solid is not uniform, the promise of the nanocrystalline approach is revealed in the fact that the results are comparable to or better than state-of-the art single crystal detectors, as will be shown next.

### 2.4.3 Spectroscopic measurements

Fig. 14A corresponds to a mixed spectrum derived from alpha particles- emerging from a *thin-film* $^{241}$Am alpha particle source and $^{133}$Ba gamma-rays impinging upon the 1 x 1 cm detector shown in Fig. 14A. The alpha source, placed 3.7 cm away from the detector surface, attenuates the 5.49 MeV particles to the energy range shown, and the $^{133}$Ba source was placed external to the box housing the detector, 5 cm distant. If one focuses on those spectral features whose appearance correlates to the presence of the $^{133}$Ba source, then we can identify the features using both the pulse amplitude and the frequency of occurrence as follows.

Although the material need not be linear, the pulse amplitudes of the full-energy gamma-ray peaks were linearly related to the energies identified in Fig. 14B. Second, the areas under the spectral features correspond to the expected values. For instance, if one takes the gamma-ray emission yield as well as the photoelectric absorption probabilities into account, the area of the 383.85 keV peak should be 12 % of that under the 356.02 keV peak, compared with a measured value of 11 %. The sizes of the x-ray escape features diminish for thicker detectors; however, the thin detector was utilized to characterize the charge-creation characteristics, minimally adulterated by charge-loss considerations, so that the inherent physical properties could be better estimated.

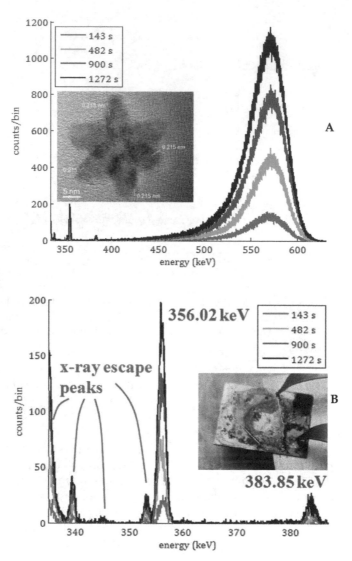

Fig. 14. (A) Energy spectrum derived from [133]Ba gamma-rays and [241]Am alpha particles, attenuated through 3.7 cm of air, impinging upon a 1 x 1 cm thin (10's of micrometers) composite assembly of para-MEH-PPV and star-shaped PbSe nanopartiles, accumulated for various durations shown in the legend. The inset shows a TEM micrograph of a typical PbSe NC in the assembly. (B) Typical [133]Ba spectrum derived from a thin detector, in which the Pb and Se x-ray escape peaks are prominent.

As reported in (Hammig & Kim, 2011), Fig. 15 shows the spectral widths compared with three state-of-art detectors, all operated at room temperature except for the high purity germanium (HPGe) detector, which was cooled with liquid nitrogen.

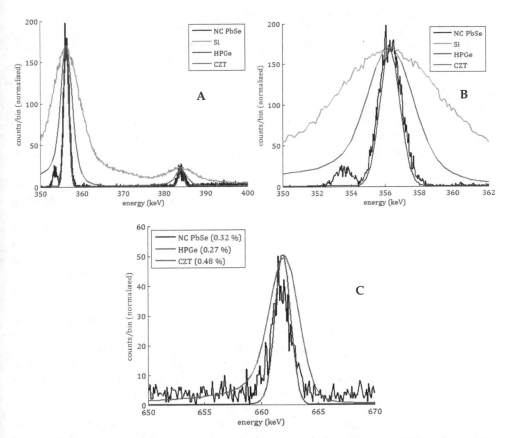

Fig. 15. (A) Spectral comparison between high-resistivity silicon (green), CZT (red), HPGe (blue), and the PbSe nanocomposite (black), when exposed to gamma-rays emitted from $^{133}$Ba. The different measurement periods (e.g. 21.2 min for NC PbSe, 24 hr for HPGe) account for the differences in curve smoothness, and the curves were normalized to the number of counts in the PbSe peak. Note that the HPGe and NC PbSe traces largely overlap except for the x-ray escape peak from the NC PbSe detector. (B) Close-up of the 356.02 keV peak. (C) Spectral comparison between CZT (red), HPGe (blue), and the PbSe nanocomposite (black), when impinged upon by gamma-rays emitted from $^{137}$Cs. The energy resolutions are shown in the legend. The measurement period for the NC PbSe spectrum was 46.4 min. (Hammig & Kim, 2011).

As shown in Figs. 15A and 15B, the energy resolution of the PbSe nanocomposite is superior to silicon and CZT and comparable to HPGe, yielding an energy resolution of 0.42 % (1.5 keV) at 356 keV, compared with a resolution of 0.96 % (3.4 keV) for CZT and 0.39 % (1.4 keV) for HPGe. Fig. 15C shows an equivalent comparison between the detectors when impinged upon by 661.7 keV gamma-rays from $^{137}$Cs. The $^{137}$Cs source was more than one hundred times less intense than the $^{133}$Ba source and the peak was therefore less well sampled. More importantly, a slight gain drift in the detector dominated the peak width shown in the figure. Nevertheless, the $^{137}$Cs results buttress those derived from $^{133}$Ba gamma-rays; namely, the nanocomposite detector provides an energy estimate with similar fidelity to that produced by HPGe. The results imply that the conductive polymers are not only accomplishing the goal of realizing good charge transport through the device, but they are not substantially participating in the charge loss of the recoil electrons.

## 2.5 Summary

Nanostructured cadmium and lead salts have been extensively studied as advanced media for optoelectronic devices. The sensitivity of their charge-transport properties to the surface conditioning of the nanoparticles and their degradation under ambient conditions has discouraged their use in a semiconductor configuration; rather, most of the efforts geared toward nuclear radiation detection have concentrated on nanostructured *scintillation* materials. Furthermore, the slight thicknesses that accompany the traditional spin-, dip-, and drop-casting methods have argued against their use as sensors of more highly-penetrating radiation.

However, we have shown that detectors with deep depletion regions can be constructed such that the deposited charge can be effectively drifted and the energy of the interacting quanta can therefore be measured. Furthermore, the accuracy of that measurement is comparable to those estimates provided by the best single-crystalline semiconducting materials, providing evidence for enhanced charge multiplication as the degree of strong confinement is increased.

Nevertheless, if one does employ thin detectors, then they are susceptible to the escape of the primary or secondary charged particles induced by the radiation's interaction, particularly for neutral particles that interact with roughly equal probability throughout the interaction depth of thin stopping media. The low energy tailing in the spectral features that are inherent to this energy loss can negatively impact the ability of the sensor to extract the physical characteristics of the impinging quantum, but more importantly, small volume detectors are inefficient for highly penetrating particles.

For neutral species, one might prefer to avoid the abrupt modality altogether, in which the particle is first converted into secondary charge particles which then confer the incident particle's information via electromagnetic interactions. If one can take advantage of the wave mechanics of the impinging quanta, then they can be guided into the detection volume such that the momentum and energy information is fully preserved, the potential of which is discussed in the next section.

## 3. Extending the bandwidth of wave-guiding media via self-assembled metamaterials

If one can inject the incident particle's energy into mechanical vibrational modes, such as the lever modes discussed in (Hammig, 20005) or the whispering-gallery modes discussed in (Zehnpfennig et al., 2011), then one can eskew the charge-conversion process altogether, as well as the inefficiencies inherent to the charge creation and collection processes. Utilizing the wave-mechanics of the radiation proceeds most simply from long wavelength electromagnetic waves, where microwave and optical light guiding and cloaking have already been demonstrated, to the shorter wavelengths that accompany ionizing radiation. The science of guiding electromagnetic waves is aided by the development of metamaterials, which permit one to exercise control over the permittivity and permeability of the transmission medium. If one can reduce the sizes of the nanoscale features in a metamaterial, then the effective medium approximation can be extended to increasingly small wavelengths such that UV, x-ray, and potentially, gamma-ray wave-guiding can be elicited. Such a development has consequnce for not only sensing applications but it also provides an alternative radiation-shielding modality to the existing materials, and provides the promise of light-weight, high efficiency shielding systems.

There has been no *fundamental* advance in the science of nuclear-radiation shielding since the dawn of the nuclear era. Existing shielding materials depend on *particle*-interaction models, in which one maximizes the number of high cross-section scattering sites that can be placed between the radiation source and the protected asset. However, the advent of nanostructured materials provides means through which novel physics can be realized. In particular, they provide a physical pathway through which the wave mechanics of electromagnetic radiation can be controlled via the spatial mapping of the permittivity and permeability. Such control can be realized over electromagnetic waves ranging from microwaves to optical photons. Extending that capability to shorter wavelengths requires either correspondingly smaller nanostructures, or new modalities for controlling the indices of refraction.

Invisibility cloaks generally operate on two operational principles. In the first, transformational optics- in which optical conformal mapping is used to design and grade a refractive index profile- can be used to render an object invisible by precisely guiding the flow of electromagnetic radiation around the object, in a push-forward mapping cloak or a conformal mapping cloak (H. Chen et al., 2010). The underlying mechanism stems from the invariance of Maxwell's equations to coordinate transformations, the effect of which is to only change constitutive parameters and field values. Thus, the space about a cloaked object can be removed from the space that is transformed by the coordinate transformation.

If one utilizes a metamaterial with a negative index of refraction, then one can also cloak an object by cancelling the wave scattering from the target. Optical negative index metamaterials (NIMs) typically require a plasmonic material, which is comprised of a metal-dielectric composite comprised of nanoscale-sized elements. For instance, Valentine et al. utilized a structure consisting of alternating layers of Ag and $MgF_2$ in a cascaded fishnet structure, in order to realize a 3D optical metamaterial with very low losses (Valentine et al., 2011). The unit cell dimensions were 860 x 565 x 265 nm and the material successfully exhibited the desired refractive index at optical wavelengths of 1200 – 1700 nm; thus, the nanostructural dimensions were approximately one-half to one third of the wavelength. Note that the regular structure of the fishnet structure, shown in Fig. 16b, was patterned using focused ion-beam (FIB) milling, which is capable of precise, nanoscale cutting with high aspect ratios.

Another example is shown in Fig. 16a, in which an optical cloak for the transverse magnetic waveguide mode is comprised of isotropic dielectric materials and delivered broadband and low-loss invisibility in the 1400-1800 wavelength range. The hole diameter shown in Fig. 16a is 110 nm, approximately one-tenth the wavelength, and was again realized with FIB milling.

In fact, most of the cloaking demonstrations to date have used a material-*removal* fabrication modality in which the composite structure is built-up via deposition procedures, and then electron beam lithography or FIB milling is used to realize the structure by etching the underlying composite. Unfortunately, both procedures are slow and costly when implemented over large areas. More importantly, they do not generally allow the precise patterning of materials to the levels required to realize wave optics with x- or gamma-rays.

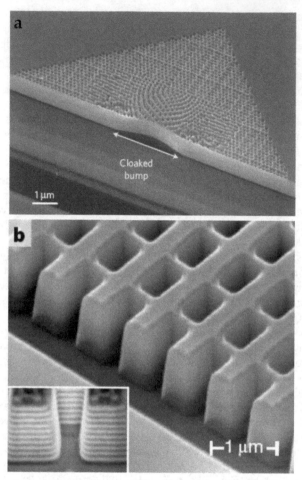

Fig. 16. (a) The carpet cloak design that transforms a mirror with a bump into a virtually flat mirror. Note the gradient index cloak above the bump, the pattern fabricated in a SOI wafer where the Si slab serves as a 2D waveguide. (b) SEM image of fabricated 21-layer fishnet structure, consisting of alternating layers of 30 nm silver and 50 nm magnesium fluoride (Valentine et al., 2011).

That is, the greatest challenge to achieving wave-guiding effects at x- or gamma-ray wavelengths, is that for existing metamaterials, the material loss and fabrication difficulties are substantial. As suggested above, the size of the nanostructure is generally sub-wavelength, so that the index of refraction of the material is governed by the effective medium approximation. The wavelength of a 1 keV x-ray is 1.23 nm; thus, a nanostructured metamaterial must have an ordered nano-crystalline structure in the nanometer range, which is beyond the capabilities of standard e-beam tools, and one must therefore deviate from the material-removal modality.

Fortunately, the desired size-scale can be realized with self-assembly, in which the current lower limit is in this nanometer scale and thus potentially applicable to x-ray cloaking. A combination of sol preparation and electrophoretic deposition has been used to synthesize a variety of oxide nanorod arrays, such as $TiO_2$, $SiO_2$, $Nb_2O_5$, $V_2O_5$, $BaTiO_3$, $Pb(Zr,Ti)O_3$, Sn-doped $In_2O_3$ (ITO), and $Sr_2Nb_2O_7$ (Kao, 2004). We have grown nanowires from CdTe and PbSe semiconductors, upon which metallic shells can be deposited. There has been a myriad of different self-assembled nanostructures, but they do not approach the process uniformity and control that accompanies batch-processing methods (cf. Ko et al., 2011) , and the deviations in the uniformity would diminish any wave-guiding achieved from a designed array of plasmonic structures of similar size. Nevertheless, some wave guiding can occur from an array of proper materials. Thus, using fabrication techniques similar to that that have been successfully employed to make high-performance charge-conversion devices, one can conceivably change the materials in the nanostructured array in order to excercise control over the wave mechanics of high energy photons.

## 4. Conclusion

Nanostructured media allow us to measure ionizing radiation with greater precision than can be achieved with single-crystal media because one has greater experimental control over the detector's governing parameters- the size and shape of the nanoparticles- than can be equivalently achieved in a nominally single-crystal media, in which defects and impurities can readily spoil the performance of the sensor. More importantly, one can yield fundamentally superior detection characteristics because one has a mechanism through which the informational processes- such as electron creation- can be controlled relative to the loss processes- such as phonon excitation. Fortuitously, this superior physics is also accompanied by a less expensive fabrication modality.

Nanostructured materials have therefore been projected as the next-generation material for the detection of ionizing radiation, the adoption of which depends on overcoming the expected difficulties associated with using a composite material comprised of innumerable interfaces at which information can be lost. As we have shown in this chapter, methods exist that can quench interface trapping so that the promise of the material can be realized today.

## 5. References

Bahl, G.; et al., Stimulated optomechanical excitation of surface acoustic waves in a microdevice. *Nat. Commun.* 2:403 doi: 10.1038/ncomms1412 (2011).

Chen, H.; C. Chan & P. Sheng. Transformation optics and metamaterials, *Nature Materials* Vol. 9, (2010).

Coe-Sullivan, S.; et al., "Large-Area Ordered Quantum-Dot Monolayers via Phase Separation during Spin Casting", *Adv. Funct. Mater.* 15 (2005).

Du, H.; et al., Optical properties of colloidal PbSe nanocrystals, *Nano Lett.*, vol. 2, no. 11, pp. 1321-1324, (2002).

Hammig, M.D; D. K. Wehe and J. A. Nees, , The measurement of sub-brownian lever deflections. *IEEE Trans. Nuc. Sci.* 52, 3005-3011 (2005).

Hammig, M.D.; G. Kim, Fine Spectroscopy of Ionizing Radiation via a Colloidal Array of Blended PbSe Nanosemiconductors (submitted, 2011).

Jacobsohn, L.G.; et al., Fluoride Nanoscintillators, Journal of Nanomaterials, Vol. 2011 (2011).

Kim, G.; J. Huang and M.D. Hammig, An Investigation of Nanocrystalline Semiconductor Assemblies as a Material Basis for Ionizing-Radiation Detectors. *IEEE Trans. Nuc. Sci.* 56 841-848 (2009).

Kim, G; M.D. Hammig, Development of Lead Chalcogenide Nanocrystalline (NC) Semiconductor Ionizing Radiation Detectors. *2009 IEEE Nuc. Sci. Sym. Conf. Record,* 1317-1320 (2009).

Klassen, N.V.; et al., Advantages and problems of nanocrystalline scintillators, IEEE Trans. on Nuc. Sci., Vol. 55, 1536-1541 (2008).

Lacy, J; et al., Boron-Coated Straw Detectors: a Novel Approach for Helium-3 Neutron Detector Replacement, *2010 IEEE Nuclear Science Symposium and Medical Imaging Conference (2010 NSS/MIC)*, p 3971-5, (2010).

Murray, C.B.; et al., Colloidal synthesis of nanocrystals and nanocrystal superlattices, *IBM J. Res. & Dev.*, vol. 45, no. 1, pp. 47-56, Jan. 2001.

Olkhovets, A.; et al., Size-dependent temperature variation of the energy gap in lead-salt quantum dots, *Phys. Rev. Lett.*, vol. 81, pp. 3539-3542, (1998).

Remacle, F. et al., Conductivity of 2-D Ag Quantum Dot Arrays: Computational Study of the Role of Size and Packing Disorder at Low Temperatures, *J. Phys. Chem.*, vol. 106, pp. 4116-4126, (2002).

Rogach, A.L; et al., II-VI semiconductor nanocrystals in thin films and colloidal crystals, *Colloids Surf. A*, vol. 202, iss. 2-3, pp. 135-144, (2002).

Schaller, R. D., et al., Breaking the phonon bottleneck in semiconductor nanocrystals via multiphonon emission induced by intrinsic nonadiabatic interactions, *Phys. Rev. Lett.*, vol. 95, 196401, (2005).

Ullom, J.N.; et al., Multiplexed microcalorimeter arrays for precision measurements from microwave to gamma-ray wavelengths. *Nucl. Inst. and Meth. A* 579 161-164 (2007).

Withers, N.; et al., Lead-iodide-based nanoscintillators for detectino of ionizing radiation, Proc. Of SPIE Vol. 7304, 73041N (2009).

Zehnpfennig, J.; et al., Surface optomechanics: calculating optically excited acoustical whispering gallery modes in microspheres, *Optics Express*, vol. 19, 14241 (2011).

# Glass as Radiation Sensor

Amany A. El-Kheshen

[1]*National Research Centre, Glass Research Department,*
[2]*Chemistry Department, Faculty of Science, Taif University,*
[1]*Egypt*
[2]*KSA*

## 1. Introduction

### 1.1 What is glass? [Noel, 1998]

Let's begin with what glass is not—it is not a crystal. The atoms in a crystal are organized in a regular, repetitive lattice so you need only locate a few atoms in order to predict where all of their neighbors are (Fig.1a). The atoms are so neatly arranged that, except for occasional crystal defects, you can predict positions for thousands or even millions of atoms in every direction. This spatial regularity is called **long-range order**.

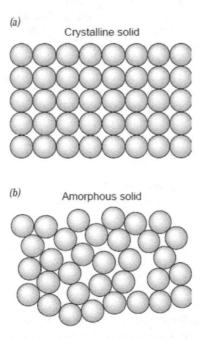

Fig. 1. (a), a crystalline solid with long-range order, (b), an amorphous solid, no long-range order.

Glass is an **amorphous solid**, a material without long-range order (Fig. 1*b*). Locating a few glass atoms tells you nothing about where to find any other atoms. The atoms in glass are arranged in the random manner of a liquid because glass is essentially a super-stiff liquid. Its atoms are jumbled together in a sloppy fashion but they can't move about to form a more orderly arrangement. Glass arrives at this peculiar amorphous state when hot liquid glass is cooled too rapidly. If molten glass were an ordinary liquid, it would begin to solidify abruptly during cooling once it reached its freezing temperature. At that point, its atoms would begin to arrange themselves in crystals that would grow in size until there was no liquid left. That's what's normally meant by freezing. However some liquids are slow to crystallize when you cool them slightly below their freezing temperatures. While they may be cold enough to grow crystals, they must get those crystals started somehow. If crystallization doesn't start, material's atoms and molecules will continue to move about and it will behave as a liquid. When that happens, the liquid is said to be **supercooled**. Supercooling is common in liquids that have difficulties forming the initial **seed crystals** on which the rest of the liquid can crystallize. Because almost all of the atoms in a seed crystal are on its surface, it has a relatively large surface tension and surface energy. Below a certain critical size, a crystal is unstable and tends to fall apart rather than grow. However once the first seed crystals manage to form, a process called nucleation, the rest of the supercooled liquid may crystallize with startling rapidity. Just below its freezing temperature, the atoms in glass don't bind to one another long enough to form complete seed crystals, and nucleation takes almost forever. At somewhat lower temperatures, seed crystals begin to nucleate, but glass's large viscosity (thickness) prevents these crystals from growing quickly. The glass remains a supercooled liquid for an unusually long time. At even lower temperatures, glass becomes so viscous that crystal growth stops altogether. The glass is then a stable supercooled liquid. At this temperature range, glass still pours fairly easily and can be stretched or molded into almost any shape.

However, when you cool the glass still further, it becomes a *glass*. Here the word *glass* refers to a physical state of the material—a type of amorphous solid. To distinguish this use of the word glass from the common building material, it is italicized. Glass, the material, becomes a *glass*, the state, at the **glass transition temperature** (Tg). Below Tg, the atoms in the glass rarely move past one another; they continue to jiggle about with thermal energy but they don't travel about the material.

## 1.2 A quick journey with glass [Gerry, M., 2002]

The production of glass has been occurring naturally for millions of years, but the discovery of manufactured glass leads us to live a luxury life.

Natural glass has existed since the beginnings of time, formed when certain types of rocks melt as a result of high-temperature phenomena such as volcanic eruptions, lightning strikes or the impact of meteorites and then cool and solidify rapidly. The earliest man-made glass objects, mainly non-transparent glass beads, are thought to date back to around 3500 B.C., which found in Egypt. Phoenician merchants and sailors spread this new art along the coasts of the Mediterranean Sea.

After 1500 B.C., Egyptian craftsmen are known to have begun developing a method for producing glass pots by dipping a core mould of compacted sand into molten glass and then turning the mould so that molten glass adhered to it. While still soft, the glass-covered mould could then be rolled on a slab of stone in order to smooth or decorate it. The earliest examples of Egyptian glassware are three vases bearing the name of the Pharaoh Thoutmosis 2I (1504-1450 B.C.).

The first instance of glass being used for windows in buildings, especially in Britain, was during the period of the Romans. In the 1st century A. D., it is known that the Romans used glass for a variety of reasons including mosaic tiles, decorating pottery and as windows. It seems that the glass window became more popular with the advent of churches and other places to worship. Most of the earliest examples of Roman window glass are colored, suggesting that they were glass windows for churches, for example. Around the world, the need for glass windows (and the manufacture of the glass) did not really take hold until the 11th century.

## 1.3 It only looks like magic (A world of glass)
### 1.3.1 First types of glass

The first types of glass invented centuries ago, scientists made that glass by mixing materials, melting and then cooling to form a solid. There are many different types of glass with different chemical and physical properties. Each can be made by a suitable adjustment to chemical compositions, but the main types of glass are:
- Commercial Glass also known as soda-lime glass
- Lead Glass
- Borosilicate Glass
- Glass Fiber

Glasses may be devised to meet almost any imaginable requirement. For many specialized applications in chemistry, pharmacy, the electrical andelectronics industries, optics, or the comparatively family of materials known as glass ceramics? Glass is a practical material for the engineer to use.

### 1.3.1.1 Commercial glass

Most of the glass we see around us in our everyday lives in the form of bottles and jars, flat glass for windows or for drinking glasses is known as commercial glass or soda-lime glass, as soda ash is used in its manufacture.

The main constituent of practically all commercial glass is sand. Sand by itself can be fused to produce glass but the temperature at which this can be achieved is $\geq 1700oC$. Adding other minerals and chemicals to sand can considerably reduce the melting temperature.

The addition of sodium carbonate ($Na_2CO_3$), known as soda ash, to produce a mixture of 75% silica ($SiO_2$) and 25% of sodium oxide ($Na_2O$), will reduce the temperature of fusion to about 800oC. However, a glass of this composition is water-soluble and is known as water glass. In order to give the glass its stability, other chemicals like calcium oxide ($CaO$) and magnesium oxide ($MgO$) are needed.

Commercial glass is normally colorless, allowing it to freely transmit light, which is what makes glass ideal for windows and many other uses. Additional chemicals have to be added to produce different colors of glass.

Most of commercial glasses have roughly similar chemical compositions of:

70wt% - 74wt% $SiO_2$ (silica)

12wt% - 16wt% $Na_2O$ (sodium oxide)

5wt% - 11wt% $CaO$ (calcium oxide)

1wt% - 3wt%$MgO$ (magnesium oxide)

1wt% - 3wt% $Al_2O_3$ (aluminium oxide)

Flat glass is similar in composition to container glass except that it contains a higher proportion of magnesium oxide.

Within these limits the composition is varied to suit a particular product and production method. The raw materials are carefully weighed and thoroughly mixed, as consistency of composition is of utmost importance in making glass.

Nowadays, recycled glass from bottle banks collections, known as cullet, is used to make new glass. Using cullet has many environmental benefits, it prevents pollution by reducing quarrying, and because cullet melts more easily, it saves energy.

Almost any proportion of cullet can be added to the mix (known as batch), but in the right condition. Although the recycled glass may come from manufacturers around the world, it can be used by any glassmaker, as the container glass compositions are very similar. It is however important, that glass colors are not mixed and that the cullet is free from impurities, especially metals and ceramics.

### 1.3.1.2 Lead glass

Commonly known as lead crystal, lead glass is used to make a wide variety of decorative glass objects.

It is made by using lead oxide instead of calcium oxide, and potassium oxide instead of all or most of the sodium oxide. The traditional English full lead crystal contains at least 30% lead oxide (PbO) but any glass containing at least 24% PbO can be described as lead crystal. Glass containing less than 24% PbO, is known simply as crystal glass. The lead is locked into the chemical structure of the glass so there is no risk to human health.

Lead glass has a high refractive index making it sparkle brightly and a relatively soft surface so that it is easy to decorate by cutting and engraving which highlights the crystal's brilliance making it popular for glasses, decanters and other decorative objects.

Glass with even higher lead oxide contents (typically 65%) may be used as radiation shielding because of the well-known ability of lead to absorb gamma rays and other forms of harmful radiation.

### 1.3.1.3 Borosilicate glass

Most of us are more familiar with this type of glass in the form of ovenware and other heat-resisting ware, better known under the trade name Pyrex.

Borosilicate glass, the third major group, is made mainly of silica (70-80wt %) and boric oxide (7-13wt %) with smaller amounts of the alkalis (sodium and potassium oxides) and aluminum oxide. This type of glass has relatively low alkali content and consequently has good chemical durability and thermal shock resistance. As a result it is widely used in the chemical industry, for laboratory apparatus, for ampoules and other pharmaceutical containers and as glass fibers for textile and plastic reinforcement.

### 1.3.1.4 Glass fiber

Glass fiber has many uses from roof insulation to medical equipment and its composition varies depending on its application.

For building insulation, the type of glass used is normally soda lime. For textiles, an alumino-borosilicate glass with very low sodium oxide content is preferred because of its good chemical durability and high softening point. This is also the type of glass fiber used in the reinforced plastics to make protective helmets, boats, piping, car chassis, ropes, car exhausts and many other items.

In recent years, great progress has been made in making optical fibers which can guide light and thus transmit images round corners. These fibers are used in endoscopes for examination of internal human organs, changeable traffic message signs now on motorways for speed restriction warnings and communications technology, without which telephones and the internet would not be possible.

## 1.3.2 Special glass

Types of special glass change according to human needs and development. Earlier, the special types of glass were:

- Vitreous silica
- Aluminosilicate glass
- Alkali-barium silicate glass
- Technical Glass
- Glass Ceramics
- Optical glass
- Sealing glass

### 1.3.2.1 Vitreous silica glass

Silica glass or vitreous silica is of considerable technical importance as it has a very low thermal expansion. This glass contains tiny holes and is used for filtration. Porous glasses of this kind are commonly known as Vycor.

### 1.3.2.2 Alumino-silicate glass

An important type of glass, aluminosilicate, contains 20wt% aluminium oxide (alumina-$Al2O3$), often including calcium oxide, magnesium oxide and boric oxide in relatively small amounts, with very small amounts of soda or potash. It is able to withstand high temperatures and thermal shock and is typically used in combustion tubes, gauge glasses for high-pressure steam boilers, and in halogen-tungsten lamps capable of operating at temperature as high as 750°C.

### 1.3.2.3 Alkali-barium silicate glass

Without this type of glass, watching TV would be very dangerous. A television produces X-rays that must be absorbed; otherwise they could in the long run cause health problems. The X-rays are absorbed by glass with minimum amounts of heavy oxides (lead, barium or strontium). Lead glass is commonly used for the funnel and neck of the TV tube, while glass containing barium is used for the screen.

### 1.3.2.4 Technical glass

Technical is the term given to a range of glasses used in the electronics industry.
Without borate glass, the computer revolution would not have been possible as it's vitally important in producing electrical components. This type of glass contains little or no silica and is used for soldering glass, metals or ceramics as it melts at the relatively low temperature (450-550oC), below that of normal glass, ceramics and many metals.
Glass of a slightly different composition is used for protecting siliconsemi-conductor components against chemical attack and mechanical damage, known as passivation glass.
Chalcogenide glass, similar semi conductor effect, is a type of glass that can be made without the presence of oxygen. Some of them have potential use as infrared transmitting materials and as switching devices in computer memories because their conductivity changes abruptly when particular voltage values are exceeded.

## 1.3.2.5 Glass ceramic

Some of these "Glass ceramics", formed typically from lithium aluminosilicate glass, are extremely resistant to thermal shock and have found several applications where this property is important, including cooking ware, mirror substrates for astronomical telescopes and missile nose cones.

An essential feature of glass is that it does not contain crystals. However, by deliberately stimulating crystal growth in glass, it is possible to produce a type of glass with a controlled amount of crystallization that can combine many of the best features of ceramics and glass.

## 1.3.2.6 Optical glass

Optical glasses are found in scientific instruments, microscopes, and fighter aircraft and most commonly in spectacles.

The most important properties are the refractive index and the dispersion. The index is a measure of how much the glass bends light. The dispersion is a measure of the way the glass splits white light into the colors of the rainbow. Glass makers use the variations in these characteristics to develop optical glasses.

## 1.3.2.7 Sealing glass

A wide variety of glass compositions are used to seal metals for electrical and electronic components. Here the available glasses may be grouped according to their thermal expansion which must be matched with the thermal expansions of the respective metals so that sealing is possible without excessive strain being induced by differing levels of expansion.

For sealing to tungsten, in making incandescent and discharge lamps, borosilicate alkaline earths or aluminous silicate glasses are suitable. Sodium borosilicate glasses may be used for sealing to molybdenum and the iron-nickel-cobalt (Fernico) alloys, are frequently employed as a substitute. The amount of sodium oxide permissible depending on the degree of electrical resistance required. With glasses designed to seal to Kovar alloy, relatively high contents of boric oxide (approximately 20%) are needed to keep the transformation temperature low and usually the preferred alkali is potassium oxide so as to ensure high electrical insulation.

Where the requirement for electrical insulation is paramount, as in many types of vacuum tube and for the encapsulation of diodes, a variety of lead glasses (typical containing between 30% and 60% lead oxide) can be used.

## 1.3.2.8 Colored glase (Weyl, 1959 )

Color is the most obvious property of a glass object. It can also be one of the most interesting and beautiful properties. Although color rarely defines the usefulness of a glass object it almost always defines its desirability.

### 1.3.2.8.1 The colored glass recipe

The earliest people who worked with glass had no control over its color. Then, through accident and experimentation glass makers learned that adding certain substances to the glass melt would produce spectacular colors in the finished product. Other substances were discovered that, when added to the melt, would remove color from the finished project.

The Egyptians and the Romans both became expert at the production of colored glass. In the eighth century, an Arab chemist known as "Gerber" recorded dozens of formulas for the production of glass in specific colors. Gerber is often known as the "father of chemistry" and he realized that the oxides of metals were the key ingredients for coloring glass.

*1.3.2.8.2 The glass color palette*

Once the methods of colored glass production was discovered, an explosion of experimentation began. The goal was to find substances that would produce specific colors in the glass. Some of the earliest objects made from glass were small cups, bottles and ornaments.

Religious organizations were among those who provided incentive to the early glass artisans. Stained glass windows became very popular additions to churches and mosques over 1000 years ago. These artists needed a full palette of colors to make a realistic stained glass scene. This search for a full palette fueled research and experimentation to produce a vast array of colors.

*1.3.2.8.3 Colors of duration*

Then, another problem was discovered. Many of the glass colors did not stand up to year-in, year-out exposure to the direct rays of the sun. The result was a stained glass scene of deteriorating beauty. Some colors darkened or changed over time, while others faded away. Research and experimentation continued in an effort to meet the need for colors of duration. Eventually a full palette of fairly stable colors was achieved.

*1.3.2.8.4 Metals used to color glass*

The recipe for producing colored glass usually involves the adding of a metal to the glass. This is often accomplished by adding some powdered oxide, sulfide or other compound of that metal to the glass while it is molten. The table below lists some of the coloring agents of glass and the colors that they produce. Manganese dioxide and sodium nitrate are also listed. They are discoloring agents - materials that neutralize the coloring impact of impurities in the glass.

| Metals Used to Impart Color to Glass | |
| --- | --- |
| Yellow | Cadmium Sulfide |
| Red | Gold Chloride |
| Blue-Violet | Cobalt Oxide |
| Purple | Manganese Dioxide |
| Violet | Nickel Oxide |
| Yellow-Amber | Sulfur |
| Emerald Green | Chromic Oxide |
| Fluorescent Yellow, Green | Uranium Oxide |
| Greens and Browns | Iron Oxide |
| Reds | Selenium Oxide |
| Amber Brown | Carbon Oxides |
| White | Antimony Oxides |
| Blue, Green, Red | Copper Compounds |
| White | Tin Compounds |
| Yellow | Lead Compounds |
| A "decoloring" agent | Manganese Dioxide |
| A "decoloring" agent | Sodium Nitrate |

Table 1.

## 1.4 What types of radiation are there?
The radiation one typically encounters is one of four types: alpha radiation, beta radiation, gamma radiation, and x radiation. Neutron radiation is also encountered in nuclear power plants and high-altitude flight and emitted from some industrial radioactive sources.

### 1.4.1 Alpha radiation
Alpha radiation is a heavy, very short-range particle and is actually an ejected helium nucleus. Some characteristics of alpha radiation are:
Most alpha radiation is not able to penetrate human skin.
Alpha-emitting materials can be harmful to humans if the materials are inhaled, swallowed, or absorbed through open wounds.
A variety of instruments has been designed to measure alpha radiation. Special training in the use of these instruments is essential for making accurate measurements.
A thin-window Geiger-Mueller (GM) probe can detect the presence of alpha radiation.
Instruments cannot detect alpha radiation through even a thin layer of water, dust, paper, or other material, because alpha radiation is not penetrating.
Alpha radiation travels only a short distance (a few inches) in air, but is not an external hazard.
Alpha radiation is not able to penetrate clothing. Examples of some alpha emitters: radium, radon, uranium, thorium.

### 1.4.2 Beta radiation
Beta radiation is a light, short-range particle and is actually an ejected electron. Some characteristics of beta radiation are:
Beta radiation may travel several feet in air and is moderately penetrating.
Beta radiation can penetrate human skin to the "germinal layer," where new skin cells are produced. If high levels of beta-emitting contaminants are allowed to remain on the skin for a prolonged period of time, they may cause skin injury.
Beta-emitting contaminants may be harmful if deposited internally.
Most beta emitters can be detected with a survey instrument and a thin-window GM probe (e.g., "pancake" type). Some beta emitters, however, produce very low-energy, poorly penetrating radiation that may be difficult or impossible to detect. Examples of these difficult-to-detect beta emitters are hydrogen-3 (tritium), carbon-14, and sulfur-35.
Clothing provides some protection against beta radiation. Examples of some pure beta emitters: strontium-90, carbon-14, tritium, and sulfur-35.

### 1.4.3 Gamma and X- radiation
Gamma radiation and x rays are highly penetrating electromagnetic radiation. Some characteristics of these radiations are:
Gamma radiation or x rays are able to travel many feet in air and many inches in human tissue. They readily penetrate most materials and are sometimes called "penetrating" radiation.
X rays are like gamma rays. X rays, too, are penetrating radiation. Sealed radioactive sources and machines that emit gamma radiation and x rays respectively constitute mainly an external hazard to humans.
Gamma radiation and x rays are electromagnetic radiation like visible light, radio waves, and ultraviolet light. These electromagnetic radiations differ only in the amount of energy they have. Gamma rays and x rays are the most energetic of these.

Dense materials are needed for shielding from gamma radiation. Clothing provides little shielding from penetrating radiation, but will prevent contamination of the skin by gamma-emitting radioactive materials.

Gamma radiation is easily detected by survey meters with a sodium iodide detector probe.

Gamma radiation and/or characteristic x rays frequently accompany the emission of alpha and beta radiation during radioactive decay.

Examples of some gamma emitters: iodine-131, cesium-137, cobalt-60, radium-226, and technetium-99m.

## 2. Effect of radiation on different types of glass

### 2.1 Silicate glass

Jiawei Shenga (2009) studied the effect of UV-laser irradiation on the soda-lime silicate glass by preparing Commercial soda-lime silicate glass substrates composed of (wt%): 73.2 $SiO_2$, 15.3 $Na_2O$, 1.3 $Al_2O_3$, and 10.2 CaO, they found that As shown in Fig. 1, the as-quenched base glass was colorless and had no measurable absorptions in the visible region. In addition, no ESR signal was observed in the blank samples. After laser irradiation, two characteristic absorption bands in the visible region with maxima at about 620, and 430 nm, respectively, were observed. More precise peak positions were determined to be 627 and 431 nm for these two bands, respectively, with the assistance of Gaussian resolution [Friebele 1991, Marshall 1997, Zhang 2007] (Fig. 2,). As a result, the glass showed slight brown. The induced color was unstable, as indicated by the decease of the peak intensity after 50 h room temperature storage following the irradiation (Fig. 2). These results were similar to the case using X-ray radiation [Sheng, 1970]. Radiation may cause the displacement of lattice atoms or electron defects that involve changes in the valence state of lattice or impurity atoms. The ionizing radiation produces electron-hole pairs in the glass structure. Accordingly, new optical absorption bands were developed [Sheng, 1970]. In general, these absorptions are associated with either oxygen deficiency or oxygen excess in the glass network. The most fundamental radiation-induced defects in glass are the nonbridging oxygen hole center (NBOHC: ^Si–O*), the E' center (^Si*), the peroxy radical (POR: ^Si–O–O*), and the trapped electrons (TE), where the notation "^" represents three bonds with other oxygen in the glass network and "*" denotes an unpaired electron. According to current knowledge, absorption bands of 627 and 431 nm in the soda-lime silicate glass were identified mainly as absorption of NBOHCs [Bishay, 1970, Wong, 1967, Sheng, 2007]. The main absorptions of the E0 center and the POR are in the far UV region, which have less effect on the glass visible color. Fig. 3 shows the ESR spectrum of the glass after the irradiation. Two distinctive signals at g ¼ 1.99w2.00 and g ¼ 1.992, respectively, were observed. As expected, the g ¼ 1.99w2.00 was identified as the defects of NBOHCs, correlating bands of 627 and 431 nm. The signal of g ¼ 1.992 might attribute to the defect of E' center. The spectra of the induced absorption with varying exposure time are presented in Fig. 4. The peak heights of the induced absorption increased with the laser irradiation time, and the peak positions absorption band were relatively constant, suggesting that only variations of their populations of defects be effected by the irradiation time. As shown in Fig. 4 (insert), the absorptions at 627 or 431 nm increased rapidly during the first 30 min radiation. The increase was then slow down and no more increase was observed after 60 min radiation. Irradiation energy is another important factor that affects the induced absorption of glass. As can be seen in Fig. 5, the induced absorption increased when the energy density was increased, while the induced absorption peak positions were relatively constant.

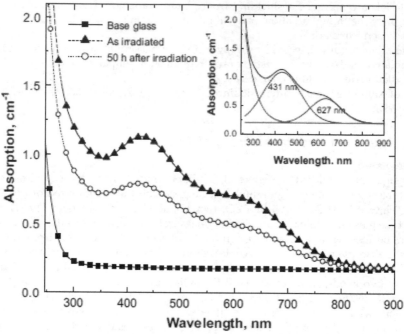

Fig. 2. Optical absorption in soda-lime silicate glass after UV-laser irradiation at 75 mJ/cm²
for 10 min (insert Gaussian resolution of induced absorption)

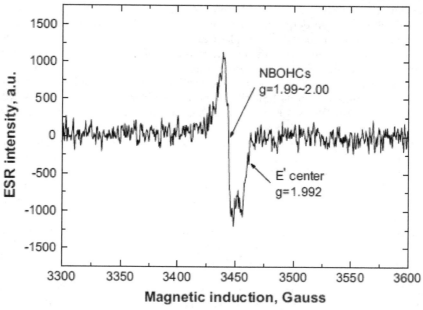

Fig. 3. ESR spectra of glass after irradiation at 75 mJ / cm² for 10 min

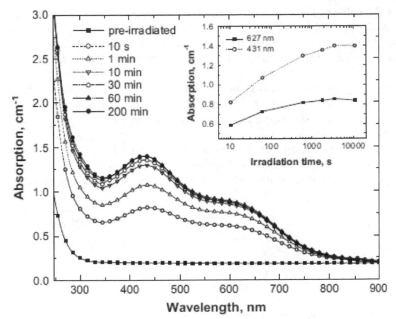

Fig. 4. Radiation induced optical Absorption with varied irradiation time (75 mJ / cm$^2$) (insert: Optical absorptions affected by irradiation time)

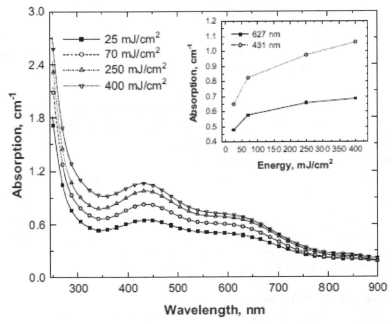

Fig. 5. Optical absorption affected by irradiation energy (100 shots) (insert: Optical absorption affected by irradiation energy).

**Gusarov et al (2010)** studied the effect of gamma irradiation on Silica-soda-lime glass with a composition $75SiO_2$–$22Na_2O$–$3CaO$, wt.% (labelled R1) and the same glass doped with 0.05 wt.% $CeO_2$ (labelled R7) were melted from high purity raw components using a technology which allowed one to keep the concentration of impurities of Fe ions at a few ppm and other transition metals below 1 ppm. Experimental samples were prepared as plates 25×25mm2, ~1mm thick, and polished for optical measurements. The glass samples contained a small number of air bubbles with diameters up to ~0.1mm, however transmission measurements at different parts on several samples showed that their influence was below the detection level. All transmission spectra were measured from 195 to 3300 nm with a commercial double-beam UV–Vis–NIR spectrophotometer. Irradiations have been performed in the Brigitte gamma irradiation facility at SCK CEN at 70, 200, and 350 °C, 25 kGy/h. They found that the addition ofCeO2 results in significant transmission degradation in the UV (Fig. 6)

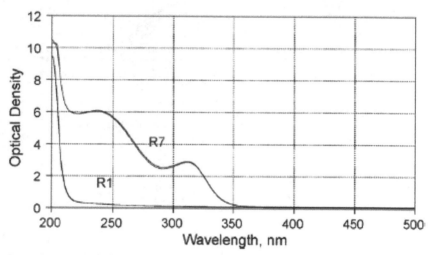

Fig. 6. Optical absorption spectra of R1-5 and R7-5 glass samples in the UV range. For each glass two spectra are plotted: before and after annealing for 24 h at 350 °C

This degradation is due to the overlapping optical absorption bands of $Ce^{3+}$ and $Ce^{4+}$ ions, with maxima at 313 and 240 nm (3.96 and 5.15 eV), respectively [**Arbuzov, 1990**]. Above 350 nm both materials show the same high transmission with no observable optical absorption bands. No detectable changes in transmission were observed after annealing R1 and R7 control glass samples for 24 h at 200 °C. Annealing for 24 at 350 °C does not change the transmission of R1. For R7 a weak band with amplitude of ~0.3cm−1 at 275 unappeased. This effect can be tentatively attributed to thermally-stimulated charge transfer between Ce and Fe ions. However, the amplitude of this thermally induced band is significantly less than the amplitude of bands produced by irradiation.

## 2.2 Phosphate glass doped with cobalt oxide
Sodium phosphate glass with different CoO percentage was prepared [El-Batal, 2010], A 60Co gamma cell (2000 Ci) was used as a c- ray source with a dose rate of 1.5 Gy/s (150 rads/s) at a room temperature of ~30 °C. The investigated glass samples were subjected to

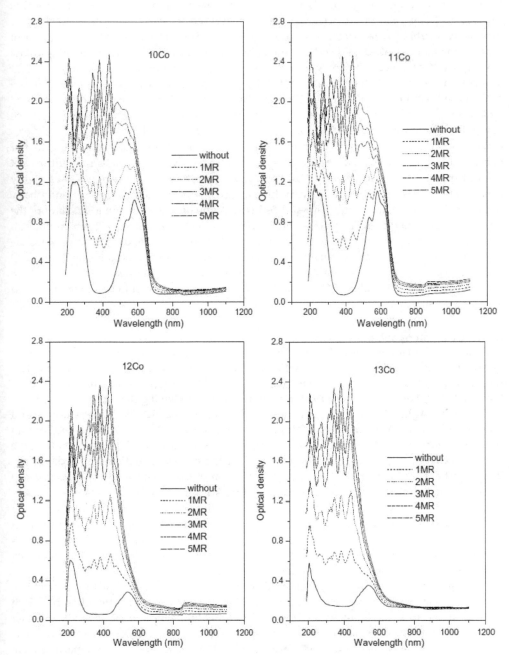

Fig. 7. Absorption spectra of CoO-doped sodium phosphate glasses with constant 0.25% CoO with additional 5% or 10% of Na$_2$O or 5% or 10% P$_2$O$_5$ before and after gamma irradiation.

the same gamma dose every time. Using a Fricke dosimeter, the absorbed dose in water was utilized in terms of dose in glass. No cavity theory corrections were made. Each glass sample was subjected to a total dose of $5 \times 104$ Gy (5MR). The UV–visible absorption of glasses containing CoO-doped sodium metaphosphate glasses. The glass containing 0.05% CoO (glass/Co) reveals before gamma irradiation a strong ultraviolet charge transfer absorption with two small peaks at about 205 and 235 nm and the visible spectrum shows a small broad band centered at about 530 nm. On subjecting this glass to successive gamma irradiation, the UV absorption progressively increases with marked splitting to three component peaks at about 205, 275, and 310 nm and followed by subsequent closely-connected four peaks at about 340, 400, 450, and 500 nm. With glass containing higher concentrations P0.5% CoO), the optical absorption spectra reveal the same strong ultraviolet absorption with two peaks at about 205 and 235 nm and the visible region shows a very broad band increasing in intensity with the increase of CoO content and exhibiting finally three small peaks at about 500, 540, and 595 nm. With continuous gamma irradiation, the absorption spectra reveal numerous connected peaks extending from 205 up to 400 nm exhibiting continuous growth with gamma irradiation and the distinct broad visible absorption is observed to be finally unaffected by progressive irradiation at high CoO content . The rate of increase with the first dose of irradiation is observed to be very high especially in the band at about 500 nm and the optical density is seen to be slowly increased with progressive irradiation.

### 2.2.1 Condition of cobalt in glass

Although cobalt can exhibit different oxidation states in many inorganic complexes but in glasses melted under normal atmospheric conditions, cobalt ion exists in the divalent state with two possible coordination forms, namely the octahedral and tetrahedral [Bates 1962, Bamford 1977, Paul 1990, Aglan 1955]; The only known ion having the 3d7 configuration is $Co^{2+}$. Early, **Aglan 1955**, and later other scientists [El-Batal,2003] have studied and interpreted the absorption spectra of glasses containing $Co^{2+}$ in terms of equilibrium between octahedral and tetrahedral coordination forms of $Co^{2+}$ ions depending on the type and composition of glass and condition of melting. The energy level diagram for d7 system in octahedral coordination predicts that the spectrum of $Co^{2+}$ in octahedral symmetry will consist essentially of three bands corresponding to spin-allowed transitions $^4\Gamma_4 \rightarrow 4C5$, $^4\Gamma_4 \rightarrow ^4\Gamma_2$ and $^4\Gamma_4 \rightarrow ^4\Gamma_4$ (P) together with several weak lines corresponding to spin-forbidden transitions. However, the band corresponding to the transition $^4\Gamma_4 \rightarrow ^4\Gamma_2$ is expected to occur only with low intensity as it corresponds to a forbidden two electron jump. **Bates, 1962** has assumed that in low alkali glasses, $Co^{2+}$ ions are present in octahedral symmetry probably with a rhombic distortion arising from the Jahn– Teller effect.

The energy diagram for d7 system in tetrahedral symmetry predicts that the spectrum of $Co^{2+}$ in four coordination consists essentially of three bands [Sreekanth, 1998] corresponding to the spin-allowed transition $^4\Gamma_4 \rightarrow ^4\Gamma_5$, $^4\Gamma_2 \rightarrow ^4\Gamma_4$ and $^4\Gamma_2 \rightarrow ^4\Gamma_5$ (P) together with several weak lines and bands corresponding to spin-forbidden transitions. The splitting of the 4Γ4 (P) bands has been attributed to L, S interactions (i.e. a departure from Russel–Sanders coupling) which is also quite large in the free ion. The high intensity of the tetrahedrally coordinated band is a consequence of the mixing of the 3d-orbitals with 4P-orbitals and ligand orbital [ElBatal, 2008].

## 2.3 Lead silicate glass

Lead silicate and borate glasses glasses [El-Kheshen, A., 2008] have been extensively studied these last decades mainly because of their presence and importance in a broad range of technological applications. These glasses have been used in the electronic and optical technologies, such as for electron multiplier [Anderson,1979], micro-channel plated [Wiza, 1979], non-linear optical and magneto-optical devises [George, 1999]. Also, high lead silicate glasses can find application as radiation shielding materials [Friebele, 1991].

A base lead silicate glass of the chemical composition PbO 60 wt%, SiO$_2$ 40 wt% (71.24 PbO mol%, 28.76 SiO$_2$ mol%) was prepared by [Azooz , 2009] .Batches containing the base glass composition to which was added 0.1% of one of the 3d-transition metal oxides: TiO$_2$, V$_2$O$_5$, Cr$_2$O$_3$, MnO$_2$, Fe$_2$O$_3$, CoO, NiO or CuO. An Indian 60Co gamma cell (2000 ci) was used as a gamma ray source with a dose rate of 1.5Gy s−1.

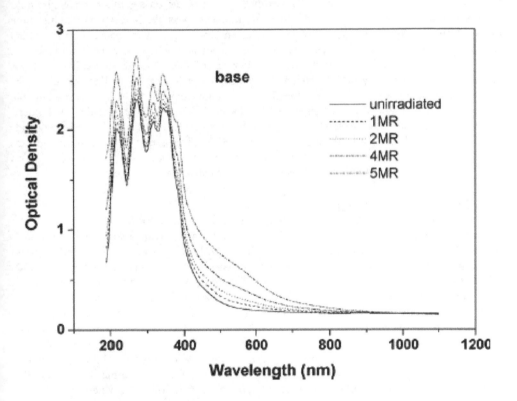

Fig. 8. The optical absorption spectra of the parent glass before and after different doses of gamma rays

Experimental results indicate that the effect of successive gamma irradiation is mainly concentrated in the changes of the intensities of the already observed UV bands which are assumed to be due to trace iron impurities (mainly $Fe^{3+}$ ions) and to the $Pb^{2+}$ ions. The main pronounced difference is the sharp growth of the kink at about 390nmto a pronounced peak and the quite resolution of a broad visible band centered at about 500 nm. It seems that the presence of high content of the heavy metal oxide (PbO), shows some shielding towards successive gamma irradiation and only limited changes are observed. This behavior is related to the presence of heavy $Pb^{2+}$ ions which retard the movement or transfer of released electrons during the irradiation process [Singh, 2004].

The realization of the effect of gamma irradiation can be interpreted by assuming that when the lead silicate host glass is subjected to ionizing radiation, electron–hole pairs are produced, which then become individually trapped at various intrinsic defect sites in the glass structure. Also, the presence of trace iron impurities in the host glass can easily trap electrons or holes and further extrinsic defects may be formed by photochemical process. It has been shown by several authors [Bishay, 1977] that the principal feature of the spectra of Pb-containing glasses is a band at 1.59 eV with a width of 0.6 eV which has been correlated with $Pb^{3+}$ in silicate glasses. [Friebele, 1977] assumed that in oxide glasses this band is observed only when the Pb concentration is less than $\approx$25 mol%. A possible explanation for this is the assumption that Pb2+ is incorporated in glass as an ionic modifier for concentration of PbO $\leq$25–30 mol%, while at higher concentrations, $PbO^{2+}$ is covalently bonded in $PbO_4$ unit [Barker, 1965] and/or $PbO_3$ unit [Fayon, 1999] as a glass former. [Friebele, 1991] further added that two bands with peak energies of 2.36 and 3.31 eV are present in high lead glasses and can be associated with Pb3+ on a network forming unit. Regarding the SiO2 as a partner in the composition, [Bishay, 1977] and [Friebele 1991] have classified the origin of the UV induced bands and attributed them to electron centers while they related to the visible induced bands to positive holes. Later, [Shkrob, 2000] have assumed that irradiation of alkali silicate glasses results in the formation of metastable spin centers such as oxygen hole centers (OHC1 and OHC2), silicon peroxy radicals and a silicon dangling electron center (E/center).

### 2.3.1 Cr doped lead silicate glass

This chromium-doped glass reveals before irradiation (Fig. 9) five strong ultraviolet absorption bands at about 205, 270, 310, 340 and 380nm followed by a strong visible band at about 440 nm and a medium band at about 480 nm and finally a very broad band centered at about 635 nm. This observed spectrum represents collective presence of absorption bands due to trace iron impurities (205, 270, 310 nm), $Pb^{2+}$ ions (340 nm) and hexavalent chromium ions $Cr^{6+}$ (380 nm) while the visible bands due to trivalent chromium bands (440 and 650 nm) and a new mixed band (470 nm). These collective specific absorption bands are identified to the respective mentioned ions in various glass systems by several authors [Ghoneim, 1983].With progressive gamma irradiation, the six successive ultraviolet and near visible absorption bands slightly increase in intensity followed by slight increase and the presence of chromium ions seems to retard the effect of gamma irradiation this mentioned region. The visible band at about 470 nm shifts to longer wavelength at 480 nm while broad band highly increases in intensity and shifts from 650 to 620 nm.

It can be assumed that some of the present $Cr^{6+}$ ions capture liberated electrons during gamma irradiation and are converted to induced $Cr^{3+}$ ions producing absorption in the same position of the trivalent chromium in the following photochemical reaction:

$$Cr^{6+} + 3e^- \rightarrow Cr^{3+}$$

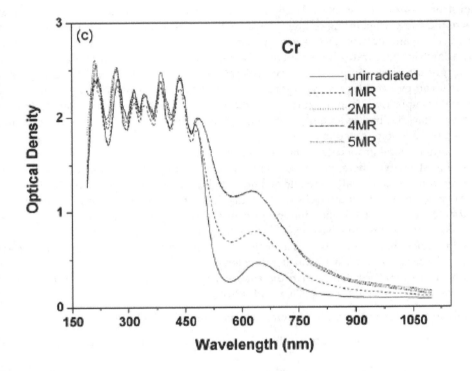

Fig. 9. The optical absorption spectra of the Cr-doped lead silicate glasses before and after different doses of gamma rays.

## 2.4 Radiation processing for glass coloration / discoloration

Glass is particularly susceptible to radiation-induced coloration/ discoloration due to its amorphous, non-crystalline structure. The nature of the optical changes varies, but usually consists of coloration in the visible light region and the formation of absorption bands in the infrared and/or ultraviolet regions. The optical density is almost always increased. It is well documented, for example, that high doses of gamma radiation turn glass various shades ranging from deep brown to pale amber [Prasil, 1991]. Clear glass discolors when exposed to gamma irradiation, such as from a Cobalt-60 source. Ordinary flint, borosilicate, and lead glass undergo a color change from clear to light amber, brownish to black, depending on the amount of energy absorbed. Milk glass, when exposed to gamma irradiation, yields a grayish color, depending on energy absorbed and any swirling effect is highlighted, probably due to concentrations of color in the glass that have not mixed uniformly. The actual mechanism of the formation of "color centers" has been described [Prasil, 1991]. The final color created is dependent on the chemical composition of the glass and can be altered by selection of additives (e.g., cerium ions can reduce browning; manganese ions induce an amethyst color). Final color is a combination of original glass color and the effects of color center formations.

Color from irradiation is considered metastable and heat, for example, is known to reverse the effect. Depending on the type of color centers formed and the hardness of the glass, varying amounts of heat or energy penetration are required to reverse color. Glass with a low diffusion rate (high softening point) is more stable. Combining all of these factors gives a very unique decorative coloration to a glass item. Very striking color contrasts can evolve, depending upon the color of the paints utilized on an item, the chemical composition, and impurities in the glass. At low doses, the color intensity increases linearly but eventually saturates at high doses. The final product is not radioactive or contaminated in any sense.

The general physical properties of colored glass include:

1.  **Color stability:** There is a slight loss of color during the first several days after processing, and thereafter the rate of fade is slight. The process can actually be reversed, causing the glass to become clear, by placing the unit in an oven at approximately 300°F for a couple of hours. Hence, for applications where the product is exposed to high heat or low heat on a routine basis (e.g., dishwashers), this may not be ideal.

2.  **Ultraviolet absorption:** There is a small amount of ultraviolet absorption characteristic which changes as the glass is further discolored.

3.  **Physical properties:** Other than the color change, the normal physical properties of glass are not affected [Dietz 1976, Prasil, 1990].

**The advantages of irradiation for decorative glass coloration include:**

1.  Processing is done in shipping carton. There is no need to unpack or handle individual pieces.

2.  The process is simple, clean, and completely safe to the end product and to the consumer.

3.  Although economics would tend to favor large volumes, smaller volumes can also be economically attractive. The irradiation process is not dependent on processing large runs.

4.  There is no waste product or breakage associated with the process

## 3. Colored glass as radiation sensor [Dunn, T.M. 1960]

This wonderful material (colored glass), with its beautiful appearance, has another use rather than decoration which is: radiation sensor. Some other instruments were used as radiation detectors but now colored glass has the same action, due to transition elements doping in glass. There are a number of properties shared by the transition elements that are not found in other elements, which results from the partially filled $d$ shell. These include:

•   The formation of compounds whose color is due to $d$ - $d$ electronic transitions

•   The formation of compounds in many oxidation states, due to the relatively low reactivity of unpaired $d$ electrons.

•   The formation of many paramagnetic compounds due to the presence of unpaired $d$ electrons. A few compounds of main group elements are also paramagnetic (e.g. nitric oxide, oxygen).

Color in transition-series metal compounds is generally due to electronic transitions of two principal types.

Charge transfer transitions. An electron may jump from a predominantly ligand orbital to a predominantly metal orbital, giving rise to a ligand-to-metal charge-transfer (LMCT) transition. These can most easily occur when the metal is in a high oxidation state. For example, the color of chromate, dichromate and permanganate ions is due to LMCT transitions. Another example is: Mercuric iodide, $HgI_2$, is red because of a LMCT transition. As this example shows, charge transfer transitions are not restricted to transition metals. A metal-to ligand charge transfer (MLCT) transition will be most likely when the metal is in a low oxidation state and the ligand is easily reduced. d-d transitions. An electron jumps from one d-orbital to another. In complexes of the transition metals the $d$ orbitals do not all have the same energy. The pattern of splitting of the $d$ orbitals can be calculated using crystal field theory. The extent of the splitting depends on the particular metal, its oxidation state and the nature of the ligands. In octahedral complexes, d-d transitions are forbidden and only occur because of coupling in which a molecular vibration occurs together with a d-d transition. Tetrahedral complexes have somewhat more intense color because mixing $d$ and $p$ orbitals is possible when there is no centre of symmetry, so transitions are not pure d-d transitions. The molar absorptivity ($\varepsilon$) of bands caused by d-d transitions are relatively low, roughly in the range 5-500 $M^{-1}cm^{-1}$ (where M = mol $dm^{-3}$). Some d-d transitions are spin forbidden. An example occurs in octahedral, high-spin complexes of manganese (2), which has a $d^5$ configuration in which all five electron has parallel spins; the color of such complexes is much weaker than in complexes with spin-allowed transitions. In fact many compounds of manganese (2) appear almost colorless. The spectrum of $[Mn(H_2O)_6]^{2+}$ shows a maximum molar absorptive of about 0.04 $M^{-1}cm^{-1}$ in the visible spectrum.

A characteristic of transition metals is that they exhibit two or more oxidation states, usually differing by one. Main group elements in groups 13 to 17 also exhibit multiple oxidation states. The "common" oxidation states of these elements typically differ by two. For example, compounds of gallium in oxidation states +1 and +3 exist in which there is a single gallium atom. No compound of Ga(2) is known: any such compound would have an unpaired electron and would behave as a free radical and be destroyed rapidly. The only compounds in which gallium has a formal oxidation state of +2 are dimeric compounds, such as $[Ga_2Cl_6]^{2-}$, which contain a Ga-Ga bond formed from the unpaired electron on each Ga atom. Thus the main difference in oxidation states, between transition elements and other elements is that oxidation states are known in which there is a single atom of the element and one or more unpaired electrons. The maximum oxidation state in the first row transition metals is equal to the number of valence electrons from titanium (+4) up to manganese (+7), but decreases in the later elements. In the second and third rows the maximum occurs with ruthenium and osmium (+8). In compounds such as $[MnO_4]^-$ and $OsO_4$ the elements achieve a stable octet by forming four covalent bonds. The lowest oxidation states are exhibited in such compounds as $Cr(CO)_6$ (oxidation state zero) and $[Fe(CO)_4]^{2-}$ (oxidation state −2). These complexes are also covalent. Ionic compounds are mostly formed with oxidation states +2 and +3. In aqueous solution the ions are hydrated by (usually) six water molecules arranged octahedrally.

## 3.1 Why do many glasses turn green, brown or black when exposed to ionizing radiation? [El-Kheshen A. A., 1999]

The phenomenon of radiation-induced color change is called 'activation of color centers'. The details are quite complex and involve an alteration of the orbital distribution of an atom's valence (outermost) electrons, causing the atom to absorb photons of a different frequency (color) after irradiation than before. What we describe as the reflected, transmitted, and emitted colors of a material is a consequence of the outermost shell of electrons, the valence electrons of atoms. In the following discussion, we will consider primarily reflected light.' White' light is actually a mixture of photons of many different frequencies (colors), and color is typically described by wavelength. The valence shell of an atom, described by classical physics as comprised of up to eight discreet electrons, is described by quantum physics to behave as a singular 'thing'. The valence shell will resonate with (it will capture, or absorb) photons of certain discrete energies. This resonance causes the valence shell to become excited, or 'pumped', to a higher energy meta-stable 'level' or state. The atom has a 'ground' (un-excited) state, and may have multiple excited states. Upon absorption of radiation of energy, electrons may be stripped from the atom, resulting in multiple degrees of ionization, each capable of multiple states of excitation. For example, a hypothetical atom with eight electrons in its valence shell has its ground state and shall have three meta-stable excited states. Upon capture of a photon of energy, one of the valence electrons is stripped resulting in the atom becoming a singly-ionized ion. This particular ion shall have two meta-stable excited states, and upon capture of a photon of energy a second of the valence electrons is stripped resulting in a doubly-ionized ion capable of four meta-stable states. And so on. For any mono-atomic atom, and for the shared valence electron orbitals of any compound, there are these discreet states capable of absorbing (and emitting) discreet quanta of energy which is shuttled around as photons. Now, let's work with a real example: Let's assume that we have a 'white' light beam composed of red, blue and yellow photons (the primary transmissive colors). We direct this beam onto a puddle of red anthracine dye. Anthracite dyes have tri-cyclic ('aromatic') structures which are capable of resonating with discreet spectra of frequencies [El-Batal, 2008]. By slight modifications of their molecular structures the valence orbitals of their constituent carbon atoms can be altered, which results in slightly different resonant characteristics. The different anthracine dyes thus appear to our eyes as having different colors. We have selected a red one. The blue and yellow photons in our white light resonate with (are captured and absorbed by) the red dye molecule, and the red photons are scattered (reflected). Some of this scattered red light is directed toward our eyes, and we thus see the puddle of dye as 'red'. If we put the puddle of red dye on a clear film, the red photons will be seen also to pass through the dye.

Now, let's examine a piece of colored glass. The materials which are added to glass as pigments are selected because their valence shell electrons are capable of selectively absorbing photons of certain frequencies, while passing and reflecting others. Cobaltic oxide for example, will absorb red and yellow but not blue, thus cobalt glass is blue. Transparent glasses in general are a mixture of alkaline and transition metal oxides, most of which are selected precisely because they *won't* interact with visible light. Put a piece of such a transparent glass in a high neutron-flux or high radiation (such as γ-ray) environment, and they will 'become colored' as the glass becomes physically altered at the atomic level [Moustafa , 2010 ].

### 3.2 Can the activation of color centers be reversed? Can the colored glass be made clear again? [Radek, 2009]

Yes. Glass, even at room temperature, is a liquid: the interatomic bonds are weak and constantly breaking and reforming. These bonds are in fact an interaction of (or, sharing of) the valence electrons of adjacent atoms. (In metals this sharing of electrons results in electrical conduction. For example, in a length of copper wire, the individual electrons at one end of the wire will, in theory, eventually migrate to the other end of the wire, due to random motion and without the application of an outside force). The inter-atomic bonding structure within the bulk of a material places physical constraints upon the valence electrons. Therefore, if you activate a color center (by any means) the alteration in the valence shell will be either stable (unchanging with time) or meta-stable (will change gradually with time). The stronger the interatomic structure (or 'lattice' in the case of a true crystal which, unlike a glass, is a solid) the more stable the change. The change can however be reversed by weakening the inter-atomic bonds which will allow the formation of new, lower energy bonds (atoms will break their initial bonds and reform bonds with other neighbors). This can be accomplished simply by the application of heat. In practice, the temperature required for complete color deactivation in an amorphous material (such as a glass) is its annealing temperature. Therefore, simply annealing a piece of glass will deactivate the color centers. Having worked extensively with researchers who conduct experiments which activate color centers by both of these means, I have frequently had need to deactivate color centers. An unexpected consequence of color center activation is the formation of strained inter-atomic bonds, and unless the piece is annealed, it may spontaneously fracture! As a rule, the more a piece is colored, the more strain the piece is under, and the more it needs to be annealed. Note that the color doesn't 'cause' the strain. Rather, color and strain are two entirely separate issues sharing the same cause. Finally, it's worth pointing out that color center activation is now so well understood that unusually colored gemstones are now being created by the process. In some cases, semiprecious irradiated stones are fraudulently sold as precious gem stones, and the difference can only be distinguished by a professional gemologist. Many materials when exposed to radiation will darken or change color. For example you can occasionally find a discarded bottle that had been left outside for many years that has become darkened due to exposure to sunlight. Co-60 emits gamma rays, which are light waves but with a much higher energy than visible light. Many hospitals use a Co-60 source to sterilize medical supplies after they have been packaged. So why does gamma radiation change the color of glass? We need to look at the structure of glass and at why certain materials absorb different colors of light to understand the effect. Glass is made of silicon and oxygen. Oxygen atoms are hungry for more electrons than they are naturally given, so each oxygen atom will share an electron from the silicon atom. This shared electron creates a very strong attraction between the oxygen and silicon, and is the reason why glass is a very hard material.

As you can tell by looking out a window, visible light goes right through glass. For light to be absorbed in a material, the light wave has to interact with an electron in the material and basically "shake it loose" from its atom. In glass the electrons are so tightly held by the silicon and oxygen atoms that visible light cannot shake one of those electrons loose, so light travels right through without any absorption. But gamma rays, which have a million times as much energy as visible light, can shake a lot of things loose in glass. When a gamma ray enters glass, it can actually knock a silicon atom out of its place, shoving it in-between other atoms, and leaving a hole, or vacancy, in the material. The electrons surrounding the hole will no longer be tightly bound to the atoms, and these "loose" electrons will be able to interact with lower energy light (visible light). When visible light is transmitted through the glass, specific colors of the light can interact and be absorbed by the electrons near the vacancy, and the transmitted light will have a color that is the complement of the absorption. These holes, or defects, are often called "color centers" because they impart a color to the glass. If you heat the irradiated glass (called annealing) the atoms will begin to shake so much that the displaced silicon atom will eventually go back to its original place and the glass will become clear again.

## 4. References

Aglan, M.A., Moore, H., J. Soc. Glass Technol. 39 (1955) 35T.

Anderson, L.P., Grusell, E., Berg, S., J. Phys. E 12 (1979) 1015.

Arbuzov, V. Belyankina, N.,"Spectroscopic and photochemical properties of cerium in silicate glass", Fiz. Khim. Stekla 16 (1990) 593–604.

A world of glass, http://www.britglass.org.uk/Index.html.

Azooz, M.A., El-Batal, F.H., " Gamma ray interaction with transition metals-doped lead silicate glasses", Materials Chemistry and Physics 117 (2009) 59–65.

Barker, R.S., McConckey, E.A., Richardson, D.A. Phys. Chem. Glasses 6 (1965) 24–29.

Bates, T., in: D. Mackenzie (Ed.), "Modern Aspects of the Vitreous State", vol. 2, Butterworth, London, 1962, p. 195.

Bamford, C.R., "Colour Generation and Control in Glass", Elsevier, New York, 1977.

Bishay A. Radiation induced color centers in multicomponent glasses. J Non-Cryst Solids 1970;3:54–114.

Bishay, A., Maklad, M. ,Phys. Chem. Glasses 7 (1966) 149–156.

Dietz, George R., "Radiation Coloring of Glass" Presentation to Society of Glass Decorators, Annual Meeting October 11- 13, 1976, Pittsburgh, PA.

Dunn, T.M., Wilkins, R.G. "Modern Coordination Chemistry. New York: Interscience., Chapter 4, Section 4, "Charge Transfer Spectra", pp. 268-273, 1960.

El-Batal, F., Ouis, M., Morsi, R. Marzouk, S.," Interaction of gamma rays with some sodium phosphate glasses containing cobalt", Journal of Non-Crystalline Solids 356 (2010) 46–55.

ElBatal, F.H., ElKheshen, A.A., Azooz, M.A., Abo-Naf, S.M., Opt. Mater. 30 (2008).

El-Batal,. Khalil, M.M.I, Nada, N., Desouky, S.A. Mater. Chem. Phys. 82 (2003) 375.

El Kheshen, A. A. El-Batal, F., Marzouk,S."UV-visible, Infrared and Raman spectroscopic and thermal Studies of tungsten doped lead borate glasses and the effect of ionizing gamma irradiation", Indian Journal of Pure & Applied Physics, 46, 4, 225-238, 2008.

El-Kheshen, A. A., "Effect of γ- rays on some glasses containing transition metals, A PhD thesis, Ain-Shams University, Egypt, 1999.

Fayon, F., Landron, C., Sakwari, K., Bessada, C., Massiot, D., J. Non-Cryst. Solids 243 (1999) 39.

Friebele EJ. Radiation effects in optical and properties of glass. New York, Westerville; 1991.

Friebele, E.J. in: D.R. Uhlmann, N.J. Kreidl (Eds.), Optical Properties of Glasses, The American Ceramic Society,Westerville, OH, 1991, p. 205.

Friebele, E.J. Proc. Int. Congr. Glass, 11th, Progue, vol. 3, 1977, pp. 87–95.

George, H.B., Vira, C., Stehle, C. , J. Meyer, S. Evers, D. Hogon, S. Feller,M. Affatigato, Phys. Chem. Glasses 40 (1999) 326.

Ghoneim, N.A., El-Batal, H.A., Zahran, A.H., Ezz ElDin, F.M. Phys. Chem. Glasses 24 (1983) 83.

Gerry, M., Alan M., "Glass: a world history, published by; University of Chicago Press", 2002.

Gusarov, A., Huysmans, S., Berghmans, F., "Induced optical absorption of silicate glasses due to gamma irradiation at high temperatures", Fusion Engineering and Design 85 (2010) 1–6.

Jiawei S, Yanfen W, Xinji Y, Jian Z., "UV-laser irradiation on the soda-lime silicate glass", International Journal of Hydrogen Energy, 34 ( 2009 ) 1123 – 1125.

Marshall CD, Speth JA, Payne SA. Induced optical absorption in gamma, neutron and ultraviolet irradiated fused quartz and silica. J Non-Cryst Solids 1997;212:59–73.

Moustaffa, F.A., El-Batal, F.H., Fayad, A.M., El-Kashef, I.M.Absorption Studies on Some Silicate and Cabal Glasses Containing NiO or $Fe_2O_3$ or Mixed NiO + $Fe_2O_3$", Acta physica polonica A, Vol. 117, 2010.

Noel, C.S., "The Glass and Glazing Handbook; Standards Australia, "SAA HB125–1998.

Paul, . A. Chemistry of Glasses, second ed., Elsevier, New York, 1990.

Prasil, Z , Marlind, T. "Two colors out of one" Glass Decoloration #125, beta-gamma, 1991, 2-3.

Prasil, Z. and Marlind, T. "Radiation Coloration of Glass-State of the Art, Glass Decoloration #249, beta-gamma 1991, 2-3.

Prasil, Z., Schweiner, E., Pesek, M. "Radiation Modification of Physical Properties of Inorganic Solids", Radiation Physics and Chemistry, Vol. 35, Nos. 4-6, pp.509-513, 1990.

Radek , P. Vojtĕch E., Viktor G., Mariana K., Martin M., Ondřej, S., Ladislav S., " A comparison of natural and experimental long-term corrosion of uranium-colored glass", J. of Non-Crystalline Solids, Volume 355, Issues 43-44, , Pages 2134-2142, 2009.

Sheng J., "Photo-induced and controlled synthesis of Ag nanocluster in soda-lime silicate glass". Int J Hydrogen Energy 2007;32(13). 2062–65.

Sheng J, Li J, Yu J. The development of silver nanoclusters in ion-exchanged soda-lime silicate glasses. Int J Hydrogen Energy 2007;32(13):2598–601).

Shkrob, I.A., Tadjikov, B.M., Trifunac, A.D. J. Non-Cryst. Solids 262 (6) (2000) 35.

Singh, N., Singh, K.J., Singh, K., Singh, H., Nucl. Instrum.Methods Phys. Res., B 255 (2004) 305.

Sreekanth , R.P., Murali, A., Lakshmara Rao, J. J. Alloy. Compd. 281(1998) 99.

Weyl, W.A "Coloured Glasses, reprinted by Dawsons of Pall Mall, London, 1959.

Wiza, J.L. Nucl. Instrum. Methods 62 (1979) 587.

Wong J, Angell CA. Glass: structure by spectroscopy. NewYork: Marcel Dekker; 1976.

Zhang J, Dong W, Qiao L, Li J, Zheng J, Sheng J. Silver nanocluster formation in the soda-lime glass by X-ray irradiation and annealing. J Cryst Growth 2007;305:278-84

# 3

# Optical Storage Phosphors and Materials for Ionizing Radiation

Hans Riesen and Zhiqiang Liu
*The University of New South Wales,*
*Australia*

## 1. Introduction

Over recent years there has been significant progress in the development of optical storage phosphors and materials for ionizing radiation, resulting in important applications in the fields of dosimetry and computed radiography. The aim of this Chapter is to provide some background information as well as an overview of a range of modern storage phosphors and materials and their applications in dosimetry and computed radiography. Dosimetry is the measurement of absorbed dose of ionizing radiation by matter or tissue and is measured in the SI unit of gray (Gy) or sievert (Sv), respectively, where 1 Gy and 1 Sv are equal to 1 joule per kilogram. It is noted here that the non-SI units of rad (absorbed radiation dose; for matter) and rem (Roentgen equivalent in man; for tissue) are still heavily used; their conversions are given by 1 Gy=100 rad and 1 Sv=100 rem. A dosimeter allows us directly or indirectly to measure exposure, kerma, absorbed or equivalent dose and associated rates of ionizing radiation. A dosimetry system consists of the actual dosimeter and an associated reader. A dosimeter is characterized by its accuracy and precision, linearity, energy response, dose and dose rate dependence, spatial resolution and directional dependence.

It is well documented that even low doses of ionizing radiation can induce cancer. Thus personal radiation monitoring, i.e. dosimetry applied to people, has become highly important in recent years as the number of potential radiation sources is ever increasing. Radiation monitoring will also become more important in medical diagnostics and in radiation therapy as regulations become stricter. Nuclear accidents, such as the recent events in Fukushima, Japan, have highlighted the need for inexpensive and effective radiation monitoring systems that can be deployed to a large population. We note here that, of course, the exposure to natural background radiation is, to a great extent, unavoidable. For example, the average annual radiation dose per capita from natural and medical sources is 2.0 mSv/year and 2.8 mSv/year in Australia and the United Kingdom, respectively. The amount of radiation exposure is a function of geography and building standards. For example, in European homes there is a relatively high level of radon as the internal walls are based on clay bricks. This is reflected in the fact that the average equivalent dose by radiation from radon progeny is 0.2 mSv/year and 1.4 mSv/year in Australia and the United Kindom, respectively.

Ionizing radiation has a wide range of applications such as in the sterilization of medical hardware and food, in medical diagnostics and radiotherapy and in technical and scientific imaging (Andersen et al., 2009; Dyk, 1999; Eichholz, 2003; Mijnheer, 2008; Podgorsak, 2005;

Riesen & Piper, 2008). For safety and quality assurance, reliable radiation monitoring systems (dosimetry) are mandatory for these applications (Mijnheer, 2008). Dosimetry systems that are currently used for evaluating dose and its distribution include (Dyk, 1999; Mijnheer, 2008; Podgorsak, 2005; Schweizer et al., 2004):

1. Ionization chambers and electrometers for the determination of radiation dose. The standard chambers are either cylindrical or parallel plates. Active volumes for cylindrical chambers are between 0.1 and 1 $cm^3$ i.e. they are relatively bulky. Ionization chambers allow an instant readout, are accurate and precise and are recommended for beam calibration. Disadvantages include the fact that they have to be connected by cables and are relatively bulky i.e. they, in general, do not allow point measurements.

2. Standard radiographic films are based on radiation sensitive silver halide emulsions in thin plastic films. Exposure to ionizing radiation leads to a latent image in the film that can be read out by measuring the optical density variations within the film. Since the films are very thin they do not perturb the beam. However, the photosensitive silver halide grains in the film are rapidly saturated with a low number of X-ray photons i.e. the dynamic range of the blackening process is very limited. Nevertheless, due to the thin film thickness which limits scattering effects, the spatial resolution of this method is still one of the highest up to date. A dark room processing facility is required to work with radiographic film and they also need proper calibration.

3. Radiochromic film is based on polymer films that contain dye molecules that polymerize upon exposure to ionizing radiation. For example, GafChromic® film yields a blue coloration due to the radiation induced polymerization to polyacetylene. The film can be read out by a standard scanner.

4. Thermoluminescence is based on thermally excited phosphorescence of a range of systems. Sensitive materials include $CaSO_4$:Dy, $CaF_2$:Dy and LiF:Mg,Ti. The most commonly used thermoluminescence (TL) dosimeters in radiotherapy include the tissue-equivalent LiF:Mg,Cu,P and LiF:Mg,Ti. TL dosimeters can come in all forms and sizes and allow high spatial resolution e.g. they are suitable for point dose measurements. However, TLD requires relatively complex reader units and hence instant readout is not possible. Moreover, the information is destroyed upon readout and can, in general, easily be lost.

5. Silicon diode and MOSFET (metal-oxide-semiconductor-field-effect-transistor) based dosimetry offer the advantages of small size, instant readout and high sensitivity but require that cables are connected. Also, their sensitivity changes with accumulated dose i.e. results are not very reproducible.

6. Optically stimulated (or photostimulated) luminescence (OSL or PSL) systems are related to TLD but require higher stimulation energy. In a typical OSL (PSL) material, such as $BaFBr/Eu^{2+}$, red light stimulates the recombination of electrons and holes that were created by exposure to ionizing radiation, yielding broad blue luminescence light. OSL materials offer high spatial resolution and sensitivity but are susceptible to ambient light and therefore data can be inadvertently lost. An OSL material that has been used in vivo fibre based dosimetry is the $Al_2O_3$:C system by Landauer. In this case, the blue emission is stimulated by green light (e.g. Nd:YAG laser at 532 nm). However, the size of the crystal required for this system is still relatively large ($2 \times 0.5 \times 0.5$ $mm^3$) and the material loses its information under ambient lighting conditions.

7. Electronic Portal Imaging Devices (EPIDs), based on amorphous silicon, are used for real time acquisition of megavolt images during patient treatment and can be, to some

extent, used to undertake dosimetry. Disadvantages of these devices include the fact that they are not linear with dose and dose rate and they are also relatively bulky and need to be connected with cables etc.

Storage phosphors can also be used as two-dimensional dosimeters, i.e. some spatial information about radiation exposure can be derived. This, of course, is based on computed radiography which is another important application of storage phosphors. The basic principle of computed radiography is illustrated in Figure 1.

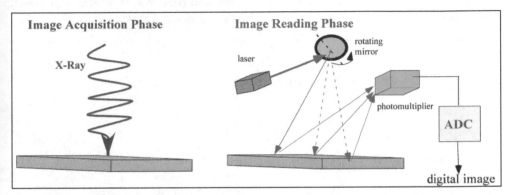

Fig. 1. Principle of computed radiography. An object on top of an imaging plate is exposed to X-rays. The storage phosphor renders a latent image of the object via the distribution of absorbed dose. The absorbed dose is measured in the reading phase and the signal is digitized. The readout for the conventional storage phosphor is undertaken by the so-called "flying-spot" method.

Since its introduction in the early 1990s, the method of computed radiography has gained significant momentum since it allows a reduction of the dose to as low as 18% in comparison with screen-film technology. In conventional computed radiography (CR), the latent image on an imaging plate (comprising of an X-ray storage phosphor) formed by exposure to ionizing radiation, is read out by photostimulated (optically stimulated) emission by using the so-called "flying-spot" method. In this method a focused red helium-neon laser beam is scanned across the imaging plate and the resulting photostimulated emission in the blue-green region of the visible spectrum as measured by a photomultiplier is converted into a digital signal pixel-by-pixel. In contrast to the screen-film method, computed radiography phosphors enable a dynamic range of up to eight orders of magnitude. However, the spatial resolution of imaging plates in CR is not yet on a par with screen-film technology due to the relatively large crystallites/grain size required in the commercially used photostimulable X-ray storage phosphor, $BaFBr(I):Eu^{2+}$ (Nakano et al., 2002; Paul et al., 2002).

In this chapter, an overview of some optical storage phosphors and materials is given and storage mechanisms and applications are briefly discussed with emphasis on a novel class of photoluminescent storage phosphors.

## 2. Optical storage phosphors and materials

We define an optical storage phosphor or material to be a system that undergoes some electronic or structural change that allows an optical readout of radiation dose. This definition encompasses thermoluminescence where glow curves are measured upon heating the thermoluminescent material that was exposed to ionizing radiation, photostimulated (optically stimulated) luminescence where the stored energy is released upon stimulation by light, radiochromic film where chemical species are created that can be read out by absorption spectroscopy, or photoluminescent or radio-photoluminescent phosphors where the radiation induced centres are metastable or stable and can be repetitively read out by photoexciation. Silver-based emulsions are also optical storage media in the broader sense, as the darkening of the film can be read out quantitatively by the change of optical density. However, such films are not discussed here in detail. Our definition of what constitutes an optical storage phosphor or material is schematically depicted in Figure 2.

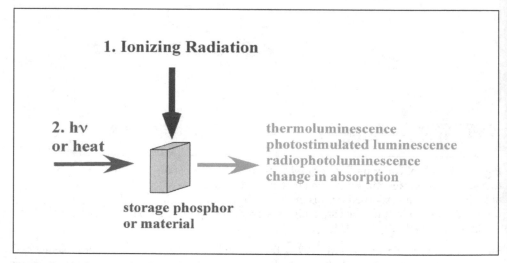

Fig. 2. General principle of an optical storage phosphor. Ionizing radiation leads to electronic and/or structural modifications in a material and the associated changes can be measured by luminescence or absorption spectroscopy.

The general mechanism of materials that exhibit thermoluminescence or photostimulated luminescence is schematically illustrated in Figure 3. Upon exposure to ionizing radiation, electron-hole pairs are created and the electrons and holes are subsequently trapped at defect sites within the crystal lattice or self-trapped such as by the creation of the $V_k$-centre. For example, electrons can be trapped at anion-vacancies such as F, Cl or I vacancies in halide crystals, forming the so-called F-centres (F stands for Farbe which is the German word for colour) which display oscillator strengths of close to unity. Holes can be self-trapped by the well-known $V_k$ centres, e.g. a hole can be shared between two adjacent halide ions, leading to significant lattice distortion, which in turn may immobilise the hole to a certain extent. In thermoluminescent materials the electrons are liberated by heating the material since the electron traps are relatively shallow and the conduction band is thermally accessible. In photostimulable (optically stimulable) materials the electron trap is deeper and higher energies are required to liberate the electron. Upon the liberation of the electron it can combine with the hole leading to emission of light, either directly through the recombination process or by transferring excitation energy to an activator centre such as $Eu^{2+}$ or the like.

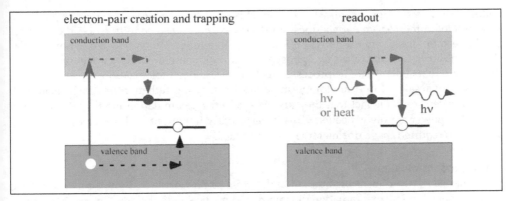

Fig. 3. Schematic diagram of the main mechanism of materials that display thermoluminescence or optically stimulated luminescence after exposure to ionizing radiation. Electrons and holes created by ionizing radiation are trapped out. Subsequently, heat or light can liberate electrons back into the conduction band and after recombination with the hole light is either emitted directly or by the transfer of the exciton energy to an activator such as a transition metal ion or a rare earth ion.

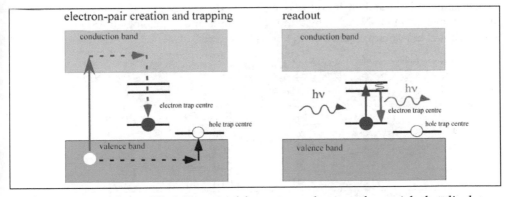

Fig. 4. Schematic and simplified diagram of the main mechanism of materials that display radiophotoluminescence after exposure to ionizing radiation. Electrons are localized in deep traps, e.g. transition metal ions, F-centres, rare earth ions. The electron trap centres can then be repetitively photoexcited from the electronic ground state to an electronically excited state. The electronically excited state is then deactivated by the emission of light and the centre induced by ionizing radiation is conserved.

In materials that display photoluminescence or radiophotoluminescence, electrons promoted to the conduction band by ionizing radiation migrate to deep traps. This is illustrated in Figure 4. These deep traps may be rare earth or transition metal ions doped as impurities into a wide bandgap host. The reduced trap can then be photoexcited without that the electron is promoted back into the conduction band. Naturally, if high light intensities are used, the electron may be promoted back into the conduction band by a two-step photoionization process. It is also possible that the excited state is close to the conduction band and hence some electrons will be liberated thermally from these excited states. However, in contrast to photostimulable materials, the signal in radiophotoluminescent materials does not decrease under ideal conditions i.e. the material can be read out for a very extended period of time to render very high signal to noise ratios. We define a photoexcitable storage phosphor as a radiophotoluminescent material that can be restored to its initial state by processes such as two-step photoionization.

A major advantage of optical storage materials for ionizing radiation in comparison with electronic devices is the fact that they are always in the accumulation mode and no battery or electric power supply is needed. This is particularly important in dosimetry applications where it is required that a dosimeter is on at all times.

## 2.1 Radiochromic film

Since the 1980s, two-dimensional radiation detectors employing radiochromic films have been widely applied in ionizing radiation dosimetry and medical radiotherapy (Soares, 2007; 2006). Radiochromic films consist of a thin polyester base impregnated with radiation-sensitive organic microcrystal monomers. Upon ionizing radiation, the film emulsions undergo a colour change through a chemical reaction (usually polymerization) and the radiation response signal can be read out by measuring the absorbance or optical density using spectrometers at certain wavelengths. Currently, the most popular radiochromic film in medical applications are the polydiacetylene based GafChromic® films designed by International Specialty Products (ISP). GafChromic® film was first introduced by Lewis (Lewis, 1986) and further developed by McLaughlin et al. at the National Institute of Standards and Technology (NIST) in the United States. Table 1 lists some current commercial GafChromic® films.

| Application | Film type | Active Layer Thickness (μm) | Dose Range (Gy) | Spatial Resolution (dpi) | Energy Range |
|---|---|---|---|---|---|
| Radiotherapy | HD-810 | 6.7 | 10 ~ 2500 | 10,000 | MV |
| | MD-55-2 | 67 | 10 ~ 100 | 5,000 | MV |
| | RTQA2 | 17 | 0.02 ~ 8 | 5,000 | MV |
| | EBT-2 | 30 | 0.01 ~ 40 | 5,000 | kV-MV |
| | EBT-3 | 30 | 0.01 ~ 40 | 5,000 | kV-MV |
| Radiology | XR-QA2 | 25 | 0.001 ~ 0.2 | 5,000 | kV |
| | XR-CT2 | 25 | 0.001 ~ 0.2 | 5,000 | kV |
| | XR-M2 | 25 | 0.001 ~ 0.2 | 5,000 | kV |
| | XR-RV3 | 17 | 0.01 ~ 30 | 5,000 | kV |

Table 1. List of commercial GafChromic® films.

With different film implementations, the GafChromic® film exhibits a variety of dose responses in the energy range from keV to MeV, targeting different areas of medical radiotherapy. For example, with only a 6.7 μm thick active layer, the transparent GafChromic® HD-810 film is predominantly employed in high dose radiation measurements with doses up to 2500 Gy. In contrast, by adding a yellow dye into the sensitive layer, the latest GafChromic® EBT film is designed mainly for radiotherapy applications. Many studies have been devoted to the characterization and performance assessment of the GafChromic® films including the dose response functions (sensitivity), film stability and image resolution (Arjomandy et al., 2010;

Rink et al., 2008; Secerov et al., 2011). In addition to GafChromic® films, other commercial radiochromic films include FWT-60 from Far West Technology Inc., the B3 dosimetry lines manufactured by the Gex Corporation and the SIRAD (Self-indicating Instant Radiation Alert Dosimeter) developed by Gordhan Patel at JP Laboratories (Mclaughlin et al., 1991). Compared with conventional silver halide based photographic films, the radiochromic film offer many advantages in medical dosimetry and radiography, such as outstanding spatial resolution (>1200 lines/mm for GafChromic® film (Secerov et al., 2011), accurate and precise dose measurements, weak energy dependence and relatively easy handling.

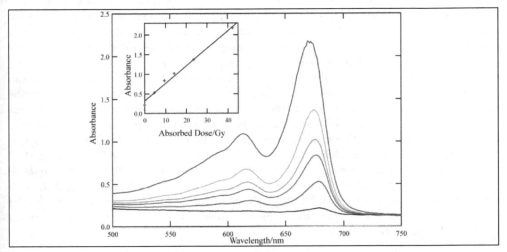

Fig. 5. Absorption spectra of GafChromic® film HD-810 as a function of high X-ray dose (40 kVp). The inset shows a plot of the absorbance at 676 nm of the film as a function of the absorbed dose. The absorption spectra were measured on a Cary-50 UV-Vis absorption spectrometer.

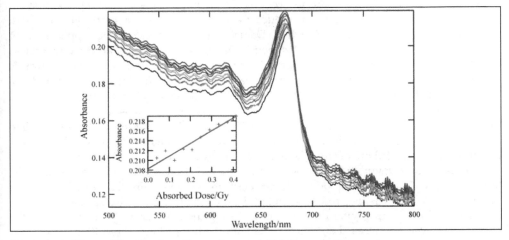

Fig. 6. As in Figure 5 but for low X-ray dose and 60 kVp radiation.

Figures 5 and 6 illustrate the X-ray dose dependence of the absorption spectra of GafChromic® film HD-810 at high and low doses, respectively. The optical absorbance at 676 nm of the film increases linearly in the dose range of 0.04 Gy to 2 Gy of X- irradiation. Weak energy dependence is another desirable property for a two-dimensional radiation detector in order to obtain accurate values of absorbed doses.

Due to the high atomic number components in silver–based film emulsions, the conventional photographic films suffer from a strong photon energy dependence of the response to dose. This has been significantly improved by radiochromic film dosimetry. Buston et al. (Butson et al., 2010) examined the energy dependence of the GafChromic® EBT2 film dose response in the X-ray range of 50 kVp to 10 MVp and a 6.5% ± 1% variation in optical density to absorbed dose response was obtained. The near energy-independence of radiochromic film has made it an ideal technology in some medical applications where a broad energy range of radiation sources is delivered. Furthermore, unlike the photographic films, the radiochromic films do not require subsequent chemical processing and can produce repetitive responses over a relatively long period of time. Despite the numerous advantages, special attention has to be paid to the environmental factors such as temperature, humidity and UV exposure in the irradiation and readout processes of the radiochromic films. Abdel-Fattah and Miller (Abdel Fattah & Miller, 1996) have investigated the effects of temperature during irradiation on the dose response of the FWT-60-00 and B3 radiochromic film dosimeters. At temperatures between 20 and 50 °C and a relative humidity between 20 and 53%, the temperature dependences of FWT-60-00 and Risø B3 dosimeters were found to be 0.25 ± 0.1% per °C and 0.5 ± 0.1% per ° C, respectively. For the GafChromic® films, it was suggested to be best stored in the dark at the temperature of ~22°C (Mclaughlin et al., 1991).

## 2.2 Thermoluminescent materials

Thermoluminescence (TL) is the emission of light produced by heating of a material. The phenomenon of thermoluminescence has been known for a long time and has been reported by Robert Boyle as early as 1663 when he observed thermoluminescence from diamond (Horowitz, 1984). At the end of the 19th century Borgman (Borgman, 1897) reported that X-rays and radioactive substances can induce thermoluminescence in particular materials. However, it was only in 1953 when Daniels et al. suggested the application of thermoluminescence in the field of dosimetry (Daniels et al., 1953). The early work was mostly involving the tissue equivalent LiF but also $Al_2O_3$. From the 1960s onwards thermoluminescence was regarded to be a viable alternative to film dosimetry and LiF became the TLD-100 standard after Harshaw purified the material and refined the crystal growth to make the properties of the material more reproducible. LiF is still in use today but the main key is to select a material for a specific monitoring application. Up to this date dosimetry based on thermoluminescence is still the main mode of personal radiation monitoring, with a plethora of suitable materials for a wide range of applications.

The mechanism of thermoluminescence induced by radiation is generally described in the following way. Upon ionizing radiation, free electron and hole pairs are created due to the photoelectric effect. The electrons and holes are then captured at different defect centres (electron and hole traps) within the material. During the subsequent heating process, the trapped electrons are released and recombine with the trapped holes, resulting in emission of light with the intensity proportional to the absorbed ionizing radiation dose (see Figure 4 above). Compared with other radiation dose measurement techniques, thermoluminescence dosimetry offers many advantages, such as a relatively small dosimeter size, high sensitivity over a wide range of surface dose levels from μGy to several Gy, long-term information storage and versatile applications in different types of radiation including X-rays, γ and β rays and neutrons.

There are a large number of materials exhibiting thermoluminescence, mostly insulators or semiconductors containing lattice defects resulted from substitutional impurities or ionizing irradiation. However, not all the materials with thermoluminescence properties are suitable for radiation dosimetry. There are several requirements for the practical use of thermoluminescence materials in radiation measurements (Kortov, 2007). For example, it is desirable that the material has a simple glow curve with the main peak occurring at around 200 °C. If thermoluminescence is also induced at lower temperatures, the thermal energy at room temperature may be high enough to liberate trapped electrons and fading can occur. On the other hand, when the glow curve is complex on the high-temperature side, the stored energy remaining after the radiation measurement has been performed, has to be dissipated (Yen et al., 2007). Figure 7 shows the glow curve of LiF:Mg, Cu, P (Bos, 2001; McKeever et al., 1993) after exposed to 0.4 Gy $^{60}$Co · -radiation. From this figure it follows that dosimeters based on this material should not be subject to temperatures above 80 °C, otherwise fading starts to occur. In addition to a simple glow curve, a linear dose response, high luminescence efficiency, low photon energy dependence and long term stability are also desirable for the application of thermoluminescent materials in radiation dosimetry (Azorin et al., 1993).

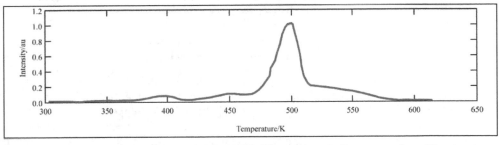

Fig. 7. Glow curve of LiF:Mg, Cu, P (TLD-100H). Thermal annealing procedure: 10 min at 240 °C with a heating rate of 2.1 °C /s followed by cooling in air. The absorbed dose was 0.4 Gy from a $^{60}$Co source. The main dosimetric peak occurs at ~200 °C.

Table 2 lists some typical commercial thermoluminescence dosimeters with their applications and useful dose ranges. These commercial dosimeters generally exhibit emission at 380 - 480 nm, which falls off in the spectral response range of common photomultipliers. The thermoluminescence peak is located from 180 °C to 260 °C, offering an easy readout process. With an effective atomic number ($Z_{eff}$) close to that of soft biological tissue, LiF-based dosimeters are the most widely used thermoluminescence phosphors in personal radiation monitoring due to its very low photon energy dependence (Azorin et al., 1993). However, the glow curves of LiF-based dosimeters consist of at least seven peaks (Satinger et al., 1999). In contrast, $CaF_2$:Mn and $Al_2O_3$:C dosimeters offer simpler glow curves and higher sensitivity (Kortov, 2007). Nakajima et al. (Nakajima et al., 1978) first introduced the LiF:Mg, Cu, P (TLD-100H) which shows a 15 times greater sensitivity than that of the LiF:Mg, Cu (TLD-100). Furthermore, to improve the sensitivity of the dosimeters to thermal neutrons, $^6$Li and $^7$Li isotope based LiF dosimeters were designed to measure thermal neutrons in a mixed radiation field (Azorin et al., 1993).

| Material | Dosimeter | $Z_{eff}$ | Application | Useful Range |
|---|---|---|---|---|
| LiF:Mg, Ti | TLD-100 | 8.2 | medical physics | 10 µGy - 10 Gy |
| CaF$_2$:Dy | TLD-200 | 16.3 | Environmental | 0.1 µGy - 10 Gy |
| CaF$_2$:Tm | TLD-300 | 16.3 | Low dose gamma rays | 0.1 µGy - 10 Gy |
| CaF$_2$:Mn | TLD-400 | 16.3 | Environmental | 0.1 µGy - 100 Gy |
| Al$_2$O$_3$:C | TLD-500 | 10.2 | Personnel | 0.05µGy - 10 Gy |
| $^6$LiF:Mg, Ti | TLD-600 | 8.2 | Neutron and gamma rays | 10 µGy - 10 Gy |
| $^7$LiF:Mg, Ti | TLD-700 | 8.2 | Gamma rays and thermal neutron | 10 µGy - 10 Gy |
| LiB$_4$O$_7$:Mn | TLD-800 | 7.4 | High dose and neutron | 450 · Gy-10$^4$ Gy |
| CaSO$_4$:Dy | TLD-900 | 15.5 | Environmental | 1 µGy – 10$^3$ Gy |
| LiF:Mg, Cu, P | TLD-100H | 8.2 | Personnel | 1 µGy - 20 Gy |
| $^6$LiF:Mg, Cu, P | TLD-600H | 8.2 | Personnel | 1 µGy - 20 Gy |
| $^7$LiF:Mg, Cu, P | TLD-700 H | 7.4 | Personnel | 1 µGy - 20 Gy |

Table 2. Examples of commercial thermoluminescence dosimeters.

One of the better performers is CaSO$_4$:Dy which is approximately 10 times more sensitive than LiF:Mg,Ti (TLD-100). This material offers a simple processing cycle and is typically suspended in a Teflon disc. Annealing of only 2 hours at a temperature of ~280ºC is required to erase the residual radiation induced thermoluminescence before the material can be reused. In contrast, LiF:Mg,Ti requires annealing of 5 hours at 300ºC, followed by 20 hours at 80ºC. However, CaF$_2$ and CaSO$_4$ based thermoluminescence materials are exceptionally light (UV) sensitive, and fading is enhanced significantly. These materials have to be handled, used and stored in opaque containers to avoid belaching of the thermoluminescence signal. Moreover, the reading accuracy depends on the heating rate and multiple readings are not possible as the stored energy is lost upon heating. Importantly, TL dosimeters cannot work in test-accumulation mode (multiple irradiation and reading). Also, the irradiation memory is dependent on the operation temperature. The precision on reuse is also lowered (±2% above 1 mSv and higher below this value) and the long term response retention is restricted (the best materials show 5% loss at room temperature for one year). A major drawback of thermoluminescence based dosimetry systems is the requirement of relatively complex reading instrumentation including a well-controlled flowing gas thermoelectric heating system (±1°C). Current thermoluminescence dosimeter manufacturers include the Solid Dosimetric and Detector Laboratory in China, Nemoto in Japan, the Institute of Nuclear Physics in Poland and Saint-Gobain Crystals and Detectors in the USA.

## 2.3 Photostimulable (optically stimulable) storage phosphors

As is illustrated in Figure 3, the mechanism of materials that show thermoluminescence and photostimulated (or optically stimulated) luminescence (PSL or OSL) is, to a great extent, the same but OSL (PSL) materials provide significantly deeper traps for the electrons and/or holes created upon exposure to ionizing radiation and hence thermal energies are not high enough to liberate the trapped electrons. Thus, in contrast to thermoluminescence, optically stimulated luminescence is an all-optical technique that does not require heating of the storage material. This allows much simpler and compact designs of reader units. The application of optically stimulable phosphors in dosimetry was first suggested in the 1950s. It was then initially mostly used for archaeological and geological dating, e.g the dating of quartz until the development of sensitive anion-deficient $Al_2O_3$:C. Photostimulable materials also became very important with the implementation of BaFBr:$Eu^{2+}$ in computed radiography.

### 2.3.1 $Al_2O_3$:C

The $Al_2O_3$:C system has been extensively covered in a book by Bøtter-Jensen et al. (Botter-Jensen et al. 2003) and an elegant overview of the main features of this material is also given in a recent article by Akselrod (Akselrod, 2010). Anion-deficient $\alpha$-$Al_2O_3$:C was firstly developed as a thermoluminescence material (Akselrod et al., 1990) but it was soon realized that it can serve as an OSL phosphor with excellent properties. In terms of its thermoluminescence properties, it is about 40-60 times more sensitive than LiF based TLD-100. However, the sensitivity is strongly dependent on the heating rate, complicating routine dosimetry measurements. In contrast, $Al_2O_3$:C appears to be an ideal material for optically stimulated luminescence. Neutral and charged oxygen vacancies (F, $F^+$) play an important part as luminescence centres. In particular, the presence of $F^+$ centres is important as they are the recombination centres for electrons, yielding excited F-centres. $Al_2O_3$:C displays a linearity of light output as a function of radiation dose over seven orders of magnitude. The long luminescence lifetime of 35 ms of the emitting F-centres allows simple time-dependent readout schemes and this has been implemented in so-called pulsed optically stimulated luminescence measurements (Akselrod et al., 1998) where the emitted light is only measured between the stimulating light pulses in order to reduce the requirement for heavy optical filtration. The system is now fully commercialized by Landauer with a suite of dosimetry products that allow static and dynamic monitoring of ionizing radiation. Landauer also offers a portable reader unit. A drawback of the $Al_2O_3$:C material is the fact that it has to be kept in light-tight containers, otherwise the stored energy is rapidly erased by visible light.

### 2.3.2 BaFBr(I):$Eu^{2+}$

Excellent and detailed reviews of this material have been published by von Seggern (von Seggern, 1999) and Schweizer (Schweizer, 2001). BaFBr(I):$Eu^{2+}$ is the most widely used storage phosphor in computed radiography, with a number of major companies that offer imaging plates and associated readers for medical imaging.

Current storage phosphors contain centres for the capture of X-ray generated electrons and holes. For the BaBrF:$Eu^{2+}$ storage phosphor both the $F^+(Br^-)$ and the $F^+(F^-)$ defects can act as electron storage centres whereas the $Eu^{2+}$ acts as the hole trap. Upon X-ray irradiation of the X-ray storage phosphor BaBrF:$Eu^{2+}$ electron-hole pairs are created. The electrons and holes are trapped at anion vacancies and the $Eu^{2+}$, respectively, creating F-centres and $Eu^{2+}$/$V_K$-centres (a $V_K$-centre is a hole shared between two neighbouring halide anions). However, some electron-hole pairs recombine spontaneously after their creation without being trapped and lead to spontaneous emission (scintillation). Upon excitation of the F-centres at 2.1 eV and 2.5 eV for the F(Br) and F(F-) centres, respectively, the electrons recombine with the holes to form excitons. When the excitons are deactivated its excitation energy is resonantly transferred to the activator, $Eu^{2+}$, which in turn, leads to the broad $4f^65d \rightarrow 4f^7$ emission at about 390 nm. It was believed initially (Takahashi et al., 1984) that the hole combines with $Eu^{2+}$ to form $Eu^{3+}$. However, there is evidence that this is not the case as follows, for example, from the fact that the $Eu^{2+}$ photoluminescence does not decrease upon X-irradiation (Hangleiter et al., 1990; Koschnick et al., 1991). The exact mechanism of the electron–hole recombination is still controversially discussed; some experiments are at variance with a recombination of the electron-hole pairs via the conduction band. An alternative mechanism proposes a spatial correlation of the electron and hole traps and it is believed that recombination occurs via tunnelling of the electrons.

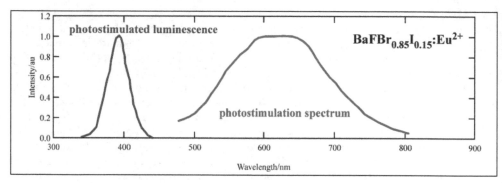

Fig. 8. Photostimulated luminescence and photostimulation spectra of $BaFBr_{0.85}I_{0.15}:Eu^{2+}$.

### 2.3.3 CsBr:Eu$^{2+}$

$CsBr:Eu^{2+}$ displays significant photostimulated luminescence after X-ray exposure (Leblans et al., 2001; Schweizer et al., 2000) and its figure-of-merit as a storage phosphor is as high as for $BaFBr(I):Eu^{2+}$ (Hackenschmied et al., 2002; Schweizer, 2001). In particular, this material can be grown in the form of needle crystals by vacuum deposition (Weidner et al., 2007), allowing an improved lateral resolution in comparison with $BaFBr(I):Eu^{2+}$. The sensitivity of this material appears to be strongly dependent on thermal treatment (Hackenschmied et al., 2003) and hydration (Appleby et al., 2009). In $CsBr:Eu^{2+}$ the holes are trapped as $V_K$-centres in the vicinity of $Eu^{2+}$, and the electrons are trapped in F-centres and hence the spectroscopy of this system is very much related to the $BaFBr:Eu^{2+}$ spectra with minor dielectric shifts of the spectra. However, poor radiation hardness appears to be a major problem with this material. In particular, the photostimulated luminescence deteriorates rapidly upon exposure to high X-ray dose and this deterioration is correlated with a reduction of the photoexcited $Eu^{2+}$ fluorescence that has been attributed to an agglomeration of $Eu^{2+}$ leading to luminescence quenching. (Zimmermann et al., 2005). This conclusion has been reached since no increase in $Eu^{3+}$ luminescence is observed.

### 2.3.4 BeO

With near tissue equivalence ($Z_{eff} = 7.1$), relatively high sensitivity and low cost, beryllium oxide (BeO) has been well-documented as a thermoluminescence dosimetry (TLD) material with a range of applications particularly in medical fields (McKeever, 1988; McKeever et al., 1995; Vij & Singh, 1997). One of the most popular forms of BeO-based TLD is Thermalox® 995. Modifications of the material can be found as BeO:Li (Yamashit et al., 1974), BeO:Na (Yamashit et al., 1974) and BeO:TiO$_2$ (Milman et al., 1996). Depending on the preparation methods and the type of radiation source, the glow curves of BeO-based TLD materials exhibit the main dosimetric peaks between 160 °C and 280 °C. Compared with other commercial thermoluminescence dosimetric systems, the main disadvantage of the BeO-based materials is the rapid light-stimulated thermoluminescence fading. Under laboratory fluorescent light, rapid fading of 50 % in 30 min for the thermoluminescence peak at 167 °C has been reported in several types of commercial BeO ceramic dosimeters (Crase & Gammage, 1975). Based on the characteristic light-sensitive thermoluminescence of BeO, Albrecht and Mandeville (Albrecht & Mandeville, 1956) first investigated optically stimulated luminescence of the X-ray irradiated BeO by using visible photons. Bulur and Göksu (Bulur & Göksu, 1998) examined the detailed dosimetric properties of Thermalox® 995 as optically stimulated luminescence dosimeter, such as the OSL signal, the stimulation spectrum, the dose dependence, multireadability and short-term fading. Upon optical stimulation in a broad band region from 420 nm to 550 nm with maximum at 435 nm (Bulur & Göksu, 1998), the stimulated luminescence of BeO is generally assumed to be in the same region as that of the thermoluminescence, i.e. UV region with main peak at around 335 nm (McKeever et al., 1995). The exact origin of the optically stimulated luminescence of BeO is not clear yet, although a possible link to the thermoluminescence peaks at 220 °C and 340 °C has been proposed (Bulur & Göksu, 1998). The linear dose response of the optically stimulated luminescence of BeO was reported from 1 µGy up to a few Gy, covering more than six orders of magnitude (Sommer et al., 2008). In practical applications of BeO in radiation dosimetry, the stimulation is usually realized by employing blue light-emitting diodes and the detection of the light can be easily achieved with a photomultiplier. Optical filters are usually used in front of the photomultiplier to discriminate the stimulation light from photostimulated luminescence. The main problem associated with BeO materials is the relatively low reproducibility but in particular the very high toxicity.

## 2.4 Photoluminescent storage phosphors

Radiophotoluminescent materials and photoluminescent storage phosphors are based on an alteration of the luminescence spectrum of a material upon exposure to ionizing radiation. The phenomenon was reported as early as 1912 by Goldstein (Goldstein, 1912) but it was only in the 1950s that it was realized that the effect could be used for dosimetry (Schulman et al., 1953; Schulman, 1950; Schulman & Etzel, 1953; Schulman et al., 1951). Schulman's system comprised of a silver-activated aluminophosphate glass. Upon exposure to ionizing radiation, electrons are promoted to the conduction band and migrate to the silver activation sites in this system, yielding a reduction of the silver atoms. This reduction results in orange luminescence that can be excited at 365 nm. The main difference to photostimulable materials is the fact that in photoluminescent storage phosphors the radiation-induced change persists upon photoexcitation. For example, if stable colour centres are induced, they can be repetitively read out by photoexcitation whereas in optically stimulable materials *one* radiation induced centre, e.g. electron-hole pair, is annihilated upon the absorption of *one* photon. A study from 1984 suggests the application of rare earth ion co-doped CaSO$_4$ to produce a radiophotoluminescent material at room temperature that also exhibits thermoluminescence i.e. a combined TL/RPL dosimetry system (Calvert & Danby, 1984). In particular, $Eu^{2+}$ ions if illuminated by ultraviolet radiation display radiophotoluminescence after exposure to ionizing radiation. However, the low sensitivity turned out to be a very limiting factor for this material.

### 2.4.1 Nanocrystalline BaFCl:Sm$^{3+}$

Nanocrystalline BaFCl:Sm$^{3+}$, as prepared by a facile co-precipitation method, is a very efficient photoluminescent storage phosphor for ionizing radiation (Liu et al., 2010; Riesen & Kaczmarek, 2007). The mechanism is based on the reduction of $Sm^{3+}$ to $Sm^{2+}$ by trapping electrons that are created upon exposure to ionizing radiation in the BaFCl host; subsequently, the $Sm^{2+}$ can be efficiently read out by measuring the intraconfigurational $f$-$f$ transitions, such as the $^5D_0$-$^7F_0$ line at around 688 nm, by excitation into the parity allowed $4f^6 \rightarrow 4f^5 5d$ transition at around 415 nm. The latter wavelength is ideal for efficient excitation by blue-violet laser diodes or LEDs as the transition is electric dipole allowed and thus very intense.

The photoluminescence and excitation spectra of nanocrystalline BaFCl:Sm$^{3+}$ after exposure to 50 mGy of 60 kVp X-ray are illustrated in Figure 9. The spectra are dominated by the $Sm^{2+}$ species that is created by the ionizing radiation. The efficiency of this class of storage phosphors can be optimized by the preparation of core-shell nanoparticles comprising of a BaFCl core and a BaFCl:Sm$^{3+}$ shell. The $Sm^{3+} \rightarrow Sm^{2+}$ conversion efficiency of these nanoparticles is drastically higher than that of microcrystals obtained by conventional high temperature (HT) sintering. This is illustrated in Figure 10 where the 415 nm excited luminescence spectra of HT and nanocrystalline BaFCl:Sm$^{3+}$ are compared.

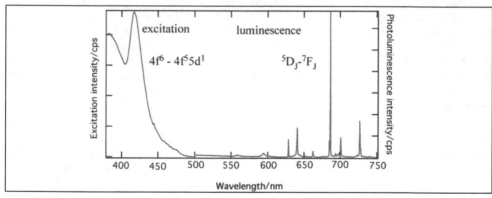

Fig. 9. Photoluminescence and excitation spectra of nanocrystalline BaFCl:Sm$^{3+}$ after exposure to 50 mGy of 60 kVp X-ray radiation. All the pronounced transitions are due to $Sm^{2+}$ which are absent before irradiation. The excitation spectrum is not corrected for the effect of a BG-18 glass filter used for the excitation light.

The HT sample was exposed to a 240,000 times higher radiation dose (12000 Gy) than the nanocrystalline material (50 mGy). It is noted here that there are significant differences in the $^4G_J \rightarrow {}^6H_J$ emission lines for the two materials, indicating a significantly different local environment for the $Sm^{3+}$ ions in the nanocrystalline material. It is also clear that both materials provide a range of sites for the $Sm^{3+}$ ions and in the case of the HT material it is straightforward to achieve high site selectivity when changing the excitation wavelength in the vicinity of 400 nm. Nevertheless, after the reduction to $Sm^{2+}$ the resulting $^5D_J \rightarrow {}^7F_J$ emission lines perfectly coincide for both materials.

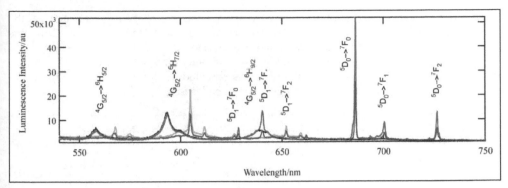

Fig. 10. Photoluminescence of high temperature (green and blue lines) and nanocrystalline (black and red lines) BaFCl:Sm$^{3+}$ before and after exposure to 12000 Gy and 50 mGy, respectively. Prominent $^4G_J \rightarrow {}^6H_J$ (Sm$^{3+}$) and $^5D_J \rightarrow {}^7F_J$ (Sm$^{2+}$) transitions are denoted. Note that the unexposed samples (green and black lines for HT and nano sample, respectively) do not exhibit any emission lines due to Sm$^{2+}$.

The much higher conversion efficiency (~500,000 times) of the nanocrystalline material manifests itself also in cathodoluminescence where the Sm$^{3+}$ and Sm$^{2+}$ peaks dominate in the spectra of the HT sintered and nanocrystalline samples, respectively (Stevens-Kalceff et al., 2010). The nanocrystalline material has been characterized with a range of techniques such as electron microscopy and high resolution synchrotron powder X-ray diffraction. The diffraction lines in the powder X-ray diffraction pattern can be indexed to the tetragonal matlockite-type structure with space group P4/nmm. The resulting lattice parameters ($a = b$ = 4.395(1) Å, and $c$ = 7.227(8) Å) indicate a small expansion of the crystal lattice in comparison with the literature values (Beck, 1976). The average volume-weighted column length of the crystallites is estimated to be 160 nm by applying the Williamson-Hall procedure (Williamson & Hall, 1953). This average size is consistent with electron microscopy micrographs and renders the nanocrystals for applications where extremely small probes for ionizing radiation are required. For example, if bonded or embedded to/in the end of an optical fibre the material can be used for remote sensing of CW or pulsed ionizing radiation; examples of such applications include real time in vivo monitoring of dose in radiotherapy and in food irradiation and the monitoring of hot labs and/or nuclear reactor environments. In such applications a selected Sm$^{2+}$ emission line, e.g. $^5D_0 \rightarrow {}^7F_0$, is monitored with a narrow bandpass filter (e.g. 1 nm FWHM) with gated detection 180$^0$ out of phase from the excitation light pulse. The latter is facilitated by the relatively long lifetime of 2 ms of the $^5D_0$ excited state and dramatically reduces the lower detection limit by elimination of excitation light, fluorescence due to other impurities and surface defects, and Raman scattered light.

The core-shell nanocrystalline storage phosphor displays a relatively linear response to ionizing radiation up to a surface dose of ca. 10 Gy in the 50 keV X-ray region as is illustrated in Figure 11. Moreover, the lower detection limits for 50 keV and 1 MeV radiation are about 100 nGy and 10 μGy, respectively. This results in an impressive dynamic range of about seven orders of magnitude for radiation detection. A range of photon-gated spectral hole-burning experiments at low temperatures has also been conducted (Liu et al., 2010). The stored energy in the phosphor material can be released by a two-step photoionization process when higher light powers are used. However, the storage phosphor can only be fully restored after exposures of up to 10-20 mGy. An example of the reproducibility of the nanocrystalline BaFCl:Sm$^{3+}$ for dosimetry applications is illustrated in Figure 12. In this example the Sm$^{2+}$ photoluminescence signal was measured as a function of accumulated dose (60 kVp X-ray irradiation) up to 1 mGy. The dosimeter material was then bleached by a two-step photoionization and the experiment was repeated. As follows from this figure, the phosphor can be fully restored in this range of absorbed doses.

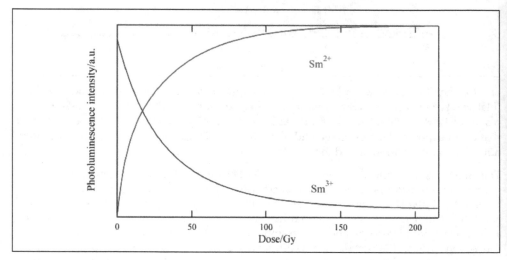

Fig. 11. Dependence of the Sm$^{2+}$ and Sm$^{3+}$ photoluminescence signals of nanocrystalline BaFCl:Sm$^{3+}$ on absorbed 60 kVp X-ray dose.

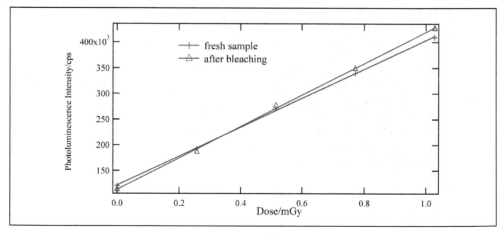

Fig. 12. Reproducibility of nanocrystalline BaFCl:Sm$^{3+}$ in dosimetry.

Figure 13 illustrates the dose dependence of the Sm$^{2+}$ photoluminescence signal upon exposure to Cs-137 γ-rays (662 keV). The phosphor's response is independent on the dose rate and fairly linear up to the maximum exposure of ca ~10 mGy.

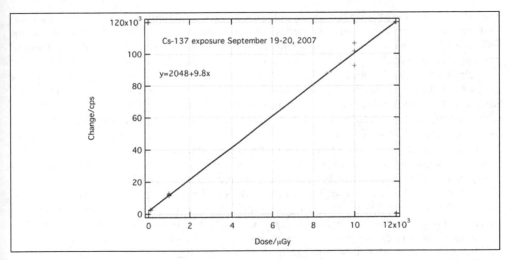

Fig. 13. Dependence of the $Sm^{2+}$ photoluminescence intensity on absorbed dose of Cs-137 γ-rays (662 keV)

As can be expected, due to the presence of heavy atoms in the $BaFCl:Sm^{3+}$ material, the phosphor's quantum efficiency is subject to a relatively strong energy dependence. This is illustrated in Figure 14. As a consequence, if used in dosimetry, filters need to be applied. However, the strong energy dependence together with a filter arrangement allows the determination of the energy of the ionizing radiation, a desirable feature.

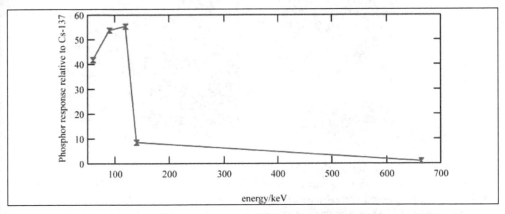

Fig. 14. Energy dependence of the conversion efficiency of nanocrystalline $BaFCl:Sm^{3+}$ relative to Cs-137 γ-rays (662 keV).

The lifetime of the excited $^5D_J$ multiplet is about 2 ms which is a severe limitation in computed radiography when the conventional "flying-spot" readout scheme is applied. However, the long lifetime allows for gated excitation and luminescence light with a phase shift of 180°. The gated readout facilitates a very sensitive detection scheme since fluorescence of impurities is blocked from reaching the light detector. A big advantage of the nanocrystalline $BaFCl:Sm^{3+}$ phosphor is the submicron particle size that increases the spatial resolution which, in the commercial BaFBr(I):Eu$^{2+}$ is limited by the large grain size that induces significant light scattering in the readout process. The nanocrystalline BaFCl:Sm$^{3+}$ also allows a higher packing density and the light scattering in the readout process can be expected to be a lesser problem. This can be tested by determining the modulation transfer function (MTF). The MTF can be determined following the procedure outlined in Figure 15 (Boone, 2001; Cunningham & Fenster, 1987; Cunningham & Reid, 1992; Samei et al., 1998; Samei et al., 2001). In particular, in the first step the edge spread function (ESF) is measured, by imaging a lead edge. The line spread function (LSF) is then calculated by differentiating the ESF and subsequently the MTF is obtained by Fast Fourier Transformation of the LSF. Thin films of nanocrystalline BaFCl:Sm$^{3+}$ were applied to a plastic substrate by mixing the phosphor with a small amount of polyvinyl acetate. A 2D reader unit that images the excited photoluminescence of the BaFCl:Sm$^{3+}$ storage plate by an electron-multiplying CCD (EMCCD) camera was employed to image the plate in parallel mode. The system relies on gated excitation-detection by pulsing the LEDs and mechanically gating the collimated luminescence light by a mechanical chopper wheel. This prevents fluorescence of impurities from reaching the detector. The 2D reader unit is illustrated in Figure 15 where the LED light sources, the film platform, the lens system, the chopper wheel and the EMCCD camera can be recognized.

Fig. 15. 2D reader for imaging plates based on nanocrystalline BaFCl:Sm$^{3+}$. The image plate is illuminated by two collimated and pulsed 405 nm mounted blue LEDs. The emitted light is collimated then passed through an aperture, that is opened with a 180º phase shift with respect to the blue LED pulses, by a mechanical light chopper wheel. The light is then refocused, filtered and detected by an EMCCD camera.

Radiographs were analyzed according to the mathematical steps shown in Figure 16. In particular, the line spread function (LSF) is obtained from the edge spread function (ESF) by differentiation and the MTF by Fast Fourier Transformation (FFT) of the LSF.

$$LSF(x) = \frac{d}{dx}ESF(x) \qquad (1)$$

$$MTF(k) = \int_{-\infty}^{\infty} LSF(x)\exp(-i2\pi kx)dx \qquad (2)$$

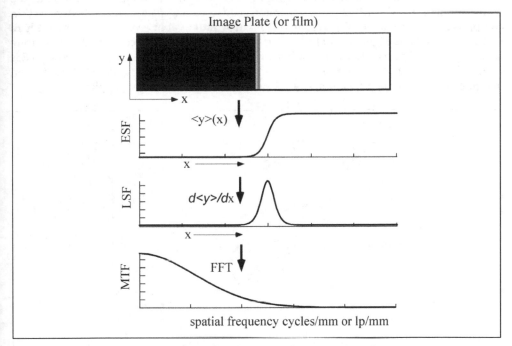

Fig. 16. Procedure to determine the MTF of an optical storage phosphor.

The resulting MTF is illustrated in Figure 17 and, for the purpose of comparison, the MTFs of the Kodak Insight dental film and Kodak GP and HR imaging plates in conjunction with Kodak Ektascan Model-400 and CR9000 computed radiography reader units are shown. The MTF points for the BaFCl:Sm$^{3+}$ imaging plates are calculated from raw data whereas for the smooth green line the ESF is based on a fit of a sigmoid to the ESF and the solid blue line is based on smoothing the ESF with the Savitzky-Golay procedure in 4th order using 37 points.

| | BaFCl:Sm$^{3+}$ imaging plate | Kodak CR9000 reader with CR-GP plate | Kodak Ektascan Model-400 reader with GP plate | Kodak Ektascan Model-400 reader with HR plate | Kodak Insight With 2D reader | Kodak Insight with plustek OpticFilm 7400 |
|---|---|---|---|---|---|---|
| **MTF=0.5** | 2.4 | 1.3 | 1.22 | 1.92 | 2.8 | 4.7 |
| **MTF=0.1** | 5.4 | 3.1 | 3.27 | 4.35 | 18 | 18 |

Table 3. Comparison of MTF values of BaFCl:Sm$^{3+}$ imaging plates with Kodak Computed radiography systems and conventional Kodak Insight dental film.

In Table 3 a comparison of MTF data is compiled, in particular the spatial frequency is summarized at MTF values of 0.5 and 0.1. It is obvious that conventional dental film exhibits by far the best values. However, the BaFCl:Sm$^{3+}$ based imaging plates are significantly better than all the commercial computed radiography systems shown in Table 3. It follows that the nanocrystalline BaFCl:Sm$^{3+}$ storage phosphor yields imaging plates with a better MTF in comparison with commercially available computed radiography systems. It is noted here that a better MTF is highly desirable as it is the main drawback of computed radiography based on currently used storage phosphors. In particular in medical imaging a good MTF is required in order to detect, for example, micro-calcifications in soft tissue. High resolution imaging is also important for a range of scientific and technical imaging applications.

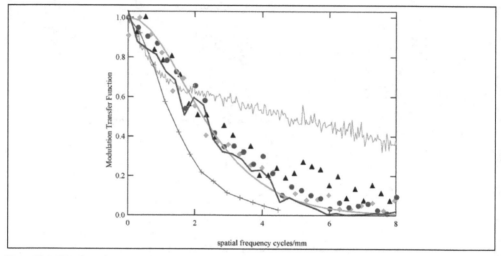

Fig. 17. MTF data for BaFCl:Sm$^{3+}$ based imaging plate (blue solid circles, black solid triangles, green diamonds and green trace). The smooth green line is based on a fit of the ESF to a sigmoid. The blue line is the result of a Savitzky-Golay smoothing procedure applied to the ESF. The red trace with markers is the MTF of the Kodak CR9000 computed radiography system. The beige trace is the MTF of Kodak Insight dental film as read out by the 2D reader unit illustrated in Figure 15.

## 3. Conclusion

Over the last 60 years an impressive range of optical storage phosphors and materials has been developed and investigated for a plethora of applications, including personal radiation monitoring, quality control in radiation therapy and food irradiation, and for applications in computed radiography both for medical and technical applications. Although full electronic devices, both for point and 2D detection, are becoming increasingly more sensitive and inexpensive, many applications for optical storage phosphors prevail, possible for many years to come, as electronic solid state devices are not flexible and hence for many applications not practical. Importantly, storage phosphor based dosimeters are always accumulating and independent of any power supply. Optical storage phosphors have a significant potential to facilitate much lower radiation doses in medical imaging, a highly desirable outcome.

## 4. Acknowledgment

The Australian Research Council is acknowledged for financial support of our research into photoluminescent X-ray storage phosphors (ARC Discovery Project DP0770415, ARC Linkage Project LP110100451). We also like to thank Dosimetry & Imaging Pty Ltd, and especially A. Ujhazy, for supporting this project. The Australian Synchrotron is acknowledged for time on the powder X-ray diffraction and X-ray absorption beamlines.

## 5. References

AbdelFattah, A.A. & Miller, A. (1996). Temperature, Humidity and Time. Combined Effects on Radiochromic Film Dosimeters. *Radiation Physics and Chemistry*, Vol. 47, No. 4, 611-621.

Akselrod, M.S., Kortov, V.S., Kravetsky, D.J. & Gotlib, V.I. (1990). Highly Sensitive Thermoluminescent Anion-Defective $\alpha$-$Al_2O_3$:C Single Crystal Detectors. *Radiation Protection Dosimetry*, Vol. 32, No. 1, 15-20.

Akselrod, M.S., Lucas, A.C., Polf, J.C. & McKeever, S.W.S. (1998). Optically Stimulated Luminescence of $Al_2O_3$. *Radiation Measurements*, Vol. 29, No. 3-4, 391-399, 1350- 4487.

Akselrod, M.S. (2011). Fundamentals of Materials, Techniques, and Instrumentation for OSL and FNTD Dosimetry, *AIP Conference Proceedings*, Sydney, Australia, September 2010.

Albrecht, H. & Mandeville, C. (1956). Storage of Energy in Beryllium Oxide. *Physical Review*, Vol. 101, No. 4, 1250-1252.

Andersen, C.E., Nielsen, S.K., Greilich, S., Helt-Hansen, J., Lindegaard, J.C. & Tanderup, K. (2009). Characterization of a Fiber-Coupled $Al_2O_3$:C Luminescence Dosimetry System for Online in Vivo Dose Verification During $^{192}Ir$ Brachytherapy. *Medical Physics*, Vol. 36, No. 3, 708-718, 0094-2405.

Appleby, G., Zimmermann, J., Hesse, S., Karg, O. & von Seggern, H. (2009). Sensitization of the Photostimulable X-ray Storage Phosphor CsBr: $Eu^{2+}$ Following Room-Temperature Hydration. *Journal of Applied Physics*, Vol. 105, No. 7, 0735111-0735115, 0021-8979.

Arjomandy, B., Tailor, R., Anand, A., Sahoo, N., Gillin, M., Prado, K. & Vicic, M. (2010). Energy Dependence and Dose Response of Gafchromic EBT2 Film Over a Wide Range of Photon, Electron, and Proton Beam Energies. *Medical Physics*, Vol. 37, No. 1942-1947.

Azorin, J., Furetta, C. & Scacco, A. (1993). Preparation and Properties of Thermoluminescent Materials. *physica status solidi (a)*, Vol. 138, No. 1, 9-46, 1521-396X.

Beck, H. (1976). A Study on Mixed Halide Compounds MFX (M= Ca, Sr, Eu, Ba; X= Cl, Br, I). *Journal of Solid State Chemistry*, Vol. 17, No. 3, 275-282, 0022-4596.

Boone, J.M. (2001). Determination of the Presampled MTF in Computed Tomography. *Medical Physics*, Vol. 28, No. 356-360.

Borgman, I.L. (1897). Thermoluminescence Provoquee par les Rayons de M. Roentgen et les Rayons de M. Becquerel. *Comptes Rendus*, Vol. 124, No. 895-896,

Bos, A. (2001). High Sensitivity Thermoluminescence Dosimetry. *Nuclear Instruments and Methods in Physics Research Section B: Beam Interactions with Materials and Atoms*, Vol. 184, No. 1-2, 3-28, 0168-583X.

Bøtter-Jensen, L., McKeever, S.W.S. & Wintle, A.G. (2003). *Optically Stimulated Luminescence Dosimetry*, Elsevier Science, Amsterdam; Boston; London.

Bulur, E. & Göksu, H. (1998). OSL From BeO Ceramics: New Observations from an Old Material. *Radiation Measurements*, Vol. 29, No. 6, 639-650, 1350-4487.

Butson, M.J., Yu, P.K.N., Cheung, T. & Alnawaf, H. (2010). Energy Response of the New EBT2 Radiochromic Film to X-ray Radiation. *Radiation Measurements*, Vol. 45, No. 7, 836-839, 1350-4487.

Calvert, R. & Danby, R. (1984). Thermoluminescence and Radiophotoluminescence From Eu- and Sm-Doped $CaSO_4$. *physica status solidi (a)*, Vol. 83, No. 2, 597-604, 1521- 396X.

Crase, K. & Gammage, R. (1975). Improvements in the Use of Ceramic BeO for TLD. *Health Physics*, Vol. 29, No. 5, 739-746, 0017-9078.

Cunningham, I. & Fenster, A. (1987). A Method for Modulation Transfer Function Determination from Edge Profiles with Correction for Finite-Element Differentiation. *Medical Physics*, Vol. 14, No. 4, 533–537.

Cunningham, I. & Reid, B. (1992). Signal and Noise in Modulation Transfer Function Determinations Using the Slit, Wire, and Edge Techniques. *Medical Physics*, Vol. 19, No. 1037-1044.

Daniels, F., Boyd, C.A. & Saunders, D.F. (1953). Thermoluminescence as a Research Tool. *Science*, Vol. 117, No. 3040, 343-349, 0036-8075.

Dyk, J.V. (1999). *The Modern Technology of Radiation Oncology: A Compendium for Medical Physicists and Radiation Oncologists*, Medical Physics Publishing, 9780944838389, Madison, Wisconsin.

Eichholz, G.G. (2003). Dosimetry for Food Irradiation. *Health Physics*, Vol. 84, No. 5, 665, 0017-9078.

Goldstein, E. (1912). Concerning the Emission Spectra of Aromatic Bonds in Ultra Violet Light, in Cathode Rays, Radium Radiation and Positive Rays. *Pysikalische Zeitschrift*, Vol. 13, No. 188-193.

Hackenschmied, P., Zeitler, G., Batentschuk, M., Winnacker, A., Schmitt, B., Fuchs, M., Hell, E. & Knüpfer, W. (2002). Storage Performance of X-ray Irradiated Doped CsBr. *Nuclear Instruments and Methods in Physics Research Section B: Beam Interactions with Materials and Atoms*, Vol. 191, No. 1-4, 163-167, 0168-583X.

Hackenschmied, P., Schierning, G., Batentschuk, M. & Winnacker, A. (2003). Precipitation- induced Photostimulated Luminescence in CsBr: $Eu^{2+}$. *Journal of Applied Physics*, Vol. 93, No. 9, 5109-5112, 0021-8979.

Hangleiter, T., Koschnick, F. & Spaeth, J. (1990). Temperature Dependence of the Photostimulated Luminescence of X-irradiated BaFBr: $Eu^{2+}$. *Journal of Physics: Condensed Matter*, Vol. 2, No. 32, 6837-6846.

Horowitz, Y.S. (1984). *Thermoluminescence and Thermoluminescent Dosimetry*, CRC Press, 9780849356650 Boca Raton, FL.

Kortov, V. (2007). Materials for Thermoluminescent Dosimetry: Current Status and Future Trends. *Radiation Measurements*, Vol. 42, No. 4-5, 576-581, 1350-4487.

Koschnick, F.K., Spaeth, J.M., Eachus, R.S., McDugle, W.G. & Nuttall, R.H.D. (1991). Experimental Evidence for the Aggregation of Photostimulable Centers in BaFBr:$Eu^{2+}$ Single Crystals by Cross Relaxation Spectroscopy. *Physical Review Letters*, Vol. 67, No. 25, 3571-3574, 0031-9007.

Leblans, P.J.R., Struye, L. & Willems, P. (2001). New Needle-Crystalline CR Detector, *Procccedings of SPIE* San Diego, CA, USA, February 2001.

Lewis, D.F. (1986). A Processless Electron Recording Medium, *Proceedings of SPSE'86 Symposium*, Arlington, VA.

Liu, Z., Massil, T. & Riesen, H. (2010). Spectral Hole-Burning Properties of $Sm^{2+}$ ions. Generated by X-rays in BaFCl: $Sm^{3+}$ Nanocrystals. *Physics Procedia*, Vol. 3, No. 4, 1539-1545, 1875-3892.

McKeever, J., Walker, F. & McKeever, S. (1993). Properties of the Thermoluminescence Emission from LiF (Mg, Cu, P). *Nuclear Tracks and Radiation Measurements*, Vol. 21, No. 1, 179-183, 0969-8078.

McKeever, S.W.S. (1988). *Thermoluminescence of Solids,* Cambridge University Press, 0521368111, Cambridge, New York.

McKeever, S.W.S., Moscovitch, M. & Townsend, P.D. (1995). *Thermoluminescence Dosimetry Materials: Properties and Uses,* Nuclear Technology Publishing, 1870965191, England.

Mclaughlin, W.L., Yundong, C., Soares, C.G., Miller, A., Vandyk, G. & Lewis, D.F. (1991). Sensitometry of the Response of a New Radiochromic Film Dosimeter to Gamma-Radiation and Electron-Beams. *Nuclear Instruments & Methods in Physics Research Section a-Accelerators Spectrometers Detectors and Associated Equipment,* Vol. 302, No. 1, 165-176, 0168-9002.

Mijnheer, B. (2008). State of the Art of in vivo Dosimetry. *Radiation Protection Dosimetry,* Vol. 131, No. 1, 117-122, 0144-8420.

Milman, I., Sjurdo, A., Kortov, V. & Lesz, J. (1996). TSEE and TL of Non-Stoichiometric BeO-$TiO_2$ Ceramics. *Radiation Protection Dosimetry,* Vol. 65, No. 1-4, 401, 0144-8420.

Nakajima, T., Murayama, Y., Matsuzawa, T. & Koyano, A. (1978). Development of a New Highly Sensitive LiF Thermoluminescence Dosimeter and its Applications. *Nuclear Instruments and Methods,* Vol. 157, No. 1, 155-162, 0029-554X.

Nakano, Y., Gido, T., Honda, S., Maezawa, A., Wakamatsu, H. & Yanagita, T. (2002). Improved Computed Radiography Image Quality from a BaFI: Eu Photostimulable Phosphor Plate. *Medical Physics,* Vol. 29, No. 592-597.

Paul, J.R.L., Willems, P. & Alaerts, L.B. (2002). New Needle-Crystalline Detector For X-Ray Computer Radiography (CR). *The e-Journal of Nondestructive Testing,* Vol. 7, No. 12, 1435-4934.

Podgorsak, E. (2005). *Radiation Oncology Physics: A Handbook for Teachers and Students,* International Atomic Energy Agency, 9201073046, Vienna.

Riesen, H. & Kaczmarek, W.A. (2005). *Radiation Storage Phosphor & Applications,* International PCT Application PCT/AU2005/001905, International Publication Number WO 2006/063409.

Riesen, H. & Kaczmarek, W.A. (2007). Efficient X-ray Generation of $Sm^{2+}$ in Nanocrystalline $BaFCl/Sm^{3+}$: A Photoluminescent X-ray Storage Phosphor. *Inorganic Chemistry,* Vol. 46, No. 18, 7235-7237, 0020-1669.

Riesen, H. & Piper, K. (2008). *Apparatus and Method for Detecting and Monitoring Radiation,* International PCT application.

Rink, A., Lewis, D.F., Varma, S., Vitkin, I.A. & Jaffray, D.A. (2008). Temperature and Hydration Effects on Absorbance Spectra and Radiation Sensitivity of a Radiochromic Medium. *Medical Physics,* Vol. 35, No. 4545-4555.

Samei, E., Flynn, M.J. & Reimann, D.A. (1998). A Method for Measuring the Presampled MTF of Digital Radiographic Systems Using an Edge Test Device. *Medical Physics,* Vol. 25, No. 102-113.

Samei, E., Seibert, J.A., Willis, C.E., Flynn, M.J., Mah, E. & Junck, K.L. (2001). Performance Evaluation of Computed Radiography Systems. *Medical Physics,* Vol. 28, No. 361- 371.

Satinger, D., Horowitz, Y.S. & Oster, L. (1999). Isothermal Decay of Isolated Peak 5 in 165degC/15 Minute Post-Irradiation Annealed LiF:Mg,Ti (TLD-100) Following Alpha Particle and Beta Ray Irradiation. *Radiation Protection Dosimetry,* Vol. 84, No. 1-4, 67-72.

Schulman, J., Shurcliff, W., Ginther, R. & Attix, F. (1953). Radiophotoluminescence Dosimetry System of the US Navy. *Nucleonics (U.S.) Ceased publication,* Vol. 11, No. 10, 52-56.

Schulman, J.H. (1950). *Dosimetry of X-rays and Gamma Rays.*, Naval Research Laboratory. Report No. 3736, Washington, DC.

Schulman, J.H., Ginther, R.J., Klick, C.C., Alger, R.S. & Levy, R.A. (1951). Dosimetry of X-Rays and Gamma-Rays by Radiophotoluminescence. *Journal of Applied Physics*, Vol. 22, No. 12, 1479-1487, 0021-8979.

Schulman, J.H. & Etzel, H.W. (1953). Small-Volume Dosimeter for X-Rays and Gamma-Rays. *Science*, Vol. 118, No. 3059, 184-186, 0036-8075.

Schweizer, S., Rogulis, U., Assmann, S. & Spaeth, J.M. RbBr and CsBr Doped with Eu$^{2+}$ As New Competitive X-ray Storage Phosphors, *Proceedings of the Fifth International Conference on Inorganic Scintillators and their Applications*, 1350-4487, Riga-Jurmala, Latvia, August 2000.

Schweizer, S. (2001). Physics and Current Understanding of X-Ray Storage Phosphors. *physica status solidi (a)*, Vol. 187, No. 2, 335-393, 1521-396X.

Schweizer, S., Secu, M., Spaeth, J.M., Hobbs, L., Edgar, A. & Williams, G. (2004). New Developments in X-ray Storage Phosphors. *Radiation Measurements*, Vol. 38, No. 4-6, 633-638, 1350-4487.

Secerov, B., Dakovic, M., Borojevic, N. & Bacic, G. (2011). Dosimetry Using HS GafChromic Films the Influence of Readout Light on Sensitivity of Dosimetry. *Nuclear Instruments & Methods in Physics Research Section a-Accelerators Spectrometers Detectors and Associated Equipment*, Vol. 633, No. 1, 66-71, 0168-9002.

Soares, C.G. (2006). New Developments in Radiochromic Film Dosimetry. *Radiation Protection Dosimetry*, Vol. 120, No. 1-4, 100-106, 0144-8420.

Soares, C.G. (2007). Radiochromic Film Dosimetry. *Radiation Measurements*, Vol. 41, No. Supplement 1, S100-S116, 1350-4487.

Sommer, M., Jahn, A. & Henniger, J. (2008). Beryllium Oxide as Optically Stimulated Luminescence Dosimeter. *Radiation Measurements*, Vol. 43, No. 2-6, 353-356, 1350- 4487.

Stevens-Kalceff, M.A., Riesen, H., Liu, Z., Badek, K. & Massil, T. (2010). Microcharacterization of Core Shell Nanocrystallites. *Microscopy and Microanalysis*, Vol. 16, No. S2, 1818-1819, 1435-8115.

Takahashi, K., Kohda, K., Miyahara, J., Kanemitsu, Y., Amitani, K. & Shionoya, S. (1984). Mechanism of Photostimulated Luminescence in BaFX:Eu$^{2+}$ (X=Cl,Br) Phosphors. *Journal of Luminescence*, Vol. 31-32, No. PART 1, 266-268.

Vij, D.R. & Singh, N. (1997). Thermoluminescence Dosimetric Properties of Beryllium Oxide. *Journal of Materials Science*, Vol. 32, No. 11, 2791-2796, 0022-2461.

von Seggern, H. (1999). Photostimulable X-ray Storage Phosphors: A Review of Present Understanding. *Brazilian journal of physics*, Vol. 29, No. 2, 254-268, 0103-9733.

Weidner, M., Batentschuk, M., Meister, F., Osvet, A., Winnacker, A., Tahon, J.P. & Leblans, P. (2007). Luminescence Spectroscopy of Eu$^{2+}$ in CsBr: Eu Needle Image Plates (NIPs). *Radiation Measurements*, Vol. 42, No. 4-5, 661-664, 1350-4487.

Williamson, G. & Hall, W. (1953). X-Ray Line Broadening From Filed Aluminium and Wolfram. *Acta Metallurgica*, Vol. 1, No. 1, 22-31, 0001-6160.

Yamashit, T. Yasuno, Y. & Ikedo, M. (1974). Beryllium-Oxide Doped with Lithium or Sodium for Thermoluminescence Dosimetry. *Health Physics*, Vol. 27, No. 2, 201-206, 0017- 9078.

Yen, W.M., Shionoya, S. & Yamamoto, H. (2007). *Phosphor Handbook, (second edition)*, CRC, 0849335647, Boca Raton, FL.

Zimmermann, J., Hesse, S., von Seggern, H., Fuchs, M. & Knüpfer, W. (2005). Radiation Hardness of CsBr: Eu$^{2+}$. *Journal of Luminescence*, Vol. 114, No. 1, 24-30, 0022-2313.

# 4

# Ionizing Radiation Induced Radicals

Ahmed M. Maghraby

*National Inst. of Standards (NIS) – Radiation Dosimetry Department, Giza, Egypt*

## 1. Introduction

When ionizing radiation passes through a material it imparts some of its energy to that material. The imparted energy may be high enough to cause a break in bonds inside the molecules or between molecules or both, in such cases, free radicals are created. The type, lifetime, extent, fate and origin of those free radicals may differ according to several factors; most of them are beyond the subject of this chapter. However, the abundance of ionizing radiation induced radicals in a material is directly proportional (unless saturated) to the amount of ionizing radiation received by that material, hence it could be a method for determination of radiation doses passively by accurate evaluation of the extent of free radicals created.

Study of radiation induced radicals is not always related to radiation dose assessment; sometimes it is of great importance to investigate what type of radicals are produced when a specific material is exposed to a specific radiation dose, and for how long those induced radicals can persist. This could be of specific interest specially when dealing with environments of high radiation levels, for example: in space environments. Behavior of radiation induced radicals may lead to further understanding of molecular interactions or molecular dynamics, or may lead to a decision on the preference of a material for a specific design from the material science point of view.

## 2. Use of EPR for studying radiation induced radicals

Electron Paramagnetic Resonance (EPR) is the major universal technique for investigation of radiation induced radicals. The first observation of an electron paramagnetic resonance peak was made in 1945 when Zavoisky detected a radio frequency absorption line from a $CuCl_2.2H_2O$ sample (Zavoisky, E., 1945), The first EPR study of radiation damage in materials of biological (organic) interest powders was made by Gordy in 1955 (Gordy, W., et al., 1955), Papers on irradiated dimethylglyoxine and α-alanine of Miyagawa and Gordy (Miyagawa, I., and Gordy, W., 1959), and malonic acid by McConell (McConell, H., et al., 1960) soon published and confirmed that it was possible to investigate radicals produced by irradiating single crystals, in a similar way to that which had been used to study paramagnetic ions. Brady et al (Brady et al, 1968) suggested using EPR dosimetry and the additive re-irradiation method to obtain dose estimates from accidental overexposures, where human teeth were used for the first time as radiation dosimeter. In 1962, irradiated L-α-alanine was suggested as a possible dosimeter material in the high-dose range by (Bradshaw, W., et al., 1962) after which great efforts were made to establish dosimetry systems based mainly on alanine/EPR systems (Deffner, U., and Regulla, F., 1980, Nette, H., et al., 1993).

## 3. Basic principles of EPR

Electron, as a rotating charge can be considered as a very tiny magnet, and hence the single electron which moves freely in absence of external magnetic field has only two orientations if placed in the field of external magnet: aligned parallel or anti-parallel. Those two cases reflect the two energy states arise after applying external magnetic field: state of lower energy when the moment of the electron, μ, is aligned with the magnetic field and a higher energy state when μ is aligned against the magnetic field.

The two states are designated by the projection of the electron spin, $m_s$, on the direction of the magnetic field. Because the electron is a spin 1/2 particle, the parallel state has $m_s = -1/2$ and the antiparallel state has $m_s = +1/2$. The energy difference between these two states, caused by the interaction between the electron spin and the applied magnetic field ($B_0$), is shown in the following relation:

$$\Delta E = g \, \mu_B \, B_0 \, \Delta m_s = g \, \mu_B \, B_0 \tag{1}$$

Where $\mu_B$ is Bohr magneton (the natural unit of the electron's magnetic moment), g is the g-factor, and $\Delta m_s = \pm 1$.

So, if this electron gains energy of ΔE transition occurs between the two spin states, if this energy is in the form of photons, then:

$$\Delta E = h. \, v = g \, \mu_B \, B_0 \tag{2}$$

Where h is Planck's constant and v is the frequency of the electromagnetic radiation.

Now, resonance may occur either by scanning frequency at a constant magnetic field or by scanning magnetic field while the frequency of the magnetic field was held constant and the later is easier from the practical point of view.

## 4. Radiation induced radicals in biological molecules

Study of radiation induced radicals in biological molecules or molecules of biological origin is of high concerns in order to understand their impacts on functional and/or structural changes of such molecules after exposure to ionizing radiation. In the following sections, there are two examples for studying radiation induced radicals using EPR in biological molecules: bovine hemoglobin as a biological molecule of animal origin, and chitosan as a biological molecule of plant origin.

### 4.1 Bovine hemoglobin
### 4.1.1 Structure
Structure of bovine Hb is shown in Figure (1). It is composed of two pairs of non-identical subunits, alpha and beta. Each alpha–beta pair is more closely associated than they are with each other, but the overall arrangement is roughly tetrahedral (http://www.bmb.uga.edu/wampler/tutorial/prot4.html) (Marta et al., 1996).

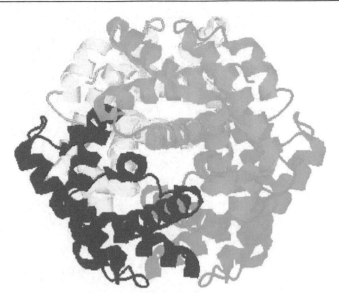

Fig. 1. Bovine Hb is composed of two pairs of non-identical subunits, alpha and beta. Each alpha–beta pair is more closely associated than they are with each other, but the overall arrangement is roughly tetrahedral.

### 4.1.2 EPR spectra of bovine hemoglobin

Fig. 2 represents the EPR spectra of unirradiated (solid line) and 743Gy gamma irradiated sample (dotted line). Major features of Bovine Hb EPR spectrum are comparable to the human one (Ikeya, 1993). About four features comprise the spectrum of g-factor equal to 5.91017, 4.27507, 2.14737 and 2.00557, respectively. The first two signals are due to Fe (III) of high spin form (S ¼ 5/2). The first signal (S1) is associated with oxidized heme iron, which clarifies its indication to methemoglobin (MetHb), in which a water molecule replaces O2 ion as a ligand of iron (Wajnberg and Bemski 1993). The second signal (S2) is corresponding to non-heme Fe(III) ions at sites endowed with rhombic symmetry, which is not associated with species involved in blood, such as transferrin which causes an EPR signal near g = 4.3 (Ślawska-Waniewska et al., 2004). From Fig. 2, it is clear that S2 is greater than S1, while this is not true in case of human Hb spectrum (Ikeya, 1993). This means higher nonheme iron content in bovine Hb than that of human. The nature of bovine Hb itself or the deterioration of its molecular structure and hence the decomposition of heme during sample preparation may give reason for the increase of non-heme iron content.

Third signal group (S3) is associated with low spin derivatives of ferrihemoglobin called "hemichrome", copper proteins and some transition-metal complexes (Rachmilewitz et al., 1971). Hemichromes are low spin derivatives of ferric Hb brought about through discrete reversible or irreversible changes of protein conformation (Venkatesh et al., 1997). The purity of our sample is about 95%, hemichromes are not easily to be separated from Hb molecules but the rest which is about 5% may contain some of these hemichromes that produce S3. Changes in normal Hb under the effect of time, pH and protein denaturants such as urea or salicylate can form different kinds of hemichromes with different endogenous ligands; hence hemichromes form the primary step to the destructive pathway for denaturation (Venkatesh et al.,1997).

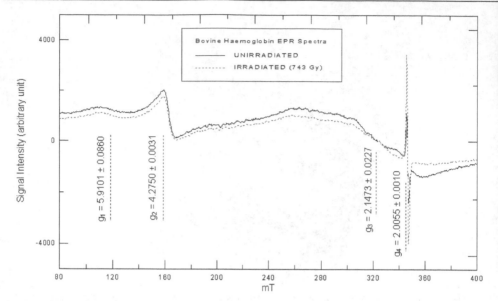

Fig. 2. EPR spectra of unirradiated and 743Gy gamma irradiated bovine Hb samples recorded at room temperature.

With regard to the fourth signal (S4), it appears as a singlet with no hyperfine structures as shown in Fig. 2. This signal is attributed to free radicals in hemoglobin formed by the degradation of blood constituents (Miki et al., 1987). Several investigators agree that at least two different kinds of radicals are formed on the protein (Kelman et al.,1994; Gunther et al., 1995). Although the formation of peroxyl radicals is well proven (Ikeya, 1993; Miki et al.,1987), this kind of radical constitutes only a fraction of the total concentration of radicals (Svistunenko et al.,1997a, b). The globin-based free radical (HB(Fe(IV)=O)) was suggested to be major contributor for S4 (Svistunenko et al., 1997a,b). Many investigations revealed that it is the tyrosine (Tyr) radical (shown in Fig. 3) (Svistunenko et al.,1997a,b, 2002, 2004; Svistunenko, 2005).

### 4.1.3 Radiation-induced changes in the bovine Hb EPR spectra

The EPR spectrum of 743 Gy gamma irradiated sample is shown in Fig. 2; from the figure it is clear that no new radicals have emerged, and no remarkable changes in the intensity of the first two signals were recorded. The unchanged intensity of the first signal (with all the irradiation doses) suggests no net change in MetHb, which may be explained as follows: The manipulated sample is in powder form, so any enzymatic reaction that can lead to change in MetHb content is excluded. The presence of oxygen is mandatory for MetHb production by other pathways such as the oxidation of heme iron by the electron transfer from Fe(II) to O2 creating Fe(III) and superoxide radicals ($O_2^-$) (Misra and Fridovich, 1972) During irradiation, wrapped samples were prevented from molecular oxygen in air. So, MetHb production through the second pathway is prevented; while the first pathway may be blocked by the removal of oxygen molecules from the bovine Hb sample during preparation in its powder form. The non-significant change in intensity of S2 by irradiation (Fig. 1), reflects the stability of non-heme iron content as radiation doses increase up to 743 Gy. It is clear from

Fig. 1 that S3 suffered apparent significant decrease upon irradiation; which ensures the decrease of the net amount of hemichromes and reflects the crosslinking processes following Hb irradiation at high doses. Results showed that the most obvious radiation-induced change in bovine Hb EPR spectrum is the significant increase in S4 even for very low dose (4.95 Gy). This may be due to the increase in the production of free radicals in Hb protein (peroxyl and tyrosyl radicals) and reflects the high sensitivity of Hb protein to irradiation.

## 4.2 Chitosan

Chitosan is a natural polysaccharide, that can be prepared on an industrial scale by deacetylation of much more abundant chitin and its chemical structure is a copolymer of β-(1-4)-D-glucosamine and N-acetyl- β-(1-4)-D-glucosamine (Jaroslaw, M., et al., 2005). Due to unique biophysical and chemical properties of polysaccharides, such as biocompatibility, biodegradability, nontoxicity and nonantigenicity (Bin Kang, et al., 2007), a broad spectrum of applications has been emerged in different modern fields: water treatment (Kurmaev E.Z. et al., 2002), chromatography, additives for cosmetics, textile treatment for antimicrobial activity (Le Hai et al., 2003), novel fibers for textiles, photographic papers, biodegradable films, biomedical devices, and microcapsule implants for controlled release in drug delivery. Also, its nano-ordered hydrogel is a potential responsive material for biochips and sensors for the development of PC- controlled biochips, and is used also in some attempts as a gene delivery system for curing of some hereditary diseases (IAEA, 2004). Tissue engineering and adsorption of metal ions as well as dyes removal are some of the many applications of chitosan (Jayakumar, R., et al., 2005).

### 4.2.1 Chitosan EPR spectral features

Figure (3) represents the EPR spectrum of chitosan-A. There are two overlapped singlet signals, the first signal (S₁) is of g-factor = 2.00725 ± 0.00018 with peak-to-peak line width

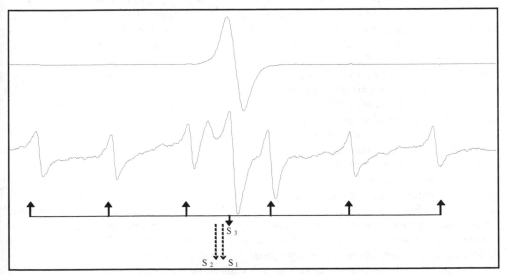

Fig. 3. EPR spectra of chitosan: lower represents the unirradiated sample spectrum, and the upper represents 30 kGy irradiated sample spectrum.

($W_{PP}$) = 1.08 ± 0.072 mT, while the second signal ($S_2$) is of g-factor = 2.00966 ± 0.00073, while its $W_{PP}$ = 5.70 ± 0.139 mT. EPR lines of Mn$^{2+}$ ($I$ = 5/2), ions are well recognized signal ($S_3$) with sextet hyperfine structure were observed ($g_{iso}$=2.01147±0.0002, $a_{av}$=9.06±0.35 mT). This signal is attributed to the presence of Mn$^{2+}$ in crab shells from which chitosan-A was extracted (Maghraby, A., 2007). No new radicals have been emerged after irradiation up to 30 kGy, while some slight broadening in line width values for $S_1$ was noticed.

## 5. EPR radiation dosimetry

Radiation dosimetry using EPR utilizes the produced radicals as a result of the passage of ionizing radiation through matter, as extent of those radicals proportionate directly to the amount of ionizing radiation. Radiation dosimetry using EPR possesses some advantages which enabled the use of EPR in different fields of ionizing radiation dosimetry.

### 5.1 Industrial applications
### 5.1.1 Food irradiation
Food irradiation is a means of microbiological contamination reductions through which, food in packages or in bulk are exposed to a high radiation dose which is enough to destroy microbiological contents (Maghraby, A., 2007, Miyagusku, L., et al., 2007). The aim of food irradiation is sterilization and preservation of food during food transfer from region to another or to increase its quality locally before use (IAEA, 1978). The use of irradiation alone as a preservation technique can play an important role in cutting losses and reducing the dependence on chemical pesticides. Each year a few hundred thousand tones of food products and ingredients are irradiated worldwide. International food stuff trade especially seafood is growing up which increases the risk of the transfer of some diseases; this led the World Health Organization to legalize radiation treatment of food (WHO, 1988, 1994). International health and safety authorities have endorsed the safety of irradiation for all foods up to a dose level of 10 kGy, however, recent evaluation of an international expert study group appointed by FAO, IAEA, and WHO showed that food treated according to good manufacturing practices (GMPs) at any dose above 10kGy is also safe for consumption(ICGFI, 1999).

Irradiated food could be identified using different methods such as Gas chromatography, mass spectrometry, thermoluminescence (Schreiber, G., et al., 1993a, Schreiber , G., et al., 1993b, Schreiber , G., et al., 1993 c, Ziegelmann, K., et al., 1999, Parlato, A. et al., 2007, D'Oca, M. C. et al., 2008), and Electron paramagnetic resonance (EPR) spectroscopy (Desrosiers, M., 1989, Sünnetcioğlu, M., et al., 1999) which is characterized by its non-destructive detection of radiation-induced radicals, hence, it is a major technique for the investigation of irradiated food and the determination of the dose delivered accurately, which results in accepting or refusing of food transfer.

Mineral nutrients and Irradiated vegetables were identified using EPR (Lakshmi Prasuna, C.P, 2008, Maghraby, A., and Maha Anwar Ali. 2007), EPR spectral investigations have been carried out on some vegetable samples and the presence of various paramagnetic metal ions in various oxidation states is indicated. In almost all the fibrous vegetable samples, the free radical signal corresponding to cellulose radical is observed. However, in some vegetables like carrot, though the free radical signal cannot be distinguished, the free radical assignments are made depending on the organic radicals. The reason is that, the organic free radicals are generated due to CO group in the organic radicals (Lakshmi Prasuna, C.P,2008).

## 5.1.2 Drugs and medical products irradiation

Use of ionizing radiation in sterilization of medicinal products, such as catheters, syringes, drug and drug raw materials, is a new technology alternative to heat and gas exposure sterilization (Jacobs, 1995; Reid, 1995; Tilquin and Rollmann, 1996; Boess and Bögl, 1996). The advantages of sterilization by irradiation include high penetrating power, low measurable residues, small temperature rise and the fact that there are fewer variables to control (Fauconnet et al., 1996; Basly et al., 1997). Thus, sterilization can be carried out on the finally packaged product and is applicable to heat-sensitive drugs.

The regulations of radiosterilization differ among countries. In other radiation sources such as X-rays, fast electrons and UV illumination. Irradiation produces new radiolytic products. To prove the safety of radiosterilization, it is important to determine physical and chemical features of the radiolytic products and elucidate the mechanism of radiolysis. Thus, it is desirable to establish a method to discriminate between irradiated and unirradiated drugs. Electron spin resonance (ESR) spectroscopy appears to be well suited for determination of free radical concentrations in complex media and so, it can be used to detect and distinguish between irradiated drugs from unirradiated ones (Gibella et al., 1993; Signoretti et al., 1994; Miyazaki et al., 1994; Onori et al., 1996).

As a cephalosporin antibiotic, duricef is used in the treatment of nose, throat, urinary tract and skin infection that are caused by specific bacteria, including staph strep and E. coli, effects of gamma radiation on cephalosporins with various substitutive groups have been reported in different papers (Miyazaki et al., 1994; Onori et al., 1996; Jacobs, 1983; Yürüs and Korkmaz, 2005; Gibella et al., 2000), but low-and high-temperature kinetic features, structure and activation energies of the radical species involved in the formation of their ESR spectra were investigated in none of these papers, except of few published work (Yürüs, S and Korkmaz, M., 2005). Eleven of the studied cephalosporins with various substitutive groups have been shown to exhibit, interestingly ESR spectra of unresolved doublet appearence (Onori et al., 1996; Yürüs and Korkmaz, 2005). This conlusion was considered as an indication relevant to the origin of radicalic species produced in irradiated cephalosporins. Species originating from substitutive groups and cephem ring are expected to be responsible from ESR spectra of irradiated cephalosporins, but the ultimate patterns of the latter are believed to be determined by the relative weights of those species. Thiamphinicol sterilization using irradiation to doses up to 25 kGy was investigated using EPR spectroscopy, and it was found that standard dose used for sterilization (25 Gy) did not produce any detectable changes in the physico-chemichal properties of Thiamfinicol.

## 5.1.3 Other applications

An electron paramagnetic resonance (EPR) investigation of a series of glasses irradiated at room temperature with $\beta$ or X radiation sources has been made in order to predict the long term behavior of glasses used in the nuclear waste disposal, in particular, no paramagnetic defects associated with aluminum ions are detected in these irradiated glasses (Bruno Boizot, et al., 1998).

Applications in polymer research and industries are numerous, for example, Poly (vinyl chloride) is one of the most used thermoplastic polymers, thanks to its low production cost, easy processing, excellent mechanical properties, high compatibility with additives and possibility to be recycled. Particularly interesting is its employment in vital single use medical devices such as catheters, cannulae, urological products and flexible tubes for extracorporeal connections. The sterilization of these products is performed by electron beam or γ irradiations, a termination step, in which the active centers are deactivated. At 25 and 50 kGy oxygen appears to saturate all radicals and the EPR spectrum shows only one component, associated with peroxyl radicals. At the 100 and 150 kGy irradiation doses the EPR spectrum shows more structure and it comprises overlapping signals from the peroxyl and polyenyl radicals not yet oxidized (Baccaro, S. et al., 2003, Costa, L. et al., 2004).

## 5.2 Medical applications
### 5.2.1 Alanine dosimetry

#### 5.2.1.1 Introduction

The natural amino acid L-α-alanine (see Figure (4)) has attracted considerable interest for use in radiation dosimetry (Bradshaw, W., etal., 1962, Regulla, D., and Deffiner, U., 1982, Nam, J., and Regulla, D., 1989, Ciesielski, B., and Wielopolski, L., 1994) and has been formally accepted as a secondary standard for high-dose and transfer dosimetry (Nett, H., ate al., 1993). The EPR powder spectrum from amorphous alanine pellets or other types of disordered alanine samples has been used for this purpose. The peak-to-peak amplitude of the central line of this powder spectrum is commonly used for monitoring the radiation dose (Eirik Malinen, et al., 2003a).

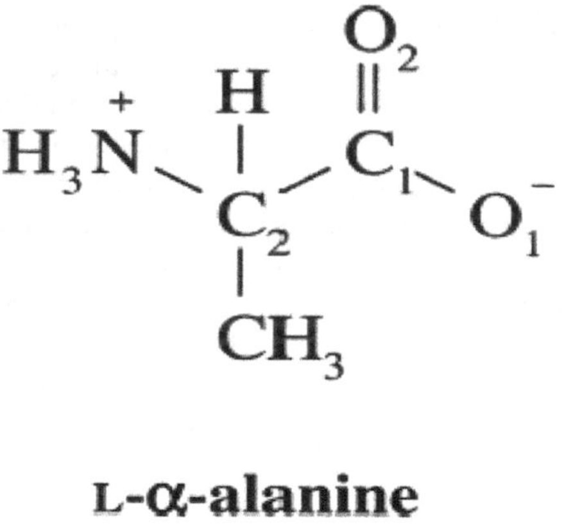

$$H_3\overset{+}{N} \diagdown \underset{\underset{\displaystyle CH_3}{|}}{\overset{\overset{\displaystyle H}{|}}{C_2}} \diagup C_1 \overset{O_2}{\diagdown} \underset{O_1^-}{}$$

## L-α-alanine

Fig. 4. Molecular structure of L-α- alanine.

It was found that at least three different radicals are formed and stabilized in alanine after X irradiation at room temperature as shown in Figure (5), (Sagstuen, E., et al., 1997a, Sagstuen, E., et al., 1997b, Mojgan Z. Heydari, et al., 2002). Two of these, the R1 radical (Structure 1) and a species formed by net hydrogen abstraction from the central carbon atom (radical R2, Structure 2), appear to occur in comparable relative amounts (55–60 and 30–35%, respectively) (Mojgan Z. Heydari, et al., 2002). The third species (denoted R3) is a minority species (5–10%), and an unambiguous structure assignment has not been made (Sagstuen, E., et al., 1997b). Structure 3 below, was suggested as a possible candidate for R3 (Eirik Malinen, et al., 2003b).

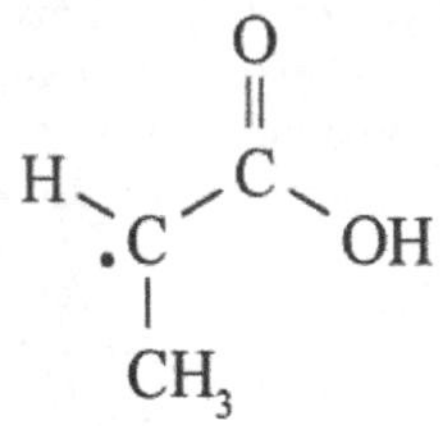

Structure 1

**R1 (SAR)**

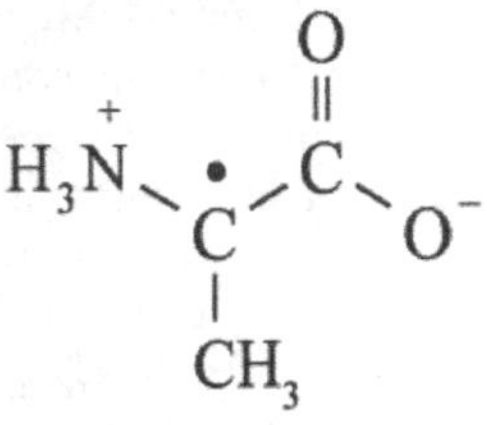

Structure 2

**R2**

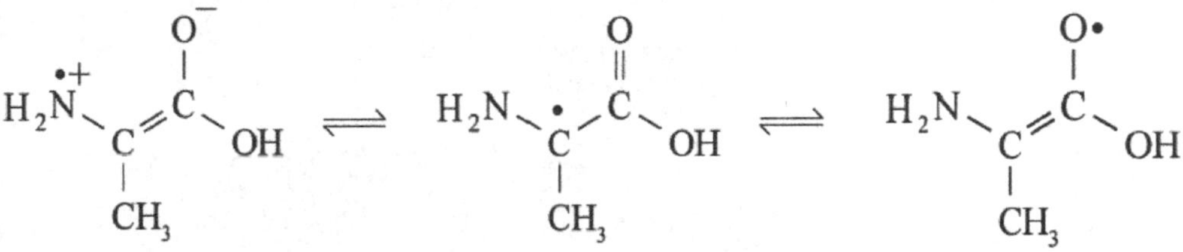

Structure 3

**R3**

Fig. 5. Three structures of alanine radicals, R1 or (SAR) is the Stable Alanine Radical, R2, and R3 in its three forms (Eirik Malinen, et al., 2003a and Eirik Malinen, et al., 2003b).

### 5.2.1.2 Dosimeters structure and shape

Alanine dosimeters often are composed of a binder and the alanine itself, the ratio and the binder type differs from type to type. The pellets are now in routine use as dosimeters for medical applications because of their robustness they are used as reference dosimeters sent by mail as a service offered by many laboratories (Maghraby, A., 2003).

Transfer alanine dosimeters which are distributed by the IAEA for intercomparison purposes were made of 70% per weight DL-α-alanine + 30% per weight polystyrene (Mehta, K., and Girzikowsky, R., 2000).

There are a lot of variations in alanine dosimeter shapes, varies according to different reasons, like the purpose of use, resolution required, and achievable sensitivity.

### 5.2.1.3 Applications

Experimental procedures of the use of alanine as dosimeter for brachytherapy purposes was described by De Angeles, C., et al., 1999. Also, a device for in vivo measurements of the rectal dose in radiotherapy for prostate cancer was developed (Daniela Wagner, et al., 2008) as shown in Figure (6). It was possible to insert this device without clinical complications and without additional rectal discomfort for the patients. For irradiations of alanine dosimeter probes under reference conditions, deviations of less than 1% in reference to the national German standard were achieved. In the absence of metallic implants, the relative deviations between measured and applied dose values at the anterior rectal wall are less than or equal to 1.5% for the in vivo measurements.

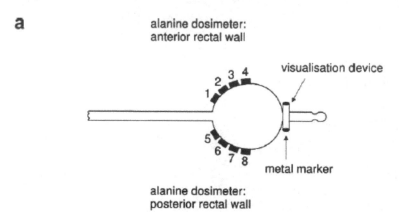

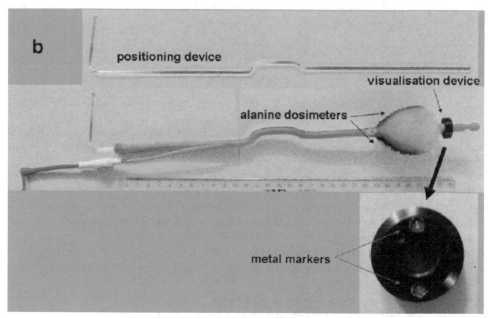

Fig. 6. (a) Sketch of the used rectal balloon equipped with alanine dosimeters. Four alanine dosimeter probes were placed at the anterior rectal wall (1–4) and: four alanine dosimeters at the posterior rectal wall (5–8). Note the visualization device on top of the balloon consisting of two metal markers. (b) Picture of one rectal balloon equipped with alanine dosimeters. For illustration, the positioning device was inserted. At the top of the rectal balloon, the visualization device with metal markers was added (Daniela Wagner, et al., 2008).

At the posterior rectal wall, relative deviations of up to 3.5% may occur. The dominant contribution to the overall uncertainty for the in vivo measurements was the positioning of the dosimeter probes in the patient's body and their corresponding localization in the Computed Tomography (CT) data. Therefore it is expected that improving the probe positioning in the patient's body ( e.g. by an increased visibility in the radiographic images ) will lead to more accurate results.

The method presented in this study turned out to be useful for in vivo quality control of the irradiations. The relative deviation between the dose determined by the ESR measurements and the planned dose determined by the treatment planning systems (TPS) was shown to be within the 5% limit recommended by the ICRU (ICRU, 1976) for doses above 0.7 Gy (Daniela Wagner, et al., 2008).

Alanine was used also for the in-phantom dose mapping in neutron capture therapy (NCT) (Baccaro, S., et al., 2004), Recoil proton dose can be measured by means of alanine detectors, after subtraction of gamma dose contribution as evaluated from the dose measured with TLD-300. To this purpose, it is necessary to study the sensitivity of alanine to proton recoils and the coefficient for converting the recoil proton dose in alanine to that in tissue.

### 5.2.1.4 Influencing factors

Time passes before starting measurements after irradiation of alanine dosimeters is a key factor for acquiring reliable measurements, it was found that there is complex dependence on the time conjugated with the received dose range (Nagy, V., and Desrseirs, M. 1996). However in one of recent studies (Anton, M., 2008), the postirradiation behavior is highly dependent on the dosimeter structure and constituents.

In addition to the time dependence, extensive studies on ambient environmental conditions effects on the alanine dosimeters sensitivity and response were performed (Nagy, V. et al., 2000, Sleptchonok, O.F., et al., 2000, Anton, M., 2005). It was found that the effect of storage conditions is dependent on the type of the dosimeter. Effect of higher temperatures also was studied (Maltar-Strmečkia, N. and Rakvin. B. 2005). Same studied parameters (relative humidity and postirradiation behavior) were carried out for a new alanine film dosimeters (Ruth M.D. Garciaa, et al., 2004).

Some other influencing factors were studied: the angular response of alanine dosimeters either in the form of pellets or powder (Jean-Michel Dolo, and Tristan Garcia. 2007), and the grain size (Tristan Garcia and Jean-Michel Dolo. 2007).

### 5.2.1.5 Energy dependence

Energy dependence of alanine dosimeters were investigated (Eva Stabell Bergstrand, 2003). It was found that the alanine response is 0.8% lower for high energy X-rays than for Co-60 gamma rays, this result indicates a small energy dependence in the alanine response for the high-energy photons relative to Co-60 which may be significant. Another recent study has been performed to reveal the response of alanine dosimeters to high energy photons 8 and 16 MV relative to the response of alanine dosimeter irradiated to the reference beam quality (Co-60), a confirmation of these results was made using EGSnrc package (Anton, M., et al., 2008). Additional study of energy dependence of Dl-alanine to 10 MV high energy photons has revealed the decrease in alanine response with respect to the Co-60 reference beam quality (Borgonove, A.F., et al., 2007).

## 5.2.1.6 Alanine derivatives

Some researchers tried to use different alanine substitute in order to take the advantages of alanine as solid-state dosimeter with minimum drawbacks, however, the positive results are not mandatory, for example: Polycrystalline phenyl-alanine and perdeuterated L-a-alanine (L-a-alanine-d4) were studied as potential high-energy radiation-sensitive materials (RSM) for solid state/EPR dosimetry. It was found that phenyl-alanine exhibits a linear dose response in the dose region 0.1–17 kGy. However, phenyl-alanine is about 10 times less sensitive to γ-irradiation than standard L-a-alanine irradiated at the same doses. Moreover, the EPR response from phenyl-alanine is unstable and, independent of the absorbed dose, decreases by about 50% within 20 days after irradiation upon storage at room temperature. γ-irradiated polycrystalline perdeuterated L-a-alanine $(CD3CD(NH2)COOH)$ has been studied at room temperature by EPR spectroscopy. By spectrum simulations, the presence of at least two radiation induced free radicals, R1 $(CH3C^*(H)COOH)$ and R2 $(H3N^+-C^*(CH3)COO-)$, was confirmed very clearly. Both these radicals were suggested previously from EPR and ENDOR studies of standard alanine crystals (Veselka Ganchevaa et al., 2006).

Also, Minidosimeters of 2-methylalanine (2MA) were prepared and tested as potential candidates for small radiation field dosimetry (Bruno T. Rossi, et al., 2005). To quantify the free radicals created by radiation, a K-Band (24 GHz) EPR spectrometer was used. X-rays provided by a 6 MV clinical linear accelerator were used to irradiate the minidosimeters in the dose range of 0.5–30 Gy. The dose–response curves for both radiation sensitive materials displayed a good linear behavior in the dose range indicated with 2MA being more radiation sensitive than L-alanine. Moreover, 2MA showed a smaller LLD (Lower Limit of Detection) value. The proposed system minidosimeter/K-Band spectrometer was able to detect 10 Gy EPR spectra with good signal-to-noise ratio (S/N). The overall uncertainty indicates that this system shows a good performance for the detection of dose values of 20 Gy and above, which are dose values typically used in radiosurgery treatments (Chen,F., et al., 2007).

Alanine-in-glass dosimeters were prepared by packing pure polycrystalline L-*a*-alanine directly as supplied by the manufacturer in glass tubes. These dosimeters exhibited a linear dose response in the dose range from 0.1 to 20 Gy. These positive properties favor the polycrystalline alanine-in-glass tube as a radiation dosimeter (Anan M. Al-Karmi, and M.A. Morsy, 2008).

A new generation of self-calibrated alanine dosimeters were developed, and a regular international intercomparison is held for evaluation of radiation dos using these dosimeters (Gancheva, V. et al., 2008), these dosimeters are consists of RSM (α-alanine, sugar, other ones), $Mn^{2+}/MgO$ as internal EPR intensity standard (IES) and a binder. Necessity to assurance of very good homogeneity of dosimeter material; and the cost of IES present in the amount of some percent in each self-calibrated dosimeter are of the main shortcomings of this technique. Also, it was found that addition of gadolinium to the alanine dosimeters in definite amounts helps to improve the Linear Energy Transfer (LET) sensitivity for γ photons (because of its high atomic number, Z = 64) and thermal neutrons as well (because of its high thermal neutron cross section) (Marrale, M., et al., 2007).

## 5.2.2 Other organics

The search for new EPR dosimeter is non stopping, the search for organic dosimeter ensures the minimum energy dependence at least at 0.5 MeV and higher photon energies, and hence tissue equivalency of the dosimeter (Maghraby, A., and Tarek, E., 2006), for example: sulfanilic acid possesses several good features of the good dosimeter and is characterized by its simple spectrum. Although its sensitivity is less than that of alanine, it could be pressed into pellets purely without need to a binder, and hence more homogeneity could be achieved. Sulfanilic acid is nearly tissue equivalent which enables its use in radiation therapy dosimetry, also it is isotropic and its detection limit is about 100±30 mGy. Sulfanilic acid EPR signal intensity shows noticeable stability for a sufficient time, which enables its use as a transfer dosimeter. Sulfanilic acid deserves further studies in order to be established as a common radiation dosimeter using EPR (Maghraby, A., and Tarek, E., 2006).

One of the most promising organic dosimeters ever since the discovery of alanine at 1962 (Bradshaw, W., et al., 1962), is the Lithium format (Gustafsson, H., et al., 2008). It is characterized by its simple spectrum (one narrow peak spectrum) and its high sensitivity (six times more than alanine), beside its obvious stability. Lithium formate also possesses tissue equivalency more than that of alanine, with suitable microwave dependence and modulation amplitude dependence (Tor Arne Vestad, et al., 2004). Lithium formate may replace alanine in the near future specially for its clinical uses.

## 5.3 Inorganic dosimeters

A wide variety of inorganic dosimeters were investigated, some of these dosimeters processed higher sensitivity than alanine with reasonable stability, and some examples are following:

## 5.3.1 Combined TL and EPR dosimeters

Beryllium oxide ceramics was investigated for TL and EPR dosimetry. Ceramics were doped with lithium and neodymium ions. TL and EPR signals associated with Li centers whose amplitude is proportional to the absorbed dose are observed. A complete anneal of the EPR signal takes place in the temperature range of the TL recordings (Kortov, V., et al., 1993b).

Onori, S., et al., have used the well known thermoluminescent material; calcium sulphate: Dy, for EPR high dose assessment (Onori, S., et al., 1998). Three EPR signals are detectable for $CaSO_4$: Dy phosphor, two out of the three signals were studied for high dose applications since one of them showed saturation at about 1 kGy. The concentration of both centers, $(SO_3)^-$ and Ca-vacancy, increases with dose at least up to $10^7$ Gy or more. In both cases, the dose-effect relationship is not linear, being supralinear for $(SO_3)^-$ center and sub-linear for $(Ca)^{2-}$ center. $(Ca)^{2-}$ center is very stable over time independently of dose, while $(SO_3)^-$ center is not stable over time and can be used for a first dose estimation soon after irradiation.

The effect of grinding on powder form of clear fused quartz was studied by EPR and TL techniques. The minimum detectable dose (MDD) for using EPR was about 2 Gy for the clear fused quartz in powder form, which is 2 times greater than the bulk form. The EPR signal for of background varied inversely with particle size and was quite high for sizes lower than 38 μm, while for a Co-60 irradiated samples (about 22 Gy), the EPR intensity of the coarse powder varied directly with particle size. Thus, the intensity of a particle size of 20-38 μm was very low (Ranjbar, A., et al., 1999).

Feasibility of reading LiF thermoluminescence dosimeters by EPR was studied by Breen, S., (Breen, S., and Battista, J., 1999). EPR signals can be observed in irradiated polycrystalline LiF rods, but only near liquid nitrogen temperatures. The magnitude of this signal was measured by two methods: direct measurement of the radiation-induced signal, and curve fitting. The direct measurement showed that the radiation-induced signal increased linearly with dose; however, the technique suffered from low sensitivity; at low doses (below 20 Gy) the signal was barely discernible above the noise in the spectrum.

The curve-fitting method isolated three peaks in the EPR spectra of TLD-100. One component increased linearly with dose, although the spectrum of unirradiated TLD-100 possessed a large contribution from this component. This behavior is similar to that observed in the peak isolated by direct measurement. A second component was uncorrelated with dose. The third component decreased with dose, although this part of the spectrum was modeled poorly, perhaps due to the presence of another (fourth) signal in the spectrum. Due to the experimental difficulty in interpreting these spectra, and the low sensitivity of the application, EPR is not recommended as a substitute for thermoluminescent dosimetry of polycrystalline LiF.

### 5.3.2 Other inorganic compounds

Sulfamic acid possesses high-sensitivity to gamma radiation and is able to detect radiation doses below 5.0 Gy which would be advantageous for a possible use in medical applications, beside other good dosimetric properties like almost energy independence, narrow line and simple spectrum of well-defined radicals, also predictable decay kinetics which enable its use as a transfer dosimeter (Maghraby, A., 2007).

Ferrous ammonium sulfate was used as a high dose dosimeter in the range from 33.5 to 546 kGy (Juárez-Calderón, J. et al., 2007). Also, Sulphur trioxide anion in $K_2CH_2(SO_3)_2$ and carbon dioxide anion in irradiated sodium formate ($NaHCO_2$) were suggested as inorganic alternatives to the EPR/alanine dosimeter by Keizer, P.N., (Keizer, P.N., et al. 1991). These two systems have a four-fold sensitivity advantage over alanine. The radicals sulphur trioxide anion and carbon dioxide anion are, moreover, found in a wide variety of matrices, and it may be possible to find one in which they are even stronger. Sulphur trioxide anion was found to be completely stable over time, while carbon dioxide anion was not so stable against decay as sulphur trioxide, for example the signal decayed by 30 % over the first six weeks when irradiated to 10 kGy. The main drawback in inorganic dosimeters is their non-water equivalence.

Dosimetric properties of magnesium sulphate were investigated (Morton, J., et al., 1993). On irradiation with Co-60 $\gamma$ rays, the stable $SO_3^-$ is produced whose EPR signal amplitude increases linearly with dose up to about $10^5$ Gy Advantages and disadvantages of the $SO_3^-$ radical system were compared with $\alpha$-alanine. The studied dose range was between 0.25 Gy and 50 Gy, focusing on the region below 10 Gy. Storage as well as measurement of the encapsulated samples was at room temperature, sample spectra are reproducible several weeks later. It was seen that the signal enhancement for magnesium sulphate is about 80%, quite significant for the lower dose range.

Magnesium oxide was suggested as a combined thermostimulated luminescence TSL-EPR detector for ionizing and ultraviolet (UV) radiations (Kortov, V., et al., 1993a).

Dose dependence of the EPR signal intensity caused by $Fe^{+3}$ ions impurities is linear from 1 to $10^4$ Gy for X irradiation. Temperature range of measurements for the EPR signal is expanded, allowing measurements of radiation dose at higher temperatures.

In Physikalisch-Technische Bundesanstalt (PTB), Schneider, C., (Schneider, C., 1994), tested the $SO_3^-$ radical in an anhydrous $MgSO_4$ matrix, it shows a three times stronger EPR signal than the alanine radical at the same dose. Because the $SO_3^-$ spectrum has negligible high frequency noise components, the pattern recognition of the derivative spectral shape out of the noise background is better. Peak-to-peak evaluation at low doses (<10 Gy) is therefore more accurate for this simple single line spectrum. Discussion of EPR dosimeter materials cannot circumvent the problem of water equivalence. At low energies (<50 keV) the photon response of a martial is proportional to $Z^4$ up to $Z^5$, so, for instance, with sulphur nuclei instead of oxygen in a dosimeter, the response will differ by a factor of more than 16. In practice, the mass collision stopping power (the quotient energy loss over beam pathlength, $(\Delta E/\Delta l)$ of alanine is about 3% lower than that of water in the electron energy range 0.1-1.0 MeV, whereas for $MgSO_4$ it is about 17% lower. Therefore, in the search for alternative EPR dosimeter materials, it is desirable that a matrix be found which combines spectral advantages (as shown by the $SO_3^-$ radical for example) with a low Z value for its constituent nuclei.

Electron paramagnetic resonance (EPR) studies have been made on lithium metaitanate ($Li_2TiO_3$) ceramics irradiated by gamma rays of Co-60 source at ambient and Liquid Nitrogen Temperatures (LNT). The EPR spectra have been found strongly dependent on irradiation dose and temperature. The radiation defects induced by gamma irradiation at LNT were found to disappear almost completely by heating up to 255 k. in contrast to this, the radiation defects produced by irradiation at ambient temperature showed tolerance of elevated temperatures up to about 600 k (Grišmanovs, V., et al., 2000).

The sulphur trioxide anion looks interesting, so, new inorganic dosimeters based on barium and strontium dithionates ($BaS_2O_6.2H_2O$ and $SrS_2O_6.4H_2O$) were developed by Bogushevich, S., (Bogushevich, S., and Ugolev, I., 2000). The minimum detection limit was found to be 0.05 Gy at the measurements uncertainty of ± 30%. The dependence for γ irradiated dosimeters was linear up to 50 kGy for barium dithionates, and 80 kGy for strontium dithionates. Radical ions $SO_3^-$ in barium and strontium dithioniates are stable at 20°C. Studies have shown that, at temperatures below 35°C, the signal is stable within ± 5% for at least two years. Similarly, Barium sulphate (BaSO4) was irradiated by γ-rays and analyzed with electron spin resonance (ESR) to study radiation induced radicals for materials as radiation dosimeter (Sharaf, M.A. and Gamal M. Hassan. 2004).

Ammonium dithionate has been investigated as a potential dosimeter material (Danilczuk, M., et al., 2008). The radical signal in irradiated polycrystalline samples is a structureless narrow line. Ammonium dithionate was found to be more sensitive an l-α-alanine by a factor of seven at he same spectrometer settings. The results indicate that the ammonium dithionate can be applied as a dosimeter for situations when a material more sensitive than l-α-alanine is needed.

### 5.4 Retrospective and emergency dosimetry
### 5.4.1 Teeth enamel dosimetry

The EPR spectrum of irradiated tooth enamel (Shin Toyoda, et al., 2008) contains a multitude of signals that can be divided into two categories, radiation-induced and radiation insensitive signals (Figure (7)). This approach is an approximation because the intensity of the so-called non-radiation sensitive EPR spectral components from tooth enamel are also slightly affected by irradiation, which is evident after irradiation with doses above one hundred Gray. However, these EPR spectral components can be considered as radiation insensitive in the application range of retrospective dosimetry.

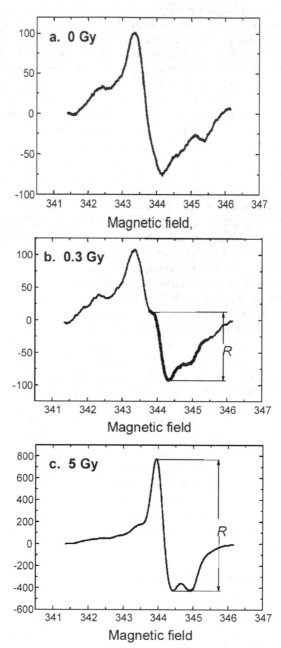

Fig. 7. EPR spectra of tooth enamel after irradiation with different doses: a- 0 Gy; b- 0.3 Gy; c- 5 Gy. The dosimetric component of the spectrum after irradiation with 0.3 Gy (in the middle – b) is in dark. R is the peak-to-peak amplitude used for EPR reconstruction (IAEA, 2002).

The majority of radiation-induced radicals in tooth enamel are carbonate derived, i.e.,$CO_2^-$, $CO_3^-$, $CO^-$, $CO_3^{3-}$, but also radicals derived from phosphate, i.e., $PO_4^{2-}$, and oxygen, i.e.,$O^-$ and $O^{3-}$ were identified. The identification of radicals was based on EPR and ENDOR (Electron Nuclear Double Resonance) measurements of irradiated synthetic hydroxyapatite doped with $^{13}C$ (Callens, F., et al., 1998). Not all radiation-induced radicals are thermally stable, e.g., the $CO_3^-$radical, with g-value of the EPR signal ranging from 2.0060 to 2.0122, decays completely at room temperature during the first two weeks after irradiation (Callens, F., et al., 1998, Cevc, P., et al., 1972, Romanyukha, A., et al., 1996). For dose reconstruction the asymmetric EPR signal with $g_\perp$=2.0018 and $g_{||}$=1.9971 (signal maximum at g=2.0032 and minimum at g=1.9971) is used. The signal is predominantly derived from stable $CO_2^-$ radical (IAEA, 2002). Two methods have been used to assess the absorbed dose of irradiated enamel by EPR: additive re-irradiation and the use of calibration curve, in the additive re-irradiation method: the sample is incrementally irradiated to construct a response curve specific to the sample in question (Pavlenko, A et al., 2007) as shown in Figure (8). The other method uses a universal calibration curve (EPR signal Intensity versus absorbed dose) generated using a large blended sample pool of enamel material designed to average the sample-to-sample variances. For doses greater than a few hundred mGy, the variation in EPR signal intensity from sample to sample for tooth enamel is about 10%, however, dose reconstruction using the universal calibration curve method is much less time-consuming and is non destructive (Desrosiers, M., and Schauer, D.A., 2001). Some other techniques are not widely used (Lanjanian, H., et al., 2008). Some consideration to be taken into account, such as the internal irradiation of human body by radioactive cesium isotopes (Borysheva, N., et al., 2007), many other influencing parameters have been investigated (El-Faramawy, N., 2008).

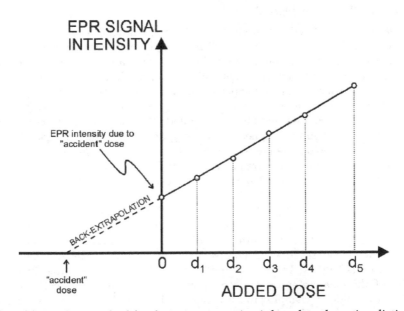

Fig. 8. The additive dose method for dose reconstruction is based on the re-irradiation ($d_1$ – $d_5$) of a tooth sample to obtain a sample-specific dose response curve, which is used to back extrapolate to the absorbed dose value (Desrosiers, M., and Schauer, D.A., 2001).

The dosimetric response of neutron irradiated human tooth enamel has been investigated (Khailov, Artem; et al., 2008). The neutron sensitivity (/Gy-100 mg) of human tooth enamel remained constant for various mean neutron energies ranging from 167 to 450 keV. Similarly, the EPR signal intensity remained independent of the neutron dose rate variation from 0.5 to 2.4 Gy/h (Rao F.H. Khan, et. al., 2004), other studies on the use of deciduous teeth have been performed (El-Faramawy, N.A., 2005, El-Faramawy, N., and Wieser, A., 2006).

## 5.4.2 Emergency dosimetry

There is growing awareness of the need for methodologies that can be used retrospectively to provide the dosimetry needed to carry out triage immediately after an event in which large numbers of people have potentially received clinically significant doses of ionizing radiation (Trompier François, et al., 2008b, Trompier François, et al., 2007a). Although some very promising approaches are being developed using biologically based parameters there also is recognition that such measurements have the potential to be confounded by other physiological and pathophysiological factors that are likely to be present in such event (Nicolalde, Roberto J; et al., 2008).

In contrast, the EPR measurements are based on physical changes in tissues whose magnitudes are not affected by the factors that can confound biologically based assessments. The EPR methods are based on the generation of stable free radicals, whose magnitude is proportional to the total dose of radiation received by the tissue, thereby allowing these tissues to be used as endogenous physical dosimeters (Nicolalde, Roberto J; et al., 2008).

### 5.4.2.1 Human subjects

#### 5.4.2.1.1 In vivo EPR measurements of teeth

*In vivo* measurements of radiation-induced EPR signals in teeth is a safe technique (Ann Barry Flood, et al., 2007) and currently it utilizes a large permanent magnet (40 mT) and, in principle, this system could be deployed in the .field using a small vehicle (Dong, Ruhong; et al., 2008). While clones of this system (see Figure (9)) would be an effective component of large deployment teams, a smaller magnet system would facilitate wider distribution of this capability. The feasibility of such magnet systems has been demonstrated (Swartz et al., 2007). These are in a form that could be incorporated into a helmet-like structure that would fit over the head. An intraoral magnet is also being developed. It is anticipated that within several years, the technology will be advanced to a point where it may be possible to obtain sufficient sensitivity with lower frequencies and thus lower the requirements for the magnetic field (Benjamin B. et al., 2008). This would further decrease the size of the magnet that is needed.

A current laboratory-based system (Benjamin B.Williams, et al., 2007) can make measurements comfortably in human subjects with a 5-min acquisition time providing dose resolution of ±0.75 Gy (1 SD) and a threshold of not more than 2.0 Gy, with the result being immediately available. There are a number of areas in which improvements should be feasible within 1–2 years. Improvements that are in process include: increasing the sensitivity of the existing types of resonators and the number of teeth in which the measurement is made by changing the size and/or shape of the resonator (see Figure (10)), improving data analysis (Demidenko, E., et al., 2007), increasing microwave power, and reducing sources of noise. Dose resolution can be improved immediately by extending the time for the measurement, with the increase being proportional to the square root of the

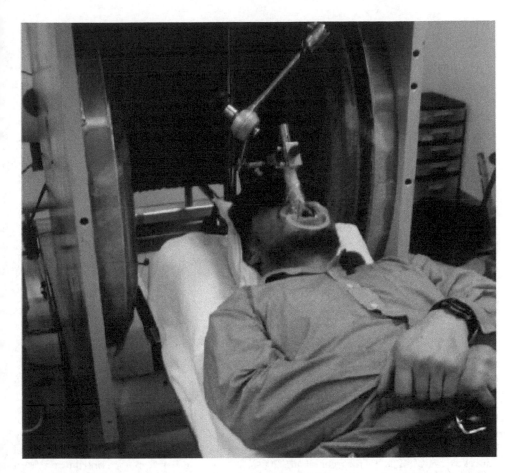

Fig. 9. Clinical spectrometer with a volunteer in it (Harold Swartz et al., 2007).

time of the measurements (i.e. increasing acquisition time from 5 to 20 min would increase the resolution by a factor of two) and by making the measurements in more than one tooth simultaneously. While the threshold, sensitivity, and accuracy can be improved further, there are some caveats that pertain to this method regardless of such improvements. The measured quantity is absorbed dose to teeth, not the critical organs of interest in radiation protection. This is not a problem if the exposure is homogeneous. In the event of an asymmetric exposure it may be feasible to utilize the Monte Carlo simulations of doses to human teeth from photon sources of eight standard irradiation geometries that have been performed and a set of dose conversion coefficients (DCCs) were calculated for 30 different tooth cells (Ulanovsky et al., 2005). DCCs were determined as ratios of tooth absorbed dose to air kerma for monoenergetic photon sources. To facilitate handling of the data set a software utility has been developed. The utility plots the DCC and computes conversion factors from enamel dose to air kerma and from enamel dose to organ dose for user-supplied discrete and continuous photon spectra.

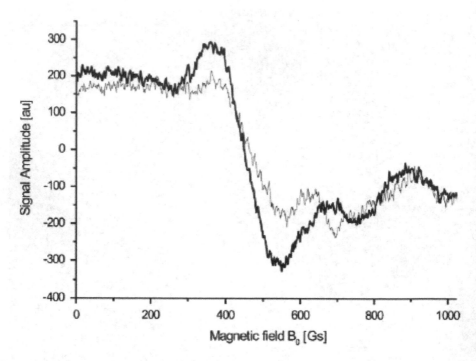

Fig. 10. Spectra from measuring two teeth (upper and lower) simultaneously (Harold Swartz et al., 2007).

The utility of EPR measurements for decision-making will depend on the homogeneity of the exposure and the type of radiation. The latter is noted because neutrons contribute very little to the EPR signal in teeth due to the low amount of hydrogen atoms in the enamel (Zdravkova et al., 2003; Trompier et al., 2004). If the dose has a major contribution from ingested or inhaled radionuclides, the dose delivered to the teeth may not closely reflect the dose to critical tissues (George A. Alexander et al., 2007).

### 5.4.2.1.2 Measurements in .fingernails (or toenails)

Although it was suggested as early as 1968 (Brady et al., 1968) that fingernails might be useful for after-the-fact dosimetry, only recently have the necessary studies been carried out to demonstrate convincingly that this approach has potential for use in the field for triage and perhaps even fairly precise determination of dose (Hongbin Li, et al., 2008). Preliminary results indicate that using simple cuttings from fingernails and X-band (9500 MHz) for the measurements, absorbed doses as of 1Gy with an uncertainty of ±0.50 Gy (1 SD) can be obtained with currently available techniques and instruments (Romanyukha et al., 2007; Trompier François et al., 2007b). If the use of fingernails for field dosimetry continues to develop, there should be no difficulty in constructing a field-deployable 9500MHz spectrometer for this purpose, which would be lightweight and automated for use by minimally trained individuals. The radiation-induced signals in fingernails are stable for at least several days (and much longer if the samples are collected within a few hours after the event and stored at low temperature). Because the measurements would be made in vitro, it should be possible to calibrate the radiation response of each sample by a simple procedure in which radiation is added to the sample. A potential advantage of measurements in fingernails, especially if combined with in vivo EPR dosimetry of teeth, include obtaining the measurement from a different location on the body (thereby providing a means to assess if there was an heterogeneous exposure).

Potential limitations to this approach may be overcome by simple modification of the collection process. For example, cutting of the fingernail can create a mechanically induced signal (MIS) that overlaps with the radiation-induced signal (RIS). However, the MIS decays rapidly and the decay is greatly accelerated by simple chemical treatment as shown in Figure (11), (Xiaoming He, et al., 2008, Dean Wilcox, et al., 2008). The influence of this MIS also can be removed by appropriate data processing because the shape is different from the RIS (Swarts Steven, et al., 2008). As is the case with any technique that requires removal of a sample from the subject, there is a potential for mislabeling the sample. This problem can be reduced by the development of automated procedures to rapidly remove any MIS and, if necessary, to calibrate the individual sample. Because only minimal manipulation of the sample is required and the measurement can be made within 5 min, it is feasible to determine the absorbed dose while the subject is still present. Finally, this method may not be applicable in children where nail volume is low (George A. Alexander, et al., 2007). Recently, doses in the range of 0.4 Gy could be detected using more sensitive EPR spectrometer with a resonator of high quality factor (Hirosuke Suzuki, 2008).

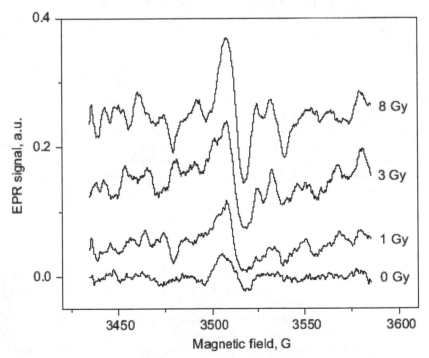

Fig. 11. EPR Spectra of fingernails treated with dithiothritol for 20 min after receiving different radiation doses (Romanyukha , a., et al., 2007).

### *5.4.2.1.3 Hai*

Hair is composed mainly of alpha keratin same like fingernails, however its use as accidental dosimeter is limited because of the melanin pigments (Trivedi, A., and Grenstock, C., 1993). Some attempts were performed to study different hair types, with different pigments, but it was found that hair EPR signal exhibit some complexities that limit its use in dose evaluation. On the other hand, recently, melanin itself (in hair) was used to indicate the presence of radiation-induced radicals (Thomas Herrling et al., 2008).

### *5.4.2.1.4 Measurements in "biopsies" of teeth using 9500 MHz EPR*

Many studies have demonstrated that retrospective measurements of dose by examination of isolated teeth with higher frequency EPR can provide very accurate estimates of dose at times ranging from immediately after the exposure to archeologically relevant times (Desrosiers and Schauer, 2001). The practical problem with this approach for acute dosimetry is the need to remove the tooth from the mouth. It now appears feasible, however, to obtain small samples from teeth rapidly and in a cosmetically acceptable manner. Small amounts can be used because of the increased sensitivity of higher frequency EPR and, there may be advantages in using frequencies even higher than 9500 MHz. Such a process could be very useful for triage and early assessment of dose to help in the determination of therapeutic intervention. Even if the technique of tooth biopsy does not fully meet the expectations, there may be situations where the value of the information that would be obtained would justify the removal of a tooth for in vitro measurement. The latter approach might be applicable in subjects for whom there are other indications of a potentially life-threatening dose and it is essential to verify the dose so that potentially risky therapies can be applied appropriately (George A. Alexander, et al., 2007).

### *5.4.2.1.5 Other tissues*

Several reports in the literature have described the effects of radiation in workers who exposed their fingers to intense radioactive sources. The radiation injuries occurring after local exposure to a high dose (20 to 100 Gy) could lead to the need for amputation. It has been investigated the use of low-frequency EPR spectroscopy to evaluate non-invasively the absorbed dose. Low-frequency microwaves are indeed less absorbed by water and penetrate more deeply into living material (~10 mm in tissues using 1 GHz spectrometers). Preliminary results obtained with baboon and human fingers compared with human dry phalanxes placed inside a surface-coil resonator. The EPR signal increased linearly with the dose. The ratio of the slopes of the dry bone to whole finger linear regression lines was around 5. The detection limit achievable with the present spectrometer and resonator is around 60 Gy, which is well within the range of accidentally exposed fingers (Zdravkova, M. et Al., 2004).

### 5.4.2.2 Non-human subjects

This item may contain everything man uses: his food, clothes, tools, constructive materials, plastics (Trompier François; et al., 2008a) and so many things that can be found in the day life. These things around the man receive same (more or less) radiation dose in case of radiation accidents, hence evaluation of radiation doses using these objects may reflect to a close relation some details. Materials have been investigated by different techniques: sugar, wall bricks, roof tiles, plastics, watch glass, ruby present in watches, medicines carried by persons and shell button (Baffa, O. 1, et al., 2008). Sweeteners based on saccharine, cyclamate, stevia and aspartame may be used.

A preliminary study of EPR signals of watch glass carried out by Wu (Wu, K., et al., 1995) and Trompier François (Trompier François, et al., 2008c) had suggested that it was an appropriate dosimeter. A large number of watch glass and display windows of mobile phone with EPR techniques to study the variability of dosimetric properties among the different types of sample. Dose response, signal stability and effects of storage conditions was presented (Bassinet, Céline; et al., 2008).
Egg shell was used as accidental EPR dosimeter, as it contains Calcium carbonate $CaCO_3$ (Da Costa, Z.M., et al., 2007), dose response is linear from the range of 3 Gy to 1kGy, and no dose rate dependence was observed.

### 5.4.3 Environmental dosimetry

Concern regarding the possibility of criminal or terrorist use of nuclear materials has led to an interest in developing the capability to measure radiation dose in a variety of natural and man-made materials. One such novel EPR dosimeter is drywall, a common construction material composed largely of gypsum (calcium sulphate dihydrate). A radiosensitive EPR signal in drywall has been observed, and suitable dose measurement protocols have been developed (Thompson, Jeroen W. et al., 2008). In other study, investigation of radiation induced radicals in coral reefs revealed that microcrystalline aragonite radicals can be used for environmental dosimetry purposes (Sharaf, M.A. and Gamal M. Hassan. 2004). Some attempts to follow the changes in sealevel were performed via the $^{230}Th/^{234}U$–ESR combined analysis for corals (Bonnie A.B. et al., 2007).

# 6. Dating

ESR has increasingly contributed to interdisciplinary research, such as dating of geological and archaeological materials (Dobosz, B., and Krzyminiewski, R. et al., 2007). Geosciences benefit from the potential of ESR spectroscopy to date minerals (e.g. carbonate deposits, calcites, silica, phosphates, etc.) (Thompson, J. and Schwarcz, H.P. 2008), and fossils (e.g. shells, corals, bones, teeth) (Ikeya, 2002). Noticeable progress in dating has, for instance, recently been reported on ESR studies (Skinner, Anne R. 2008) of limestones from Cretaceous–Tertiary boundary on the extinction theory of the dinosaurs 65 million years ago (Griscom and Beltrán-Lopez, 2002), and on ESR dating of the most ancient human settlements in Europe by ESR spectroscopy of fossil herbivorous teeth, coupled with U-series measurements, to model the uptake of uranium (Falguéres et al., 2002). The very long lifetime of some signals allows ESR to provide information over the entire span of human evolution. A survey on the present state in geological dating using ESR is given by Skinner (2000) and Skinner et al. (2002) (Dieter F. Regulla, 2005).
Dating of megafauna can contribute to the better knowledge of megafauna presence in this region as well as to the events associated to the extinction of these species. In some cases ESR is an interesting possibility of dating, since no remains of compounds containing C-14 can be found (Oliveira, L., et al., 2008).
Also, dating of barnacles may reflect past sealevel and sea level changes (Skinner, Anne, et al., 2008). Knowing the dates of these changes can constrain the periods of regional hominid occupation and provide a better understanding of local and global environments. Also, dating of teeth may lead to the accurate determination of sedimentation rates, improving the age estimates for humids and other Paleolithic cultural deposits (Anne R. Skinner, et al., 2008).

## 7. Uncertainties

One of the problems in evaluation of uncertainty in EPR tooth dosimetry is that different factors that influence the total uncertainty cannot be separated. Another problem is that blank samples are absent resulting into a lack of calibration standards. Each tooth extracted for medical indications has its own age that starts from hydroxiapatite formation. Due to natural radiation exposure there are not any tooth enamel with zero dose. Therefore, each tooth accumulates some unknown dose doe to natural background (Shishkina E., et al., 2008).

For the reasons above, the accuracy of EPR dosimetry is not a trivial task. A new approach for estimation of the uncertainty using combination of physical and numerical (Monte Carlo simulation) experiments has been developed. The exp erimental basis of that work was the results of intercomparison of three EPR laboratories: GSF (Munich, Germany); ISS (Rome, Italy) and IMP (Ekaterinburg, Russia).

Results of the study show that EPR measurement uncertainty depends on both amplitude of signal and sample mass. Moreover, statistical modeling of EPR signal demonstrates the presence of not only a random error of signal evaluation but also a systematic bias of spectra processing that depend on amplitude (Shishkina E., et al., 2008).

For radiation therapy, a small uncertainty of the applied dose is required. The uncertainty budget for the alanine/ESR dosimetry system of the Physikalisch-Technische Bundesanstalt (PTB) was determined, which relies on the use of a reference sample. A method is also presented which allows a reduction of the influence of fading or other changes of the ESR amplitude of irradiated alanine probes with time. If certain conditions are met which are described in detail, a relative uncertainty of less than 0.5% can be reached for probes irradiated with [60]Co in the 5–25 Gy dose range, including the uncertainty of the primary standard (Krauss, A. 2006). First results for dose values between 2 Gy and 10 Gy are presented as well (Mathias Anton, 2006). More methods of evaluation of uncertainties of alanine dosimetry can be found elsewhere (Bartolotta, A., et al., 1993, Bergstrand, S., et al., 1998, Kojima, T., et al., 1999, Nagy, V., et al., 2002).

## 8. References

Anan M. Al-Karmi, M.A. Morsy. 2008. " EPR of gamma-irradiated polycrystalline alanine-in-glass dosimeter". Radiation Measurements 43, 1315 – 1318.

Ann Barry Flood, Shayan Bhattacharyya, Javier Nicolalde, R., Harold M. Swartz. 2007. "Implementing EPR dosimetry for life-threatening incidents: Factors beyond technical performance". Radiation Measurements 42, 1099 – 1109.

Baccaro, S., Brunella, V., Cecilia, A., Costa, L. 2003. "γ irradiation of poly(vinyl chloride) for medical applications". Nuclear Instruments and Methods in Physics Research B 208, 195–198

Baccaro, S., Cemmi, A., Colombi, C., Fiocca, M, Gambarini, G., Lietti, B., Rosi, G. 2004. "In-phantom dose mapping in neutron capture therapy by means of solid state detectors". Nuclear Instruments and Methods in Physics Research B 213, 666–669.

Baffa, O. 1, Oliveira, P.B.1 José, F.A.1 and Kinoshita, A. 2008. "An attempt to use sweeteners as a material for accident dosimetry". BioDos 2008, Dartmouth College, Hanover, NH., USA (7-11 September 2008).

Bartolotta, A., Fattibene, P., Onori, S., Pantaloni, M., Pettti, E. 1993. "Sources of uncertainity in therapy level alanine dosimetry". Appl. Radiat. Isot. 44 (1\2), 13-17.

Basly, J.P., Longy, I., Bernard, M., 1997. In.uence of radiation treatment on two antibacterial agents and four antiprotozoal agents: ESR dosimetry. Int. J. Pharm. 154, 109–113.

Bassinet, Céline; Trompier, François; Clairand, Isabelle. 2008. "Radiation accident dosimetry on glass by tl and epr spectrometry". BioDos 2008, Dartmouth College, Hanover, NH., USA (7-11 September 2008).

Benjamin B. Williams, Ruhong Dong, Maciej Kmiec, Greg Burke, Bruce Corliss, Eugene Demidenko, Oleg Grinberg, Javier Nicolalde, Jenna Pollack, Tim Raynolds, Ildar Salikhov, Artur Sucheta, Piotr Lesniewski, Harold Swartz. 2008. "In Vivo EPR Tooth Dosimetry". BioDos 2008, Dartmouth College, Hanover, NH., USA (7-11 September 2008).

Benjamin B.Williams, Artur Sucheta, Ruhong Dong,Yasuko Sakata, Akinori Iwasaki, Gregory Burke, Oleg Grinberg, Piotr Lesniewski, Maciej Kmiec, Harold M. Swartz. 2007. " Experimental procedures for sensitive and reproducible in situ EPR tooth dosimetry". Radiation Measurements 42, 1094 – 1098.

Bergstrand, S., Hole, E., Sagstuen, E. 1998. "A simple method for estimating dose uncertainty in ESR\ alanine dosimetry". Appl. Radiat. Isot. 49 (7), 845-854.

Bin Kang, Chang S, Dai Y, Chen D (2007b) Radiation synthesis and magnetic properties of novel $Co_{0.7}Fe_{0.3}$/chitosan compound nanoparticles for targeted drug carrier. Rad Phys Chem 76: 968–973

Boess, C., Bo¨ gl, K.W., 1996. In.uence of radiation treatment on pharmaceuticals: a review. Alkaloids, morphine derivatives and antibiotics. Drug Dev. Ind. Pharm. 22, 495–529.

Bogushevich, S., and Ugolev, I., 2000. "Inorganic EPR dosimeter for medical radiology". Applied radiation and Isotopes. 52, 1217-1219.

Bonnie A.B. Blackwell, Steve J.T. Teng, Joyce A. Lundberg, Joel I.B. Blickstein, Anne R. Skinner. 2007. "Coupled $^{230}Th/^{234}U$–ESR analyses for corals: A new method to assess sealevel change". Radiation Measurements 42, 1250 – 1255.

Borgonove, A.F., Kinoshita, A., Chen, F., Nicolucci, P., Baffa, O. 2007. "Energy dependence of different materials in ESR dosimetry for clinical X-ray 10MV beam". Radiation Measurements 42, 1227 – 1232.

Borysheva, N., Ivannikov, A., Tikunov, D., Orlenko, S., Skvortsov, V., Stepanenko, V., Hoshi, M. 2007. "Taking into account absorbed doses in tooth enamel due to internal irradiation of human body by radioactive cesium isotopes at analysis EPR dosimetry data: Calculation by Monte-Carlo method". Radiation Measurements 42, 1190 – 1195.

Bradshaw, W., Cadena, E., Crawford, and Spetzler, H., 1962. "The use of alanine as a solid dosimeter". Rad.Res., 17, 11-21.

Brady, J., Aarestad, N., Swartz, H., 1968. "In vivo dosimetry by electron spin resonance spectroscopy". Health Phys. 15, 43–47.

Breen, S., and Battista, J., 1999. "Feasibility of reading LiF thermoluminscent dosimeters by electron spin resonance". Phys. Med. Boil., 44, 2063-2069 . UK.

Bruno Boizot, Guillaume Petite, Dominique Ghaleb, Georges Calas. 1998. " Radiation induced paramagnetic centres in nuclear glasses by EPR spectroscopy". Nuclear Instruments and Methods in Physics Research B 141, 580±584.

Bruno T. Rossi, Felipe Chen, Oswaldo Baffa. 2005. "A new 2 methylalanine-PVCESR dosimeter". Applied Radiation and Isotopes 62, 287–291

Callens, F. Vanhaelwyn, G., Mathys, P., Boesman, E. 1998. "EPR of carbonated derived radicals: Applications in dosimetry, dating and detection of irradiated food". Applied Magneic Resonance, 14, 235-254.

Cevc, P., Schara, M., Ravnik, N., 1972. "Electron paramagnetic resonance study of irradiated tooth enamel". Radiation Research. 51, 581.

Chen,F., Grae, C.F.O Baffa, O. 2007. "Response of L-alanine and 2-methylalanine minidosimeters for K-Band (24 GHz) EPR dosimetry". Nuclear Instruments and Methods in Physics Research B 264, 277–281

Ciesielski, B., and Wielopolski, L. 1994. "The effects of dose and radiation quality on the shape and power saturation of the EPR signal in alanine". Radiat. Res. 140, 105–111.

Costa, L. Brunella, V. Paganini, M., Baccaro, S., Cecilia, A. 2004. "Radical formation induced by γ radiation in poly(vinyl chloride) powder". Nuclear Instruments and Methods in Physics Research B 215 (2004) 471–478.

D'Oca M.C., Bartolotta A., Cammilleri C., Giuffrida S., Parlato A., Di Stefano V. 2008. "The additive dose method for dose estimation in irradiated oregano by thermoluminescence technique". Food Control. In Press.

Da Costa, Z.M., Pontuschka, W.M., Ludwig, V., Giehl, J.M., Da Costa, C.R., Duarte. E.L. 2007. "A study based on ESR, XRD and SEM of signal induced by gamma irradiation in eggshell". Radiation Measurements 42, 1233 – 1236.

Daniela Wagner, Mathias Anton, Hilke Vorwerk, Tammo Gsänger, Hans Christiansen, Bjoern Poppe, Clemens Friedrich Hess, Robert Michael Hermann. 2008. "In vivo alanine/electron spin resonance (ESR) dosimetry in radiotherapy of prostate cancer: A feasibility study". Radiotherapy and Oncology 88, 140–147.

Danilczuk, M., Gustafsson, H., Sastry, M.D., Lund, E., Lund A., 2008. "Ammonium dithionate − A new material for highly sensitive EPR dosimetry". Spectrochimica Acta Part A 69, 18–21.

Dean Wilcox, Xiaoming He, Jiang Gui, Andres Ruuge, Honbin Li, Benjamin Williams, Harold Swartz. 2008. " Dosimetry Based on EPR Spectral Analysis of Fingernail Clippings". BioDos 2008, Dartmouth College, Hanover, NH., USA (7-11 September 2008).

Deffner, U., and Regulla, F., 1980. "Influences of physical parameters on high-level amino acid dosimetry". Nuclear instruments and methods. 175, 134-135.

Demidenko, E., Williams, B.B., Sucheta, A., Dong, R., Swartz, H.M. 2007. "Radiation dose reconstruction from L-band in vivo EPR spectroscopy of intact teeth: Comparison of methods". Radiation Measurements 42, 1089 – 1093.

Desrosiers, M., 1989. "Gamma-irradiated seafood: identification and dosimeter by electron paramagnetic resonance spectroscopy" J. Agric. Food. Chem. 37, 96-100.

Desrosiers, M., Schauer, D.A., 2001. Electron paramagnetic resonance (EPR) biodosimetry. Nucl. Instrum. Meth. B 184 (1–2), 219–228.

Dobosz, B., and Krzyminiewski, R. 2007. "Linear transformation of EPR spectra as a method proposed for improving identification of paramagnetic species in ceramic". Applied Radiation and Isotopes 65, 392–396

Dong, Ruhong; Williams, Benjamin B.; Kmiec, Maciej; Demidenko, Eugene; Sucheta, Artur; Raynolds, Timothy; Nicolalde, Jean P.; Lesniewski, Piotr; Burke, Gregory; Swartz, Harold M. 2008. "Estimation of Radiation Doses of Teeth Using In Vivo EPR

Spectroscopy". BioDos 2008, Dartmouth College, Hanover, NH., USA (7-11 September 2008).

Eirik Malinen, Elin A. Hult, Eli O. Holea and Einar Sagstuen. 2003b. "Alanine Radicals, Part 4: Relative Amounts of Radical Species in Alanine Dosimeters after Exposure to 6-19 MeV Electrons and 10 kV-15 MV Photons" Radiation Research. 159, 149-153.

Eirik Malinen, Mojgan Z. Heydari, Einar Sagstuen and Eli O. Hole. 2003a. "Alanine Radicals, Part 3: Properties of the Components Contributing to the EPR Spectrum of X-Irradiated Alanine Dosimeters". RADIATION RESEARCH 159, 23-32.

El-Faramawy, N., 2008. "Investigation of some parameters influencing the sensitivity of human tooth enamel to gamma radiation using electron paramagnetic resonance". J. Radiat. Res. 49, 305-312.

El-Faramawy, N., and Wieser, A., 2006. 'The use of deciduous molars in EPR dose reconstruction". Radiat. Environ. Biophys. 44, 273-277.

El-Faramawy, N.A., 2005. "Comparison of γ and UV-light induced EPR spectra of enamel from deciduous molar teeth". Appl. Radiat. Isot. 62, 191-195.

Eva Stabell Bergstrand, Ken R Shortt, CarlKRoss and Eli Olaug Hole. 2003." An investigation of the photon energy dependence of the EPR alanine dosimetry system". Phys. Med. Biol. 48, 1753-1771.

Falguéres, C., Voinchet, P., Bahain, J.J., 2002. "ESR dating as a contributor to the chronology of the earliest humans in Europe". Adv. ESR Appl. 18, 67-76.

Fauconnet, A.L., Basly, J.P., Bernard, M., 1996. Gamma radiation induced effects on isoproterenol. Int. J. Pharm. 144, 123-125.

Gancheva, V. Yordanov, N.D., Callens, F., Vanhaelewyn, G., Raf., J. Bortolin, E., Onori, S., Malinen, E., Sagstuen, E., Fabisiak, S., Peimel-Stuglik, Z. 2008. " An international intercomparison on "self-calibrated" alanine EPR dosimeters". Radiation Physics and Chemistry 77, 357-364

George A. Alexander, Harold M. Swartz, Sally A. Amundson,William F. Blakely, Brooke Buddemeier, Bernard Gallez , Nicholas Dainiak, Ronald E. Goans, Robert B. Hayes, Patrick C. Lowry, Michael A. Noska, Paul Okunieff, Andrew L. Salner, David A. Schauer, Francois Trompier, KennethW. Turteltaub, Phillipe Voisin, Albert L.Wiley Jr., RuthWilkins. 2007. "BiodosEPR-2006 Meeting: Acute dosimetry consensus committee recommendations on biodosimetry applications in events involving uses of radiation by terrorists and radiation accidents". Radiation Measurements 42 (2007) 972 - 996.

Gibella, M., Crucq, A.S., Tilquin, B., 1993. De´ tection RPE de l'irradiation de me´dicaments. Journal de Chimie Physique et de Physico-Chimie Biologique 90, 1041-1053.

Gibella, M., Crucq, A.S., Tilquin, B., Stoker, P., Lesgards, G., Raffi, J., 2000. Electron spin resonance studies of some irradiated pharmaceuticals. Radiation Physics and chemistry, 58(1), 69-76.

Gordy. W., Ard, W., and Shields, H., 1955. "Microwave spectroscopy of biological substances. I: paramagnetic resonance in x-irradiated amino acids and proteins". Proc.Nat.Acad.Sci. USA.

Griscom, D.L., Beltra´ n-Lopez, V., 2002. "ESR spectra of limestones from the Cretaceous-Tertiary boundary: traces of a catastrophe". Int. Adv. ESR Appl. 18, 57-64.

Grišmanovs, V., Kumada, T., Tanifuji, T., and Nakagawa, T., 2000. "ESR spectroscopy of γ-irradiated $Li_2TiO_3$ ceramics". Radiation physics and chemistry. 58, 113-117.

Gunther, M.R., Kelman, D.J., Corbett, J.T., and Mason, R.P. (1995) Self-peroxidation of methmyoglobin results in formation of an oxygen-reactive tryptophan-centered radical. J. Biol. Chem. 270, 16075-16081.

Gustafsson, H., Lund, E., Olsson, S. 2008. "Lithium formate EPR dosimetry for verifications of planned dose distributions prior to intensity-modulated radiation therapy". Phys. Med. Biol. 53 4667-4682.

Hirosuke Suzuki, Katsuhisa Kato, Kenji Tamukai, Naoki Yoshida, Hiroaki Ohya, Kazunori Anzai, H. M. Swartz. 2008. "0.4 Gy and more sensitive Compact ESR system for irradiated nail". BioDos 2008, Dartmouth College, Hanover, NH., USA (7-11 September 2008).

Hongbin Li, Xiaoming He, Andres Ruuge, Jiang Gui, Ben Williams, Theodore E. MacVeagh, Dean E. Wilcox, Harold M. Swartz. 2008. "The Impact of Hydration and Microwave Power on EPR Signals in Fingernails/Toenails". BioDos 2008, Dartmouth College, Hanover, NH., USA (7-11 September 2008).

IAEA 2002 "Use of electron paramagnetic resonance dosimetry with tooth enamel for retrospective dose assessment" International Atomic Energy Authority, Vienna. IAEA-TECDOC-1331.

IAEA, (2004) 'Emerging applications of radiation in nanotechnology'. Proceedings of a consultants meeting held in Bologna, Italy, 22–25 March 2004, IAEA-TECDOC-1438

IAEA, 1978 "Food preservation by irradiation". International Atomic Energy Agency, Vienna, Austria, V. I and II.

ICGFI, 1999. "Facts about food irradiation" International Consultative Group on food irradiation.

ICRU, 1976. "Determination of absorbed dose in a patient irradiated by beams of x or gamma rays in radiotherapy procedures". Report 24, International Commission on Radiation Units and Measurements, Bethesda, MD.

Ikeya, M., 1993. "New applications of electron spin resonance-dating, dosimetry and microscopy". World scientific, Singapore.

Ikeya, M., 2002. "New prospects of ESR dosimetry and dating in 21th century". Adv. ESR Appl. 18, 3-7.

Jacobs, G.O., 1995. A review of the effects of gamma-radiation on pharmaceutical materials. J. Biomater. Appl. 10, 59–96.

Jacobs, G.P., 1983. Stability of cefazolin and other new cephalosporins following gamma irradiation. Int. J. Pharm. 17, 29–38.

Jaroslaw M. Wasikiewicz, Fumio Yoshii, Naotsugu Nagasawa, Radoslaw A. Wach, Hiroshi Mitomo. 'Degradation of chitosan and sodium alginate by gamma radiation, sonochemical and ultraviolet methods'. 2005. Radiation Physics and Chemistry 73 (2005) 287-295.

Jayakumara R, Prabaharana M, Reis RL, Mano JF (2005) Graft copolymerized chitosan – present status and applications. Carboh Polym 62: 142–158.

Jean-Michel Dolo, and Tristan Garcia. 2007. "Angular response of alanine samples: From powder to pellet". Radiation Measurements 42, 1201 – 1206.

Juárez-Calderón, J. Negrón-Mendoza, A., Ramos-Bernal, S. 2007. "Irradiation of ferrous ammonium sulfate for its use as high absorbed dose and low-temperature dosimeter". Radiation Physics and Chemistry 76, 1829–1832.

Kelman, D., J., DeGraz, J. A., and Mason, R.P. (1994) Reaction of myoglobin with hydrogen peroxide forms a peroxyl radical which oxidizes substrates. J. Biol. Chem. 269, 7458-7463.

Khailov, Artem; Ivannikov, Alexander; Skvortsov, Valeri; Stepanenko, Valeri; Tsyb, Anatoli; Hoshi, Masaharu. 2008. "The neutron dose conversion coefficients calculation forhuman tooth enamel in antropomorphic phantom". BioDos 2008, Dartmouth College, Hanover, NH., USA (7-11 September 2008).

Kizer, P.N., Morton, J.R., and Preston, K.F., 1991. "Electron paramagnetic resonance radiation dosimetry: possible inorganic alternatives to the EPR/alanine dosimeter". J. Chem. Soc. Faraday Trans. 87-19, 3147-3149.

Kojima, T., Tachibana, H., Haneda, N., Kawashima, I., Sharpe. 1999. "Uncertainity estimation in Co-60 gamma-ray dosimetry at JAERI involving a two-way dose range intercomparison study with NPL in the dose range 1-50 kGy". Radiation Physics and chemistry. 5, 619-626.

Kortov, V., Milman, I., Monakhov, A., and Sleasarev, A., 1993a. "Combined TSL-ESR MgO detectors for ionizing and UV radiations". Radiation protection dosimetry. 47, ¼, 273-276.

Kortov, V., Milman, I., Sleasarev, A., and Kijko, V., 1993b. "New BeO ceramics for TL ESR dosimetry". Radiation protection dosimetry. 47, ¼, 267-270.

Krauss, A. 2006. "The PTB water calorimeter for the absolute determination ofr absorbed dose to water in Co-60 radiation".Metrologia 43, 259.

Kurmaev EZ, Shin S, Watanabe M, Eguchi R, Ishiwata Y, Takeuchi TA, Moewes A, Ederer DL, Gao Y, Iwami M, Yanagihara M (2002) Probing oxygen and nitrogen bonding sites in chitosan by X-ray emission. J Electron Spectro Rel Phen 125: 133-138.

Lakshmi Prasuna, C.P., Chakradhar, R.P.S., Raoa,, J.L., Gopal, N.O. 2008. "EPR as an analytical tool in assessing the mineral nutrients and irradiated food products– vegetables". Spectrochimica Acta Part A, in press.

Lanjanian, H., Ziaie, F., Modarresi, M., Nikzad, M., Shahvar, A., Durrani, S.A. 2008. "Atechnique to measure the absorbed dose in human tooth enamel using EPR method". Radiation Measurements 43, S648–S650.

Le Hai, Diep TB, Nagasawa N, Yoshii F, Kume T (2003) Radiation depolymerization of chitosan to prepare oligomers. Nucl Inst Meth Phys Res B 208: 466–470

Maghraby, A., 2003. "Characterization and calibration of some organic compounds for use in EPR dosimetry". Ph.D. thesis, Cairo University, Giza, Egypt.

Maghraby, A., 2007 "Identification of Irradiated Crab using EPR", radiation measurement,V. 42, 2, 220-224.

Maghraby, A., 2007. "A sensitive EPR dosimetry system based on sulfamic acid". Nuclear Instruments and Methods in Physics Research B 262, 46–50.

Maghraby, A., and Maha Anwar Ali. 2007. " Spectroscopic study of gamma irradiated bovine hemoglobin". Radiation Physics and Chemistry 76, 1600–1605.

Maghraby, A., Tarek, E., 2006. "A new EPR dosimeter based on sulfanilic acid". Radiation Measurements 41, 170– 176.

Maltar-Strmečkia, N. and Rakvin. B. 2005. "Thermal stability of radiation-induced free radicals in γ-irradiated L-alanine single crystals". Applied Radiation and Isotopes 63, 375–380.

Marrale, M., Braia, M., Gennaro, G., Triolo, A., Bartolotta, A. 2007. "Improvement of the LET sensitivity in ESR dosimetry for γ-photons and thermal neutrons through gadolinium addition". Radiation Measurements 42, 1217 – 1221.

Marta M, Patamia M, Lupi A, Antenucci M, et al (1996) Bovine hemoglobin cross-linked through the β-chains, functional and structural aspects The Journal of Biological Chemistry, 271(13), 7473–7478.

Mathias Anton 2008. "Postirradiation effects in alanine dosimeter probes of two different suppliers" Phys. Med. Biol. 53, 1241-1258.

Mathias Anton, 2005. "Development of a secondary standard for the absorbed dose to water based on the alanine EPR dosimetry system". Applied Radiation and Isotopes 62, 779–795.

Mathias Anton. 2006. "Uncertainties in alanine/ESR dosimetry at the Physikalisch-Technische Bundesanstalt". Phys. Med. Biol. 51 5419-5440.

McConnell, H., Heller, H., Cole, T., and Fessenden, R., 1960. J. Am. Chem. Soc., 82, 766.

Mehta, K., and Girzikowsky, R., 2000. "IAEA high-dose intercomparison in Co-60 field". Applied Radiation and Isotopes. 52, 1179-1184.

Miki T, Kai A and Ikeya M (1987) Electron spin resonance of blood stains and its application to the estimation of time after bleeding Forensic Science International, 35, 149-158.

Misra H and Fridovich I (1972) The generation of peroxide radical during the autoxidation of hemoglobin. J. Biol. Chem. 247, 6960-6962.

Miyagawa, I., and Gordy, W., 1959. J. Chem. Phys., 30, 159, 1960, ibids, 32, 255.

Miyagusku, L., Chen, F., Kuaye, A., Castilho, C.J.C., Baffa, O. 2007. "Irradiation dose control of chicken meat processing with alanine/ESR dosimetric system". Radiation Measurements 42, 1222 – 1226.

Miyazaki, T., Kaneko, T., Yoshimura, T., Crucq, A.S., Tilquin, B., 1994. Electron spin resonance study of radiosterilization of antibiotics: ceftazidime. J. Pharm. Sci. 83, 68–71.

Mojgan Z. Heydari, Eirik Malinen, Eli O. Hole, and Einar Sagstuen. 2002. "Alanine Radicals. 2. The Composite Polycrystalline Alanine EPR Spectrum Studied by ENDOR, Thermal Annealing, and Spectrum Simulations". J. Phys. Chem. A. 106, 8971-8977.

Morton, J., Ahlers, F., and Schneider, C., 1993. "ESR dosimetry with magnesium sulfate". Applied radiation and isotopes. 40, 851-857.

Nagy, V., Sholom, S., Chumak, V., Desroseirs, M. 2002. "Uncertainities in alanine dosimetry in the therapeutic dose range". Appl. Radiat. Isot. 56, 917-929.

Nagy,V., and Desrosiers, M., 1996. "Complex time dependence of the EPR signal of irradiated L-α-alanine". Applied radiation and isotopes. 47, 789-793.

Nam, J. and Regulla, D. 1989. "The significance of the international dose assurance service for radiation processing". Appl. Radiat. Isot. 40, 953–956.

Nette, H.P., Onori, S., Fattibene, Regulla, D., and Wieser, A., 1993. "Coordinated research efforts for establishing an international radiotherapy dose intercomparison service based on the alanine/ESR system". Appl. Rad. Isot. 44, 7.

Nicolalde, Roberto J; 1, Gougelet, Robert M; 2, Swartz, Harold M. 2008. "EPR Spectroscopy for the Process of Triaging Mass Casualties after a Catastrophic Nuclear Event: A Simulated Exercise". BioDos 2008, Dartmouth College, Hanover, NH., USA (7-11 September 2008).

Oliveira, L., Kinoshita, A., Lopes, R., Figueiredo, A., Baffa, O., 2008. "Dating of brazilian southern megafauna by ESR spectroscopy". BioDos 2008, Dartmouth College, Hanover, NH., USA (7-11 September 2008).

Onori, S., Bortolin, E., Lavalle, M., and Fuochi, P., 1998. "CaSO$_4$: Dy phosphor as a suitable material for EPR high dose assessment". Radiation physics and chemistry. 52, 1-6, 549-553.

Onori, S., Pantalon, M., Fattibene, P., Signoretti, E.C., Valvo, L., Santucci, M., 1996. ESR identi.cation of irradiated antibiotics: cephalosporins. Appl. Radiat. Isotop. 47, 1569-1572.

Parlato, A. Calderaro, E. Bartolotta, A._, D'Oca, M.C. Giuffrida, S.A. Brai,L. M. Tranchina, Agozzino, P. Avellone, G. Ferrugia, M. Di Noto, A.M. Caracappa. S. 2007. "Gas chromatographic/mass spectrometric and microbiological analyses on irradiated chicken". Radiation Physics and Chemistry 76 1463-1465.

Pavlenko, A Mironova-Ulmane, N Polakov D Riekstina, M. 2007. "Investigation of EPR signals on tooth enamel". J. Phys.: Conf. Ser. 93, 12047-12052.

Rachmilewitz E, Peisach J and Blumberg W (1971) Studies on the stability of oxyhaemoglobin A and its constituent chains and their derivatives J. Biol. Chem., 246, 3356-3366.

Ranjbar, A., Durrani, S., and Randle, K., 1999. "Electron spin resonance and thermoluminescence in powder of clear fused quartz: effect of grinding". Radiation measurements. 30, 73-81.

Rao F.H. Khan, Aslam, Rink, W.J., Boreham, D.R. 2004. "Electron paramagnetic resonance dose response studies for neutron irradiated human teeth". Nuclear Instruments and Methods in Physics Research B225, 528-534.

Regulla, D. F. 2005. "ESR spectrometry: a future-oriented tool for dosimetry and dating". Applied Radiation and Isotopes, 62, 117-127

Regulla, D., and Deffner, U. 1982. "Dosimetry by ESR spectroscopy of alanine". Appl. Radiat. Isot. 33, 1101-1114.

Reid, B.D., 1995. Gamma processing technology: an alternative technology for terminal sterilization of parenterals. PDA J. Pharm. Sci. Technol. 49, 83-89.

Romanyukha, A., Trompier, F., LeBlanc, B., Calas, C., Clairand,I., Mitchell, C., Smirniotopoulos, J.G., Swartz, H., 2007. "EPR dosimetry in chemically treated .fingernails". Radiat. Meas., 42, 1110-1113.

Romanyukha, A., Wieser, A., Regulla, D., 1996. "EPR dosimetry with different biological and synthetic carbonated materials". Radiation Protection Dosimetry, 65, 389-392.

Ruth M.D. Garciaa, Marc F. Desrosiers, John G. Attwood, David Steklenski, James Griggs, Andrea Ainsworth, Arthur Heiss, Paul Mellor, Deepak Patil, Jason Meiner. 2004. " Characterization of a new alanine .lm dosimeter: relative humidity and post-irradiation stability". Radiation Physics and Chemistry 71, 373-377.

Sagstuen, E., Eli O. Hole, Sølvi R. Haugedal, and William H. Nelson. 1997b. "Alanine Radicals: Structure Determination by EPR and ENDOR of Single Crystals X-Irradiated at 295 K". J. Phys. Chem. A 101, 9763-9772.

Sagstuen, E., Hole, E., Haugedal, S., Lund, A., Eid, O., and Erickson, R., 1997a. "EPR and ENDOR analysis of x-irradiated l-alanine and NaHC2O4. H2O. Simulation of microwave power dependence of satellite lines". Nukleonika 42, 353-372.

Schneider, C., 1994. "Electron spin resonance (ESR) spectroscopy applied to radiation dosimetry and other fields". Physikalisch-Technische Bundesanstalt (PTB-Bericht E-51), Braunschweig, Germany.

Schreiber, G., Wagner, U., Helle, N., Ammon, J., Buchholtz, H., Delinceé, H., Estendorfer, S., Von Grabowski, H., Kruspe, W., Mainczyk, K., Munz, H., Schleich, C., Vreden, N.,

Wiezorek., C., Bögl, K., 1993a. "Thermoluminescence analysis to detect irradiated fruit and vegetables-an intercomparaison study". Bericht des Instituts Für Sozialmedizin und Epidemiologie des Bundesgesundheitsamtes, Berlin, SozEp-Heft 3/1993.

Schreiber, G., Wagner, U., Leffke, N., Helle, N., Ammon, J., Buchholtz, H., Delinceé, H., Estendorfer, S., Fuchs, K., Von Grabowski, H., Kruspe, W., Mainczyk, K., Munz, H., Nootenboom, H., Schleich, C., Vreden, N., Wiezorek., C., Bögl, K., 1993b. "Thermoluminescence analysis to detect irradiated spices, herbs, and spice and herb mixtures – an intercomparison study". Bericht des Instituts Für Sozialmedizin und Epidemiologie des Bundesgesundheitsamtes, Berlin, SozEp-Heft 2/1993.

Schreiber, G., Ziegelmann, B., Quitzsch, G., Helle, N.,Bögl, K.,1993c. "Luminescence techniques to identify the treatment of foods by ionizing radiation". Food Structure, 12, 385.

Sharaf, M.A. and Gamal M. Hassan. 2004. " Radiation induced radical in barium sulphate for ESR dosimetry: a preliminary study". Nuclear Instruments and Methods in Physics Research B 225, 521–527.

Sharaf, M.A. and Gamal M. Hassan. 2004. "ESR dosimetric properties of modern coral reef". Nuclear Instruments and Methods in Physics Research B 217, 603–610.

Shin Toyoda, Eldana Tieliewuhan, Satoru Endo, Ken'ichi Tanaka, Kunio Shiraishi, Chuzo Miyazawa, Alexandre Ivannikov, Masaharu Hoshi, Kenzo Fujimoto, Makoto Akashi. 2008. "ESR dosimetry of enamel and dentin taken from victims of JCO accident". BioDos 2008, Dartmouth College, Hanover, NH., USA (7-11 September 2008).

Shishkina E., Ivanov D., Wieser A., Fattibene P. 2008. "Application of experimental and numerical methods for estimation of uncertainties in EPR tooth dosimetry". BioDos 2008, Dartmouth College, Hanover, NH., USA (7-11 September 2008).

Signoretti, E.C., Valvo, L., Fattibene, P., Onori, S., Pantalon, M., 1994. "Gamma radiation affects on cefuroxime and cefotaxime. Investigation on degradation and syn-anti izomerization. Drug Dev. Ind. Pharm. 20, 2493–2508.

Skinner, A.F., Blackwell, B.A.B., Chasteen, N.D., Brassard, P., 2002. "New clues to limits on ESR dating". Intern. Adv. ESR Appl. 18, 77–82.

Skinner, A.R., 2000. "ESR dating: is it still an 'experimental' technique?". Appl. Radiat. Isot. 52, 1311–1316.

Skinner, Anne R. 2008. "An overview of the relationship between ESR dating and other forms of dosimetry". BioDos 2008, Dartmouth College, Hanover, NH., USA (7-11 September 2008).

Skinner, Anne R. Blackwell, Bonnie A. B., Gong, Jane J. J., Blais-Stevens, Andrée. 2008. "A new dating proposal: electron spin resonance dating with pleistocene barnacles". BioDos 2008, Dartmouth College, Hanover, NH., USA (7-11 September 2008).

Ślawska-Waniewska A, Msiniewicz-Szablewska E, Nedelko N, Galazka-Friedman J and Friedman A, (2004) Magnetic studies of iron-entities in human tissues J. of magnetism and magnetic materials, 272-276, 2417-2419.

Sünnetcioğlu, M., Dadayli, D., Celik, S., Koksel, H., 1999. "Use of EPR spin probe technique for detection of irradiated wheat". Appl. Radiat. Isot. 50, 557-560.

Svistunenko, D A, Dunne J, Fryer M, Nicholls P, Reeder B, Wilson M, Bigott M, Cutruzzolà F and Cooper C (2002) Comparative study of tyrosine radicals in hemoglobin and myoglobins treated with hydrogen peroxide Biophysical Journal, 49, 2845-2855.

Svistunenko, D A. (2005) Reaction of haem containing proteins and enzymes with hydroperoxides: the radical view. Biochem. Biophys. Acta (bioenergetics). 1707, 127-155.

Svistunenko, D A. Chris E., Cooper (2004) A new method of identifying the site of tyrosyl radicals in proteins Biophysical journal, 87, 582 595.

Svistunenko, D A. Nathan A., Davies, Michael T. Wilson, Raz P. Stidwill, Mervyn Singer, and Chris E., Cooper (1997a) Free radical in blood: A measure of haemoglobin autoxidation *in vivo* J.Chem.Soc., Perkin trans. 2. 2539-2543.

Svistunenko, D A., Rakesh P., Sergey V., Voloshchenko, and Michael T. Wilson (1997b) The globin-based free radical of ferryl hemoglobin is detected in normal human blood. J. Biol. Chem. 272, 7114-7121.

Swarts, Steven; Black, Paul; and Bernhard, William. 2008. "Ex vivo analysis of irradiated finger nails: quantifying the radiation-induced signal in fingernails in the presence of an interfering signal from mechanically-induced radicals". BioDos 2008, Dartmouth College, Hanover, NH., USA (7-11 September 2008).

Swartz, H. Greg Burke, M. Coey, Eugene Demidenko, Ruhong Dong, Oleg Grinberg, James Hilton, Akinori Iwasaki, Piotr Lesniewski, Maciej Kmiec, Kai-Ming Lo, R. Javier Nicolalde, Andres Ruuge,Yasuko Sakata, Artur Sucheta, TadeuszWalczak, Benjamin B.Williams, Chad A. Mitchell, Alex Romanyukha, David A. Schauer. 2007. "In vivo EPR for dosimetry". Radiation Measurements 42, 1075 – 1084

Swartz, H.M., Burke, G., Coey, M., Demidenko, E., Dong, R., Grinberg, O., Hilton, J., Iwasaki, A., Lesniewski, P., Kmiec, M., Lo, K.-M., Nicolalde, R.J., Ruuge, A., Sakata, Y., Sucheta, A., Walczak, T., Williams, B.B., Mitchell, C., Romanyukha, A., Schauer, D.A., 2007. "In vivo EPR for Dosimetry", Radiat. Meas., 42, .5. 23.

Thomas Herrling, Katinka Jung , Jürgen Fuchs. 2008. " The role of melanin as protector against free radicals in skin and its role as free radical indicator in hair". Spectrochimica Acta Part A 69, 1429–1435.

Thompson, J. and Schwarcz, H.P. 2008." Electron paramagnetic resonance dosimetry and dating potential of whewellite (calcium oxalate monohydrate)". Radiation Measurements 43, 1219 – 1225.

Thompson, Jeroen W.; Abu Atiya, Ibrahim; Boreham, Doug R. 2008. "Electron paramagnetic resonance (EPR) dosimetry of drywall". BioDos 2008, Dartmouth College, Hanover, NH., USA (7-11 September 2008).

Tilquin, B., Rollmann, B., 1996. Recherches a` conseiller pour l'application de la sterilization ionisante des me´ dicaments. Journal de Chimie Physique et de Physico-Chimie Biologique 93, 224–230.

Tor Arne Vestad, Eirik Malinen, Dag Rune Olsen, Eli Olaug Hole, Einar Sagstuen. 2004. "Electron paramagnetic resonance (EPR) dosimetry using lithium formate in radiotherapy: comparison with thermoluminescence (TL) dosimetry using lithium fluoride rods". Phys. Med. Biol. 49, 4701-4715.

Tristan Garcia and Jean-Michel Dolo. 2007. " Study of the in.uence of grain size on the ESR angular response in alanine radicals". Radiation Measurements 42, 1207 – 1212.

Trivedi, A., and Grenstock, C., 1993. "Use of sugars and hair for ESR emergency dosimetry". Applied Radiation and Isotopes. 44 (1-2), 85-90.

Trompier, F., Fattibene, P., Tikunov, D., Bartolotta, A., Carosi, A., Doca, M.C., 2004. EPR dosimetry in a mixed neutron and gamma radiation .field. Radiat. Prot. Dosim. 110, 437–442.

Trompier, F., Kornak, L., Calas, C., Romanyukha, A., LeBlanc, B., Clairand, I., Mitchell C.A., Swartz, H., 2007b. "Protocol for emergency EPR dosimetry in. fingernails". Radiat. Meas. 42, 1085-1088.

Trompier, F., Sadlo, J., Michalik, J., Stachowicz, W., Mazal, A. Clairand, I., Rostkowska, J., Bulski , W., Kulakowski, A., Sluszniak, J., Gozdz, S., Wojcik, A. 2007a. "EPR dosimetry for actual and suspected overexposures during radiotherapy treatments in Poland". Radiation Measurements 42, 1025 - 1028.

Trompier, François; Bassinet, Céline; Clairand, Isabelle. 2008a. "Radiation accident dosimetry on plastics by EPR spectrometry". BioDos 2008, Dartmouth College, Hanover, NH., USA (7-11 September 2008).

Trompier, François; Bassinet, Céline; Romanyukha, Alex; Clairand, Isabelle. 2008c. "Overview of physical and biophysical techniques for radiation accident dosimetry". BioDos 2008, Dartmouth College, Hanover, NH., USA (7-11 September 2008).

Ulanovsky, A., Wieser, A., Zankl, M., Jacob, P., 2005. "Photon dose conversion coefficients for human teeth in standard irradiation geometries". Health Phys. 9, 645–659.

Venkatesh B, Ramasamy S, Asokan R and Manoharan P (1997) Hemichromes in hemoglobin – An EPR study J. of Inorganic Biochemistry, 67, 121.

Veselka Ganchevaa, Einar Sagstuenb, Nicola D. Yordanov, 2006. "Study on the EPR/dosimetric properties of some substituted alanines". Radiation Physics and Chemistry 75, 329–335

Wajnberg E and Bemski G (1993) Electron spin resonance measurements of erythrocytes and hemoglobin stored at 77 K Naturwissenschaften, 80, 472-473.

WHO, 1988. "Food irradiation: a technique for preserving and improving safety of food" World Health Organization, Geneva, Switzerland.

WHO, 1994. "Safety and nutritional adequacy of irradiated food" World Health Organization, Geneva, Switzerland.

Wu, K., Sun, C.P., Shi, Y.M. 1995. "Dosimetric properties of watch glass: a potential practical ESR dosemeter for nuclear accidents". Radiat. Prot. Dosim., 59, 223-225.

Xiaoming He, Jiang Gui, Hongbin Li, Andres Ruuge, Dean Wilcox, Benjamin Williams, Harold Swartz. 2008. "Investigation of the Mechanically Induced EPR Signals in Fingernails/Toenails". BioDos 2008, Dartmouth College, Hanover, NH., USA (7-11 September 2008).

Yūrūs , S., Korkmaz, M., 2005. "Kinetics of radiation-induced radicals in gamma irradiated solid cefazolin sodium". Radiat. Eff. Defect. Solid. 160, 11–22.

Zavoisky, E., 1945. "Spin-magnetic resonance in paramagnetics". J. Phys., USSR, 9, 211.

Zdravkova, M Crokart, N Trompier, F Beghein, N Gallez, B Debuyst, R. 2004. "Non-invasive determination of the irradiation dose in fingers using low-frequency EPR". Phys. Med. Biol. 49, 2891-2898.

Zdravkova, M., Crokart, N., Trompier, F., Asselineau, B., Gallez, B., Gaillard- Lecanu, E., Debuyst, R., 2003. "Retrospective dosimetry after criticality accidents using low-frequency EPR: a study of whole human teeth irradiated in a mixed neutron and gamma-radiation .field" Radiat. Res. 160, 168–173.

Ziegelmann, B., Bögl, K., Schreiber, G., 1999. "TL and ESR signals of mollusk shells-correlations and suitability for the detection of irradiated foods". Radiation physics and chemistry, 54, 413-423.

# 5

# Formation and Decay of Colour Centres in a Silicate Glasses Exposed to Gamma Radiation: Application to High-Dose Dosimetry

K. Farah[1,2], A. Mejri[1], F. Hosni[1],
A. H. Hamzaoui[3] and B. Boizot[4]
[1]*National Center for Science and Nuclear Technology, Sidi-Thabet,*
[2]*ISTLS, 12, University of Sousse,*
[3]*National Centre for Research in Materials Science, Borj Cedria, Hammam-Lif,*
[4]*Laboratory of Irradiated Solids, UMR 7642 CEA-CNRS -
Polytechnic School, Route de Saclay, Palaiseau,*
[1,2,3]*Tunisia*
[4]*France*

## 1. Introduction

The interactions of ionizing radiation with glass matrix produce ionization, excitation, and atomic displacement. The main modification induced during $\gamma$, X rays, or electron irradiation on the glass structure is the creation of stable defects and the changes of the valence state of lattice atoms or of the incorporated impurities in glass. Some of the modified electronic configurations or defects cause preferential light absorption. Thus, glass becomes coloured and consequently these defects are called "colour centres". These centres are of many types and depending on the glass composition (Yokota, 1954, 1956) and are associated with optical absorption bands and EPR signals.

The change of optical properties of glasses when subjected to ionizing radiation has been investigated by many authors due to wide applications of this kind of material. The earlier studies were focused on ways to prevent the darkening in glasses used in reactor or hot-cell windows and optical devices (Friebele, 1991). Recently, many studies have been concentrated on the application of the irradiation induced colour to develop recyclable colour glasses which is of great interest in the glass industry from the economical and environmental point of view (Sheng et al., 2002a).

In addition to the applications mentioned above, the ionizing radiation induced colour centres in some glasses have been found wide application in radiation dosimetry (Farah et al., 2010; Fuochi et al., 2008, 2009; Mejri et al., 2008). This material is very interesting for dosimetry and very useful for many applications such as food irradiation, sterilization of medical devices, radiation treatment of industrial and municipal waste-water and radiation processing of materials.

In the present work, we investigated the effect of the irradiation dose and thermal annealing on the formation and the decay of the induced colour centres in gamma irradiated silicate glass. The activation energies of the two colour centres were calculated from Arrhenius equation. In order to evaluate its potential as radiation-sensitive material in high dose dosimetry, the main dosimetric properties were also studied in details.

## 2. Materials and methods

### 2.1 Glass composition

The glass samples were obtained from the same glass sheets purchased from the local market and were cut into pieces of 11 x 30 x 3 mm$^3$ dimensions for optical measurements. The chemical composition of the glass samples were determined by the Prompt Gamma Activation Analysis technique (Anderson et al., 2004) in the Budapest Neutron Centre (wt%: 68.52 SiO$_2$, 13.77 Na$_2$O, 8.19 CaO, 4.34 MgO, 1.003 Al$_2$O$_3$, 0.588 K$_2$O, 0.105 Fe$_2$O$_3$ and about 3.5 % of other components). To avoid grease contamination on glass surface, which may affect the absorbance measurements, the samples were carefully cleaned with ethyl alcohol. A thermal treatment at 300°C for 1 h was used to eliminate any spurious optical signal. Then the samples were wrapped in aluminium papers and stored in a dust free dark place to avoid any possible light effect.

### 2.2 Irradiation sources and procedures

Irradiations of the glass samples were done at the Tunisian semi-industrial $^{60}$Co gamma irradiation facility at the dose rate of about 2 kGy/h (Farah et al. 2006). Dosimetry was done using Fricke and Ethanol-ChloroBenzene chemical dosimeters and the traceability was established with alanine/EPR dosimetry system in terms of absorbed dose to water traceable to Aérial Secondary Standard Dosimetry Laboratory (SSDL), Strasbourg-France (Aérial, 2011). All the irradiated samples were stored in the dark in a room where the temperature was maintained between 20 and 25°C and humidity 45-60% R.H.

### 2.3 Optical absorption measurements

When exposed to gamma radiation, glass turns to a brown colour in a quantifiable and reproducible manner. Measurement of the change in absorbance with calibrated spectrophotometers at specified wavelengths provides a method for accurately determining absorbed dose.

Optical absorption spectra were taken with a Perkin Elmer spectrophotometer Lambda 20 in the range 350-800 nm. The optical spectra of non irradiated samples were measured with reference to air. All optical spectra of the samples after irradiation were measured against to non irradiated sample in order to obtain the net induced changes of absorption. Genesis 5 spectrophotometer and Käfer MFT 30 thickness gauge were used to measure the specific absorbance changes produced in glass (i.e. absorbance divided by dosimeter thickness). An electrical furnace and a freezer were used to reach the desired temperatures.

### 2.4 Temperature control during irradiation

To control and maintain the glass samples at the desired temperature during gamma irradiation, Julabo refrigerating circulator type F25-EC, with ultra purified water for the temperature range 5-90°C or a mixture of water and ethylene glycol for the range –25°C to +50°C as coolant liquid, was used. Samples were placed inside an aluminium cylinder in which the coolant liquid was circulating thus maintaining the desired temperature. Before irradiation, the set-up was kept for 20 minutes to reach the equilibrium temperature. Temperature fluctuation inside the aluminium cylinder during irradiation was within ±1°C.

## 2.5 Method and conditions of calibration

In order to minimize the contribution of influence quantities to the overall uncertainty and to ensure similar irradiation conditions both for calibration and routine dosimetry during the production run, full in-plant calibration of the glass samples was performed by irradiating them at the Tunisian semi-industrial cobalt-60 gamma irradiation facility together with transfer standard dosimeters in the Risø HDRL calibration phantoms (Sharpe&Miller, 1999). Four glass samples were placed inside the phantom, together with four alanine transfer standard dosimeters in their small 3 mm thick polyacetal holders as shown. The phantoms containing the alanine pellets and the glass samples were then taped on cartons of simulated product (dummy product box) placed on the test site facing the cobalt source. Care was taken to minimize self shielding effects between the dosimeters. The distance of the dummy product box from the source pencils was such as to reduce the spatial variations in the radiation field over the surface area of the dosimeter package to negligible small values.

The alanine pellets were sent back after irradiation for evaluation to the Aérial SSDL. The absorbed doses, as measured by the alanine dosimeters, were then used to establish the calibration curves for the glass.

Temperature strips were placed in the dosimeter packages during the calibration irradiations to record maximum temperature. Temperatures were found to vary around a mean value of 26°C with a maximum variation of +3°C.

## 3. Results and discussion

### 3.1 Effect of gamma radiation dose

Before gamma irradiation, the glass used in this study was transparent. When irradiated, two induced bands have been observed at 410 and 600 nm leading to the development of the brown colour. The intensities of the overall absorption spectra are observed to increase progressively with increasing doses between 1 and 1200 kGy (Fig.1). It is obvious that the broad band at 600 nm is less sensitive to radiation than that at 410 nm.

The induced optical absorption by gamma irradiation in the visible range of this silicate glass is due to the generation of two Non Bridging Oxygen Hole Centres (NBOHCs) ( $\equiv$Si-O°): HC1 at 410 nm and HC2 at 600 nm (Griscom, 1984).

Fig.2 shows the two bands separated in the region between 375 and 800 nm of the absorption spectrum of glass irradiated to 10 kGy. The induced absorption spectra were well modeled through a Gaussian shape with correlation coefficients ($R^2$) better than 0.99.

Table 1 shows the results of the best fit with two Gaussian bands (correlation coefficients ($R^2$) better than 0.99) of the induced absorption spectra in the region between 380-800 nm. It was found that the gamma radiation dose had no influence on the absorption band position. The band peak positions the FWHM were relatively constant and only the heights and the area under the bands changed, suggesting that only an increase of the number of induced colour centres were affected by the increasing of the irradiation dose.

These results show that this glass is radiation-sensitive material and the induced colour centres may be used for dose determination in large dose range.

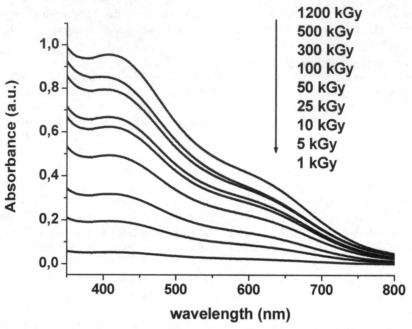

Fig. 1. Optical absorption spectra for the gamma irradiated glass samples in the range 1-1200 kGy

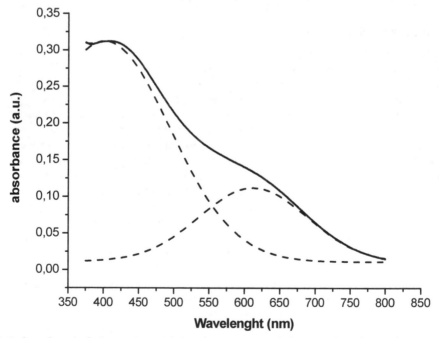

Fig. 2. Induced optical absorption with band separation of glass irradiated at 10 kGy.

Formation and Decay of Colour Centres in a Silicate Glasses Exposed to Gamma Radiation: Application
to High-Dose Dosimetry

111

| Dose (kGy) | λ (nm) | FWHM (nm) | A |
|---|---|---|---|
| 1 | 402 | 163.93 | 10.57 |
| | 598 | 174.93 | 4.50 |
| 10 | 405 | 162.42 | 62.31 |
| | 599 | 182.78 | 29.04 |
| 100 | 403 | 156.12 | 140.48 |
| | 602 | 174.82 | 43.93 |
| 1200 | 403 | 165.50 | 200.88 |
| | 603 | 182.07 | 60.31 |

Table 1. Results of band separation for the induced absorbance spectra

*Where* λ is the peak wavelength of band (nm), FWHM is the full width at half-maximum of
the band (nm) and A is the area of the band

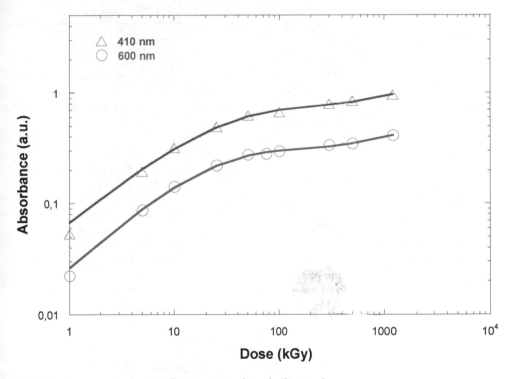

Fig. 3. Dose response curves of gamma irradiated silicate glass

The dose response curves shown in Fig.3 are successfully fitted following Mashkov equation
(Eq.1) (Mashkov et al., ). The fitting parameters are given in Table 2.

$$Q(D) = Q_c (kD)^b + Q_a [1 - \exp[-(kD)^b]] \qquad (1)$$

Where $Q(D)$, $Q_a(D)$ et $Q_c(D)$ represent the measured quantities (Optical Absorption)
proportional respectively to the total concentration of the colour centres for an accumulated

dose D, the concentration of the colour centres created by an extrinsic process of activation and the concentration of the colour centres created during an intrinsic process of the rupture, k is a rate constant and b is a number between 0 and 1. The nonlinear dose dependence of the specific absorbance as function of the dose can be interpreted in terms of two different processes involved in the creation of colour centres: the creation of colour centres induced by activation of precursory defects which saturate with the dose because their concentration in glass is limited, and an "unlimited" creation of new colour centres during an intrinsic process of Si-O-Si bond rupture (Boizot, 1997).

| Fitting parameters | Wavelengths | |
|---|---|---|
| | 410 nm | 600 nm |
| R2 | 0.995 | 0.996 |
| k (kGy$^{-1}$) | $0.049 \pm 0.007$ | $0.052 \pm 0.005$ |
| b | $0.72 \pm 0.06$ | $0.86 \pm 0.07$ |

Table 2. Fitting parameters for Figure 3

The results of fitting our experimental data by Mashkov Equation showed that the values obtained for the k parameter (defects formation rate or rate constant) are equal to 0.049 and 0.052 kGy$^{-1}$ for the two bands at 410 and 600 nm respectively. The similarity of this parameter for the two bands suggests that the related optical transitions correspond to the same type of NBOHCs in different configurations.

## 3.2 Activation energy

Figures 4 and 5 show the kinetics, for the absorption band at 410 nm, of temperature annealing performed between (-20 °C) and 150 °C of the glass samples irradiated at 30 kGy. The annealing process can be described by a sum of two first order decay kinetic functions ($\exp(-t/\tau)$) where t is the annealing time and $\tau$ is an appropriate time constant.

The activation energy characteristic of the annealing process was calculated from the Arrhenius equation. The obtained values for the fast and slow components of the 410 nm and 600 nm bands are presented in Table 3. The similarity of the activation energy values for both bands may suggest that the related optical transitions correspond to the same type of NBOHC's in different configurations (Griscom, 1984). The HC1 centre is a hole trapped in the 2p orbital of one non-bridging oxygen (NBO), analogue to NBOHC in silica glass, to which is correlated the absorption band at 410 nm. The HC2 centre is a hole trapped on two or three NBO's bonded to the same silicon to which is correlated the absorption band at 600 nm (Suszynska& Macalik, 2001).

By the mean of real-space multigrid electronic structure calculations, Jin and Chang proposed a diffusion mechanism of interstitial oxygen ions generated from $O_2$ under the UV irradiation with activation energies of 0.27 eV for $O^-$ and 0.11 eV for $O^{2-}$ (Jin & Chang, 2001). The value of 0.27 eV corresponding to the $O^-$ diffusion activation energy in the glass network under UV irradiation is identical, within experimental error, to our values of activation energies calculated from Arrhenius plots of the slow component $\tau$ corresponding to the long-time isothermal annealing. Approximately the same value of activation energy was found by Tsai and al. for the long-time slow thermal annealing of radiolytic atomic hydrogen in OH containing amorphous silica (Tsai et al., 1989).

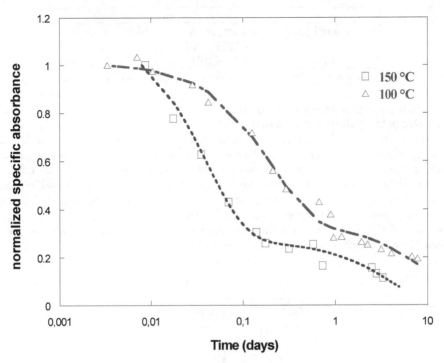

Fig. 4. Isothermal annealing of glass samples at high temperatures.

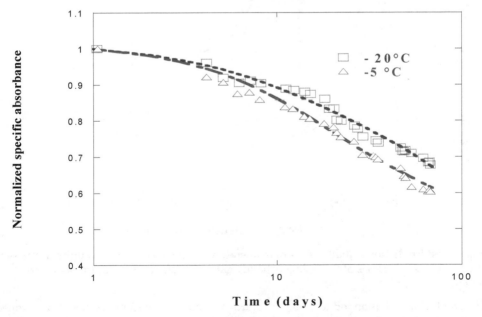

Fig. 5. Isothermal annealing of glass samples at low temperatures.

| | Activation energy (eV) | |
|---|---|---|
| Wavelength(nm) | Fast component ($\tau_1$) | Slow component ($\tau_2$) |
| 410 | 0.251± 0.022 | 0.310 ± 0.015 |
| 600 | 0.261± 0.022 | 0.353 ± 0.030 |

Table 3. Activation energy for the fast and slow components of both bands.

### 3.3 Application to high dose dosimetry
### 3.3.1 Room temperature fading behaviour

Five replicate glass dosimeters were irradiated with $^{60}$Co gamma rays to 30 kGy. The changes of absorbance were followed up to 535 days. After each measurement glass samples were stored in the dark at room temperature. Fig. 6 presents the fading behaviour of the 410 nm absorption band at room temperature respectively for the long-term and the short-term period. A strong fading can be observed in the first 9 days followed by a slow fading up to 535 days. The 600 nm absorption band showed similar behaviour.

The data in Fig. 6 were fitted using first-order kinetics based on the data after the first 10 days. The coefficient of correlation ($R^2$) was 0.99.

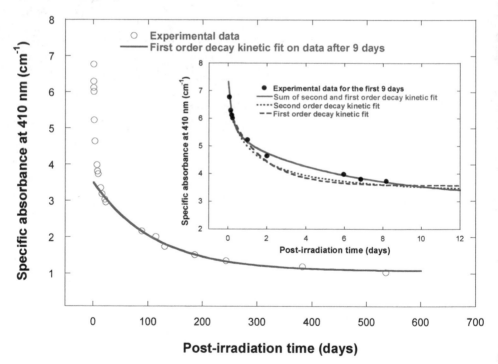

Fig. 6. Post-irradiation stability at room temperature of response of glass samples irradiated at 30 kGy. Inset: kinetics of the fading behaviour at room temperature for the short-term period.

The stability of radiation induced colour centres at room temperature was mainly controlled by the initial strong fading process. This initial fading process seems to follow neither a

simple first-order nor second-order kinetics, but it can be well described by a sum of a first
order decay kinetics function (exp (-t/τ)) and second order decay kinetics function
[1+ k (t/κ)]$^{-1}$, where t is the annealing time, τ and κ are appropriate time constants and k is a
fitting parameter.

The thermal decay of colour centres after irradiation is mainly controlled by the diffusion of
species in glass and it is generally attributed to diffusion - limited processes (Agnello & Boizot,
2003; Sheng et al., 2002c). Agnello and Boizot speculate that reaction of mobile species induced
by irradiation with the induced colour centres is in the origin of their thermal decay (Agnello&
Boizot, 2003). These mobile species are for example oxygen ions or anions and cations
impurities released by broken bonds. In fact, a large number of studies have demonstrated that
alkalis ions are movable and diffusible under irradiation. The association with NBO confines
the alkali ion to local motion, whereas the absence of a co-ordinating NBO allows the alkali ion
to explore more easily its environment (Ojovan & Lee, 2004).

By the combination of (MAS NMR) and (XPS) spectroscopy investigations, (Boizot and al.,
2000) provided additional evidence in favour of a long range migration of sodium in glass.

Sheng et al., 2002c), gave evidence that the long-term fading process of colour centres
induced by X-rays in soda-lime silicate glass was dominated by a first order kinetics, while,
both the first and the second order kinetics played role in the short-term fading process
(Sheng et al., 2002). Indeed, each NBOHC induced by gamma irradiation is surrounded by
electrons and other NBOHCs. They assumed that the recombination of NBOHCs is
controlled by the diffusion of electrons in the glass network (reaction 4) and/or by reaction
of NBOHC with neighboring NBOHCs (reactions 2 and 3):

$$\equiv Si - O° + e^- \rightarrow \equiv Si - O^- \tag{2}$$

$$2(\equiv Si - O°) \rightarrow \equiv Si - O - O - Si \equiv \tag{3}$$

It is obvious that the reaction (2) dominate the recombination process at short-term range
because of the small initial distance between NBOHCs (Waite, 1957). Sheng and al.
demonstrated that in addition to reaction (2), reaction (3) played also role in the short-term
recombination process, while the long-term recombination process was dominated only by
reaction (2).

### 3.3.2 Effect of post irradiation heat treatments on the room temperature fading

The effect of post irradiation heat treatments on the glass response fading was studied in the
temperature range of 60-150 °C using sets of three glass samples. After gamma irradiation
with 30 kGy absorbed dose, dosimeter sets were immediately submitted to the different heat
treatments for 20 minutes, which was found to be the best treatment time, and stored after
irradiation in the dark at room temperature. Optical absorbance measurements were carried
out up to two months. The specific absorbance values were normalized to the first
measurements taken 5 minutes after the heat treatments. The results are presented in Fig. 7
and Table 4. The best results have been obtained with heat treatments at 150 °C (20 min).
This procedure is very effective for the removal of unstable entities responsible for the initial
strong fading. The standard deviation of glass dosimeters response measurements is about
0.5 % (1σ) within the first two hours after irradiation. The response decay of irradiated glass
dosimeters is about 8 % between the first 24 h and 20 days. This means that glass dosimeters
can be evaluated either within the first two hours or just after one day after irradiation and
heating.

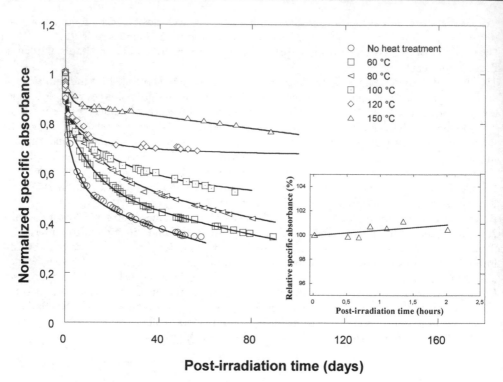

**Post-irradiation time (days)**

Fig. 7. Post-irradiation stability at room temperature of glass samples after 30 kGy irradiation and different heat treatments for 20 minutes. Inset: behaviour of glass samples, heated at 150 °C for 20 min, within the first two hours after the heat treatment.

| Storage time | Reduction of relative specific absorbance (%) | | | | | | | |
|---|---|---|---|---|---|---|---|---|
| | Room temperature | -5 °C | -20 °C | 60 °C | 80 °C | 100 °C | 120 °C | 150 °C |
| 24 h | 25 | 5 | 5 | 17 | 16 | 15 | 12 | 7 |
| 20 days | 54 | 40 | 39 | 49 | 45 | 35 | 29 | 15 |
| 60 days | 65 | 66 | 65 | 60 | 54 | 45 | 31 | 19 |

Table 4. Effect of thermal treatments on fading of glass samples after an irradiation of 30 kGy with gamma rays

### 3.3.3 Dose response curves

In order to find out the useful dose range for this silicate glass, the response curve (specific absorbance versus dose) was measured in the dose range 0.5 kGy– 87 kGy. Irradiations were carried out at the 60Co Gamma cell 220 excel of the Egyptian Radiation Technology Centre with the dose rate of 4.78 kGy/h at controlled temperature of 34°C. All data for the dose response curve were taken 24 h after irradiation. The reported specific absorbance was measured at 410 nm (Figure 8). The specific absorbance shows a rapid growth up to 40 kGy.

At higher doses the specific absorbance continued to grow slowly up to 87 kGy which was the upper dose level of the present experiments. The glass response had not yet reached saturation at this dose level. The dose response curve taken after irradiation and heat treatment at 150°C for 20 min for the dose range of 0.5–87 kGy is shown as inset of the Fig.8. The dose response curves shown in Fig.8 are successfully fitted following Mashkov equation.

| T °C | First-order kinetics τ (h) | Second order kinetics κ (h) |
|---|---|---|
| Room temperature | 189.6 | 30.9 |
| -5 | 1128 | 60 |
| -20 | 2184 | 117 |
| 60 | 169.5 | 7.5 |
| 80 | 168 | 0.43 |
| 100 | 110.4 | 0.024 |
| 130 | 6.25 | 0.0006 |
| 150 | 3.15 | 0 |

Table 5. Fading time constants for temperature range from -20 to 150 °C

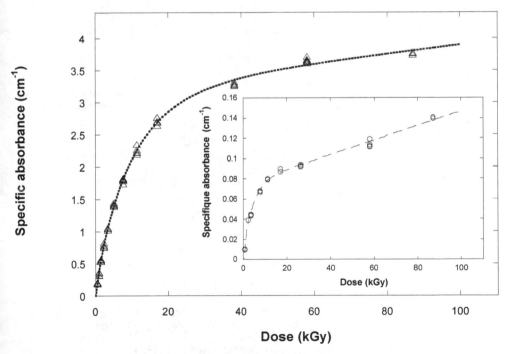

Fig. 8. Response curve for gamma irradiated glass in the dose range of 0.5-87 kGy (dose to water), absorbance at λ = 410 nm, dose rate: 4.78 kGy/h. Inset: response curve after heat treatment at 150°C for 20 min.

### 3.3.4 Effect of temperature during irradiation

The response of the majority of dosimeters is affected by the temperature during irradiation (Abdel-Fattah and Miller, 1996, Farah et al., 2004). In large gamma ray irradiators, the dosimeter temperature can reach 60°C for high doses. This effect should be carefully investigated especially if the dosimeters are going to be used for dose measurement at temperatures different than that for which they were calibrated.

In order to study the effect of temperature during irradiation on response of the glass dosimeters, samples were irradiated to an absorbed dose of 5 kGy at temperatures in the range from –3°C to +80°C, temperature range used during the usual irradiation processing. After irradiation, dosimeters were stored for 24 h in dark at room conditions (25 ± 3°C, 40-60% R.H.) then specific absorbance was measured at 410 nm. Figure 9 shows the specific absorbance, normalized at -3°C, plotted as a function of temperature during irradiation. The response of glass dosimeter decreases from –3°C to +80° C and the temperature coefficient of optical absorption was negative corresponding to a mean value of (-0.53 ± 0.02) %°C-1 and (-0.45 ± 0.04) % °C-1 respectively for 410 nm and 600 nm. These results extend the published data of (Zheng et al., 1988) and (Zheng, 1996). These authors observed, for glass dosimeters irradiated with 3 kGy, a negligible variation of absorbance at the wavelength of 500 nm between 0 and 50 °C followed by a fast decrease between 50 and 80 °C.

These differences can be explained by the fact that the colour centre studied by (Zheng et al., 1988) is probably due to free electrons trapped at some imperfection in the glass structure. While the induced absorption bands observed in our glass can be attributed to a trapped holes (NBOHCs colour centres).

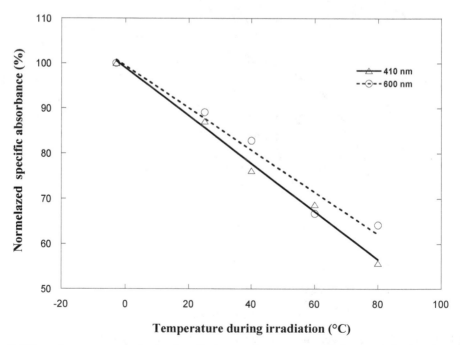

Fig. 9. Effect of temperature during irradiation on the response of glass samples irradiated at 5kGy and measured at 410 and 600 nm absorption bands

Formation and Decay of Colour Centres in a Silicate Glasses Exposed to Gamma Radiation: Application
to High-Dose Dosimetry

119

### 3.3.5 Relative humidity effect during post-irradiation storage

In order to investigate the effect of Relative Humidity (R.H.) during post-irradiation storage on the response of glass dosimeter, sets of three replicate glass samples were irradiated using 60Co gamma rays at a dose of 7 kGy. Following irradiation, glass samples were stored in the dark under different extreme R.H. conditions, dried condition (about 0%), moist condition (in water) and room condition (between 40-60%). The specific absorbance was measured at 410 and 600 nm up to 22 days. As seen in Figs. 10 and 11, small differences of specific absorbance have been observed between the glass dosimeters stored in the two extreme R.H. conditions (0 and 100%). Relative differences of specific absorbance compared to the ambient conditions of glass dosimeters stored in extreme R.H. conditions (0 and 100%) are significant for the two wavelengths 410 and 600 nm.

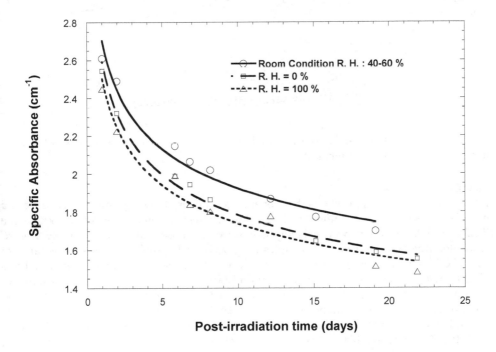

Fig. 10. Post-irradiation stability of irradiated glass to 7 kGy and stored at a different humidity conditions (410 nm band).

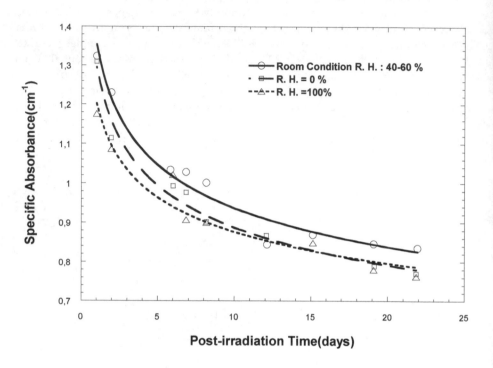

**Post-irradiation Time(days)**

Fig. 11. Post-irradiation stability of irradiated glass to 7 kGy and stored at a different humidity conditions (600 nm band).

### 3.3.6 Dose rate effect

In the semi-industrial applications that depending on plant design, products may be treated in different positions with different dose rates, whereas, dosimeters were calibrated at a fixed dose rate. In order to investigate the effect of the dose rate on the response of glass dosimeter, several groups of glass samples were irradiated to different doses in the range of 0.5-50 kGy at the following dose rates: 1 and 6 kGy/h. Glass samples were submitted to a thermal treatment at 150°C for 20 minutes immediately after irradiation. Specific absorbance measurements were carried out after 24 hours. As seen in Fig. 12, the dose rate effects did not appear to be significant in the dose range of 0.5-20 kGy. Only a weak change of response within 1-2% was observed. This result is in good agreement with published data (Engin et al., 2006; Zheng et al., 1988). Above 20 kGy, glass samples exhibit significant dose rate dependence within 8-11%. In fact, the absorbance increases within 8-11% when the dose rate decreases. This observation is in agreement with the data published by (Ezz-Eldin et al., 2008). For low dose rate the rate of electrons production is low, thus the ejected electrons has enough time to annihilate a glass defect or forming non-bridging oxygen centers which causes an increase in the glass absorbance.

For the high dose rate, the rate of electrons production become high, which gives a better chance for their fast recombination rather than for annihilation resulting in less absorbance.

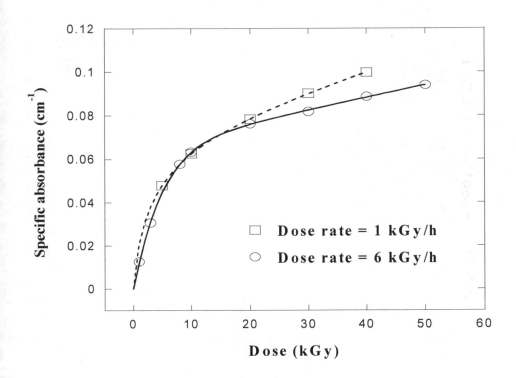

Fig. 12. Dose rate effect on the response of glass samples irradiated at 1 and 6 kGy/h and
measured at 410 nm absorption band.

### 3.3.7 Reuse and reproducibility of glass dosimeter

Few investigations were found in the literature (Sheng, 2002b) about the reuse of glass
dosimeters, by thermal bleaching of the radiation induced coloration.

In order to study the possibility of re-using glass dosimeter, three sets of glass samples, each
containing three, were irradiated, at room temperature, with 60Co gamma rays at three
different doses 1, 5 and 20 kGy.

Measurement of the specific absorbance was taken 30 min after irradiation. Then the
irradiated samples were submitted to a heat treatment of 300°C for 30 min, sufficient to
remove the radiation induced colour centres, before performing a new irradiation. This
procedure was repeated six times giving good reproducibility as it can be seen from the
results shown in Fig. 13. The standard deviation of the measurements was found to be lower
than 4% (1σ).

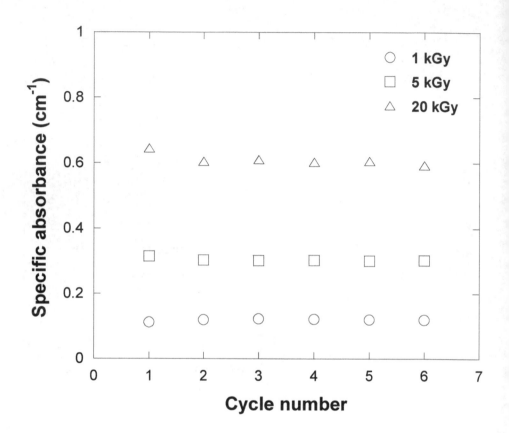

Fig. 13. Reproducibility of the results by reuse of glass dosimeter.

### 3.4 In-plant calibration of glass dosimeter for gamma irradiation
### 3.4.1 Calibration conditions
The in-plant calibration of glass dosimeters was done in the Tunisian semi-industrial cobalt-60 gamma irradiation facility rigorously following the procedure described in section 2.5 Method and condition of calibration. The phantoms, containing the alanine transfer standard dosimeters and the glass samples, were fixed on dummy product boxes and irradiated to nominal doses from 0.5 to 17 kGy at the dose rate of 1 kGy/h.

All dosimeters were read 24 hours after irradiation. Figs. 14 and 15 represent the response functions where the specific absorbance of the irradiated glasses vs dose to water, as measured by the alanine dosimeters, is reported. Here the mean values of the specific absorbance for each group of glass samples are plotted.

Formation and Decay of Colour Centres in a Silicate Glasses Exposed to Gamma Radiation: Application
to High-Dose Dosimetry

123

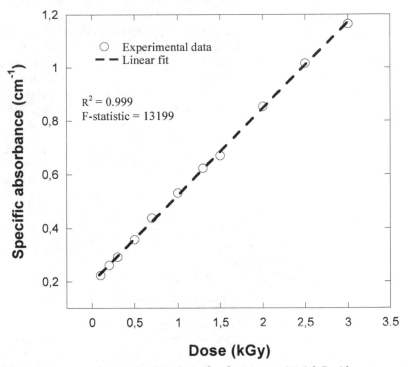

Fig. 14. Calibration curve of glass dosimeter in the dose range 0.1-3 kGy (dose to water).
Y = 0.1982 +0.3245.X equation is used for linear fit.

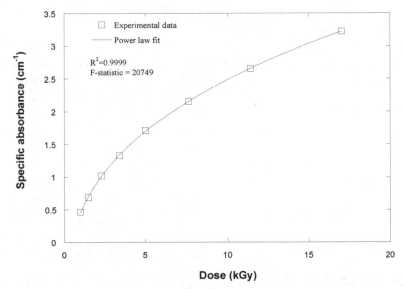

Fig. 15. Calibration curve of glass dosimeter in the dose range of 1-17 kGy (dose to water).
Y = - 1.262 + 1.717.X 0.34 equation is used for power law fit.

## 3.5 Estimation of uncertainty

The procedure outlined in ISO/ASTM Standard 51707 (ISO/ASTM, 2002) was followed to estimate the overall uncertainty in the measurement of absorbed dose from the glass dosimeter. Three components of uncertainty were taken into account. These included, (1) uncertainty in absorbed doses reported by the Aérial, (2) uncertainty in the spectrophotométric analysis of dosimeter response and measurement of dosimeter thickness (Eq. 4), and (Eq. 5) goodness of fit of the power law function to the calibration data.

$$CV_{overall}(\%) = \sqrt{\frac{\sum\limits_{i}(n_i - 1)S_{i-1}^2/\overline{k}^2}{\sum_i(n_i - 1)}} \times 100\% \tag{4}$$

Where:
C.V. (%) = overall coefficient of variation
Si-1= sample standard deviation for ith set of measurement,
(ni-1) degrees of freedom for ith set of data,
$\overline{k}$ = average value of specific absorbance for ith set of measurement,
ni = number of replicate measurements for ith set of data.

$$u_f = \sqrt{\frac{\sum\limits_{i=1}^{i=n}(y_i - f(x_i))^2}{n - m}} \tag{5}$$

Where:
uf = fit standard uncertainty
yi = y data value
f (xi) = predicted y value using the power law function
n = number of data points
m = number of coefficient fitted

Results of this analysis are given in Table 6. The components of uncertainty combined in quadrature gave an estimate of overall uncertainty at a 95% confidence level of ± 7.58% (2σ) which is in basic agreement with the expected uncertainty for the routine use of these glass dosimeters and acceptable for the intended applications.

| Uncertainty | Type A (%) | Type B (%) |
|---|---|---|
| Reference Dose (Aérial) | | 2.50 |
| Uncertainty in calibration curve | 2.43 | .48 |
| Overall uncertainty (%) | | 3.79 |
| Expanded overall uncertainty (%) (k=2) | | 7.58 |

Table 6. Components of uncertainty.

### 3.6 Field trial irradiations

In order to check the applicability and the performance of glass for routine application, field trial irradiations were carried out with the same irradiation arrangement for used the in-plant calibration. Five sets of three glass dosimeters and five sets of three Gammachrome dosimeters were irradiated, together in the polystyrene phantom in close contact, without shielding each other, to nominal doses of 0.2, 1, 1.5, 2.5, and 3 kGy. Four sets of three glass dosimeters and four sets of three Amber Perspex dosimeters were irradiated, together in the polystyrene phantom in close contact, without shielding each other, for the nominal doses: 1, 3, 7, and 15 kGy.

Dose determination for the irradiated glass was done using the established in-plant calibration curves of Figs.14 and 15. The evaluation of the absorbed dose with Harwell Perpsex dosimeters was done using our own calibration function traceable to the Aérial SSDL.

Linear regressions were applied to compare the absorbed dose results obtained in the gamma plant, measured by glass dosimeters and by the Gammachrome and Amber Perspex dosimeters (Figs. 16 and 17). The values of slope lines and correlation coefficients calculated from the data of Figs. 16 and 17 indicate the good agreement.

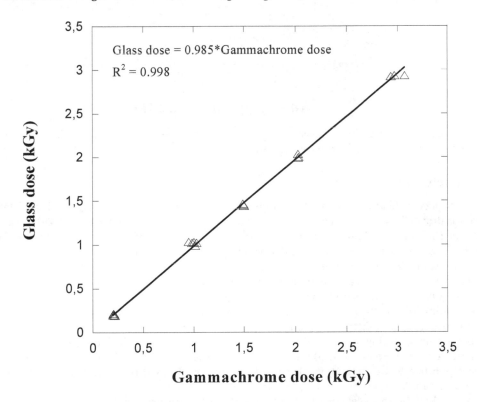

Fig. 16. Comparison of doses measured by Gammachrome dosimeters and glasses irradiated together during production runs at the gamma plant.

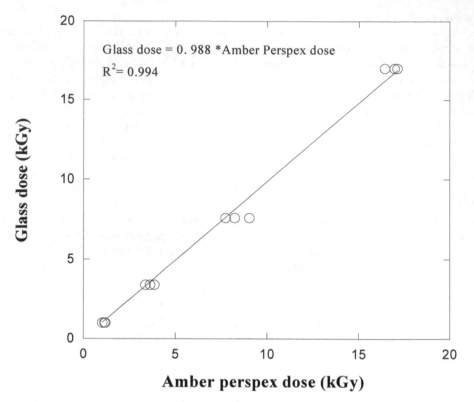

Fig. 17. Comparison of doses measured by Amber Perspex dosimeters and glasses irradiated together during production runs at the gamma plant.

## 4. Conclusion

The induced optical absorption by gamma irradiation in the visible range of this silicate glass is due to the generation of two Non Bridging Oxygen Hole Centres: $HC_1$ at 410 nm and $HC_2$ at 600 nm. The similarity of the activation energy values of the annealing process for both bands suggests that the related optical transitions correspond to the same type of NBOHCs in different configurations. The increase of gamma irradiation dose had no influence on the band peak positions and the FWHM of the induced colour centres and only the heights and the areas under the bands changed.

The nonlinear growth of the absorbance with increasing of dose can be explained by the competition of two different colour centres creation processes. The first is an intrinsic process corresponding to the rupture of the bond $\equiv$ Si-O-Si $\equiv$ in which the concentration of colour centres increases linearly with the dose. The second one is an extrinsic process corresponding to the activation of a fixed number of pre-existing precursors which must therefore saturate.

These colour centres were unstable at room temperature or at heating conditions. Most induced colour disappeared after being treated at 300 °C for 20 minutes and completely disappeared at 500 °C. The initial strong post-irradiation decay can be explained by reaction of NBOHC with neighboring NBOHCs because of the small initial distance between them. At long-term stage the slow decay is only controlled by the diffusion of impurities and electrons in the glass network.

The present study reveals that the optical absorbance response of this type of glass to γ can be quantified reasonably well in the 0.1-50 kGy range. The tests conducted on this type of glass show its good performances in the production irradiators and if the glass samples are calibrated in the plant they are reliable dosimeter material that can be used as a routine dosimeter for measuring doses between 0.1-17 kGy ensuring that the environment conditions are carefully controlled during the use of this dosimeter. The effect of influence quantities on the glass response could be minimized by performing the in-plant calibration. Even if the fading effect can give rise to significant errors during the routine use of these dosimeters, it is not of great importance if calibration curves and routine dose evaluation are done at the same time interval.

These results indicate that this material is very interesting for dose evaluation in many radiation processing applications.

## 5. References

Abdel-Fattah, A., Miller, A. (1996). Temperature, Humidity and Time. Combined effects on Radiochromic film dosimeters. *Radiat. Phys. Chem.* 47, 611–621.

Aérial. (2011). Available from http://www.aerial-crt.com

Agnello, S. and Boizot, B. (2003). Transient visible-UV absorption in beta irradiated silica. *J. Non-Cryst. Solids* 332, 84.

Anderson, D.L., Belgya, T., Firestone, R.B., Kasztovsky, Zs., Lindstom, R.M., Molnár, G.L., Révay, Zs., and Yonezawa, C. (2004). Handbook of Prompt Gamma Activation Analysis (Kluwer Academic Publishers, Dordrecht), The Netherlands.

Boizot, B. (1997). Défauts d'irradiation dans la silice amorphe a-SiO₂, 1997. Rapport Bibliographique CEA-R-5749.

Boizot, B., Petite,G., Ghaleb, D., Pellerin, N., Fayon, F., Reynard, B., Calas, G. (2000). Migration and segregation of sodium under β-irradiation in nuclear glasses. *Nucl. Instrum. and Meth. B* 166-167, pp 500-504.

Engin, B., Aydas, C., Demirtas, H. (2006). ESR dosimetric properties of window glass. *Nucl. Instrum. and Meth. B* 243, 149-155.

Ezz-Eldin, F.M., Mahmoud, H.H., Abd-Elaziz, T.D., El-Alaily, N.A. (2008) Response of commercial window glass to gamma doses. *Physica B* 403, 576.

Farah,K., Kuntz, F., Kadri,O., Ghedira, L. (2004). Investigation of the effect of some irradiation parameters on the response of various types of dosimeters to electron irradiation. *Radiat. Phys. Chem.* 71, 337–341.

Farah, K., Jerbi, T., Kuntz, F., and Kovacs, A. (2006). Dose measurements for characterization of a semi-industrial cobalt-60 gamma-irradiation facility. Radiat. Meas. 41, 201.

Farah K, Mejri A, Hosni, F., Ben Ouada , H., Fuochi P G, Lavalle, M, Kovács, A. (2010). Characterization of a silicate glass as a high dose dosimeter. Nuclear Instruments and Methods in Physics Research A, 614, pp. 137–144.

Friebele, E.J. (1991). Radiation effects. In: Optical Properties of Glass, Uhlmann. , D.R., Kreidl, N.J. (Eds.), 205, American Ceramic Society.

Fuochi P G, Corda, U, Lavalle, M, Kovács, A, Baranyai, M, Mejri, A and Farah, K (2008). Commercial window glass tested as possible high dose dosimeter, Electron and gamma irradiation. Proc. 10th Int. Conf. on Astroparticle, Particle, Space Physics, Detectors and Medical Physics Applications Ed M Barone et al (Villa Olmo, Italy, 8-12 October 2007) pp 70-74.

Fuochi P G, Corda, U, Lavalle, M, Kovács, A, Baranyai, M, Mejri, A and Farah, K. (2009). Dosimetric properties of gamma-and electron-irradiated commercial window glasses Nukleonika, 54, pp. 39-43.

Griscom, D.L. (1984). Electron spin resonance studies of trapped hole centers in irradiated alkali silicate glasses: A critical comment to current models for HC1 and HC2. J. Non-Cryst. Solids 64, 229.

ISO/ASTM Standard. (2002). Guide for Estimating of Uncertainties in Dosimetry for Radiation Processing, ISO/ASTM Standard 51707. American Society for Testing and Materials, Philadelphia, PA.

Jin, Y.G. and Chang, K.J. (2001). Mechanism for the Enhanced Diffusion of Charged Oxygen ions in $SiO_2$. Phys. Rev. Lett. 86 (9), 1793.

Mashkov, V.A., Austin, W. R., Zhang, L., and Leisure, R.G. (1996). Fundamental role of creation and activation in radiation-induced defect production in high-purity amorphous $SiO_2$. Phys. Rev. Lett., 76 (6), 2926.

Mejri A, Farah K, Eleuch H, and Ben Ouada, H. ( 2008). Application of commercial glass in gamma radiation processing Radiat. Meas. 43 1372 – 76.

Ojovan, M.I. and Lee, W.E. (2004). Alkali ion exchange in γ-irradiated glasses. J. Nucl. Mater. 335, 425.

Sharpe, P.H.G. and Miller, A. (1999). Guidelines for the calibration of routine dosimetry systems for use in radiation processing. NPL report CIRM 29, National Physical Laboratory, (Teddington, UK).

Sheng, J. (2002a). Easily recyclable coloured glass by X-Ray irradiation induced coloration. Glass Technology, 43, 238.

Sheng, J, Kadono, K., Utagawa, Y., Yazawa,T. (2002b). X-ray irradiation on the soda-lime container glass. Appl. Radiat. Isotop. 56, 61.

Sheng, J., Kadona, K., Yazawa, T. (2002c). Fading behaviour of X-ray induced color centers in soda-lime silicate glass. Appl. Radiat.Isotop. 57, 813.

Suszynska,M. and Macalik, B. (2001). Optical studies in gamma irradiated commercial soda-lime silicate glasses. Nucl. Instrum. and Meth. B 179, 383.

Tsai, T.E., Griscom, D.L. and Friebele, E.J. (1989). Medium-range order and fractal annealing kinetics of radiolytic atomic hydrogen in high-purity silica. Phys. Rev. B 40 (9) 6374.

Waite, T.R. (1957). Diffusion-limited annealing of radiation damage in germanium. Phys. Rev. 107 (21), 471.

Yokota, R. (1954). Colour centres in alkali silicate and borate glasses. Phys. Rev. 95, 1145.

Yokota, R. (1956). Colour centres in alkali silicate glasses containing alkaline earth ions. Phys. Rev. 101, 522.

Zheng, Z., Honggui, D., Jie, F., Daochuan, Y. (1988). Window glass as a routine dosimeter for radiation processing. Radiat. Phys. Chem. 31, 419-423.

Zheng, Z. (1996). Study on the possibility of reading two kinds of data from one glass detector. Radiat. Phys. Chem. 50, 303-305.

# 6

# Atmospheric Ionizing Radiation from Galactic and Solar Cosmic Rays

Christopher J. Mertens[1], Brian T. Kress[2], Michael Wiltberger[3], W. Kent Tobiska[4], Barbara Grajewski[5] and Xiaojing Xu[6]

[1]*NASA Langley Research Center, Hampton, Virginia*
[2]*Dartmouth College, Hanover, New Hampshire*
[3] *High Altitude Observatory, National Center for Atmospheric Research, Boulder, Colorado*
[4] *Space Environment Technologies, Pacific Palisades, California*
[5]*National Institute for Occupational Safety and Health, Cincinnati, Ohio*
[6]*Science Systems and Applications, Inc.*
*USA*

## 1. Introduction

An important atmospheric state variable, driven by space weather phenomena, is the ionizing radiation field. The two sources of atmospheric ionizing radiation are: (1) the ever-present, background galactic cosmic rays (GCR), with origins outside the solar system, and (2) the transient solar energetic particle (SEP) events (or solar cosmic rays), which are associated with eruptions on the Sun's surface lasting for several hours to days with widely varying intensity. Quantifying the levels of atmospheric ionizing radiation is of particular interest to the aviation industry since it is the primary source of human exposure to high-linear energy transfer (LET) radiation. High-LET radiation is effective at directly breaking DNA strands in biological tissue, or producing chemically active radicals in tissue that alter the cell function, both of which can lead to cancer or other adverse health effects (Wilson et al., 2003; 2005b). Studies of flight attendants have suggested adverse reproductive health outcomes (Aspholm et al., 1999; Lauria et al., 2006; Waters et al., 2000). The International Commission on Radiological Protection (ICRP) classify crews of commercial aircraft as radiation workers (ICRP, 1991). The US National Council on Radiation Protection and Measurements (NCRP) reported that among radiation workers monitored with recordable dose, the largest average effective dose in 2006 (3.07 mSv) was found in flight crew. In contrast, the average for the workers with the second largest effective dose, commercial nuclear power workers, was 1.87 mSv (NCRP, 2009). However, aircrew are the only occupational group exposed to unquantified and undocumented levels of radiation. Furthermore, the current guidelines for maximum public and prenatal exposure can be exceeded during a single solar storm event for commercial passengers on intercontinental or cross-polar routes, or by frequent use ($\sim$ 10-20 flights per year) of these high-latitude routes even during background conditions (AMS, 2007; Copeland et al., 2008; Dyer et al., 2009).

There is an important national need to understand and to predict the real-time radiation levels for the commercial aviation industry and the flying public, which has broad societal, public health, and economic benefits. NASA has met this need by developing the first-ever, real-time, global, physics-based, data-driven model for the prediction of biologically hazardous atmospheric radiation exposure. The model is called Nowcast of Atmospheric Ionizing Radiation for Aviation Safety (NAIRAS). In this chapter the underlying physics of the NAIRAS model is reviewed and some of the key results and applications of the model are presented. More specifically, the materials reviewed are the latest understanding in the physics and transport of atmospheric ionizing radiation from galactic and solar cosmic rays, the influence of space weather on the atmospheric ionizing radiation field, basic radiation dosimetry applied to atmospheric exposure, and the latest epidemiological understanding of radiation effects on pilots and aircrew.

## 2. Scientific and historical overview

### 2.1 Basics on cosmic rays and matter interactions

GCR consist of roughly 90% protons and 8% helium nuclei with the remainder being heavier nuclei and electrons (Gaisser, 1990). When these particles penetrate the magnetic fields of the solar system and the Earth and reach the Earth's atmosphere, they collide with air molecules and create cascades of secondary radiations of every kind (Reitz et al., 1993). The collisions are primarily due to Coulomb interactions of the GCR particle with orbital electrons of the air molecules, delivering small amounts of energy to the orbital electrons and leaving behind electron-ion pairs (Wilson et al., 1991). The ejected electrons usually have sufficient energy to undergo similar ionizing events. The cosmic ray ions lose a small fraction of their energy and must suffer many of these atomic collisions before slowing down. On rare occasions the cosmic ray ion will collide with the nucleus of an air molecule in which large energies are exchanged and the ion and nucleus are dramatically changed by the violence of the event. The remnant nucleus is highly disfigured and unstable, emitting further air nuclear constituents and decaying through the usual radioactivity channels (Wilson et al., 1991). One of the most important secondary particles created in GCR-air interactions is the neutron. Because of its charge neutrality, the neutron penetrates deep into the atmosphere, causing further ionization events along its path and contributing over half the atmospheric radiation exposure at typical commercial airline altitudes (Wilson et al., 2003). Furthermore, neutron exposures pose a relatively high health risk, since the massive low-energy ions resulting from neutron interactions always produce copious ions in the struck cell and repair is less efficient for these events (Wilson, 2000).

The intensity of the atmospheric radiations, composed of GCR primary and secondary particles, their energy distribution, and their effects on aircraft occupants vary with altitude, location in the geomagnetic field, and the time in the sun's magnetic activity (solar) cycle (Heinrich et al., 1999; Reitz et al., 1993; Wilson, 2000). The atmosphere provides material shielding, which depends on the overhead atmospheric depth. The geomagnetic field provides a different kind of shielding, by deflecting low-momentum charged particles back to space. Because of the orientation of the geomagnetic field, which is predominately dipolar in nature, the polar regions and high latitudes are susceptible to penetrating GCR (and SEP) particles. At each geographic location, the minimum momentum per unit charge (magnetic rigidity) a vertically incident particle can have and still reach a given location above the earth is called the vertical geomagnetic cutoff rigidity. The local flux of incident GCR at a given time varies widely with geomagnetic location and the solar modulation level. When solar activity is high, GCR flux is low, and vice versa. The dynamical balance between outward convective flux of solar wind and the inward diffusive flux of GCR is responsible for the anti-correlation between the incident GCR and the modulation level of solar cycle activity (Clem et al., 1996; Parker, 1965).

It is now generally understood that SEP events arise from coronal mass ejections (CME) from active regions on the solar surface (Kahler, 2001; Wilson et al., 2005b). The CME propagates through interplanetary space carrying along with it the local surface magnetic field frozen into the ejected mass. There is a transition (shock) region between the normal sectored magnetic structure of interplanetary space and the fields frozen into the ejected mass, where the interplanetary gas is accelerated forming the SEP. As the accelerated region passes an observation point, the flux intensity is observed to increase dramatically, and no upper limit in intensity is known within the shock region. The SEP energy spectrum obtained in the acceleration process is related to the plasma density and CME velocity. During a solar storm CME event, the number flux distribution incident at Earth's atmosphere is a combination of the GCR and SEP distributions. The SEP-air interaction mechanisms are the same as GCR-air interactions described above. The atmospheric radiations caused by a SEP also vary with altitude and geomagnetic field.

## 2.2 Commercial aircraft radiation exposure

GCR radiations that penetrate the atmosphere and reach the ground are low in intensity. However, the intensities are more than two orders of magnitude greater at commercial aircraft altitudes. At the higher altitudes of High Speed Civil Transport (HSCT), the GCR intensity is another two orders of magnitude higher (Wilson et al., 2003). When the possibility of high-altitude supersonic commercial aviation was first seriously proposed (The Supersonic Transport program proposed in 1961), Foelsche brought to light a number of concerns about associated atmospheric radiation exposure due to GCR and SEP, including the secondary radiations (Foelsche, 1961; Foelsche & Graul, 1962). Subsequently, Foelsche et al. (1974) conducted a detailed study of atmospheric ionizing radiation at high altitudes from 1965 to 1971 at the NASA Langley Research Center (LaRC). The study included a comprehensive flight program in addition to theoretical investigations. The measured data and theoretical calculations were integrated into a parametric Atmospheric Ionizing Radiation (AIR) model (Wilson et al., 1991). Prior to that study the role of atmospheric neutrons in radiation exposure was generally regarded as negligible (Upton et al., 1966). The LaRC studies revealed neutron radiation to be a major contributor to aircraft GCR exposure. Still the exposure levels were comfortably below allowable exposure limits for the block hours typical of airline crews of that time, except during a possible SEP event (less than 500 block hours were typical of the 1960's, although regulation allowed up to 1000 hours). Assessments of radiation exposure extending back to the 1930's for former Pan Am flight attendants were conducted by Waters et al. (2009) and Anderson et al. (2011).

There have been a number of significant changes since the original work of Foelsche (Wilson et al., 2003). A partial list of these changes, relevant to the development of the NAIRAS model, are: (1) the highly ionizing components of atmospheric radiations are found to be more biologically damaging than previously assumed and the associated relative biological effectiveness for fatal cancer has been increased (ICRP, 1991; ICRU, 1986); (2) recent animal and human studies indicate large relative biological effectiveness requiring protection for reproductive exposure (BEIR V, 1990; Chen et al., 2005; Fanton & Gold, 1991; Jiang et al., 1994; Ogilvy-Stuart & Shalet, 1993); (3) recent epidemiological studies (especially the data on solid tumors) and more recent atom-bomb survivor dosimetry have resulted in higher radiation risk coefficients for gamma rays (ICRP, 1991; NAS/NRC, 1980; UNSCEAR, 1988), resulting in lower proposed permissible limits (ICRP, 1991; NCRP, 1993); (4) subsequent to deregulation of the airline industry, flight crews are logging greatly increased hours (Barish, 1990; Bramlitt, 1985; Friedberg et al., 1989; Grajewski et al., 2011; Wilson & Townsend, 1988); and (5) airline crew members are now classified by some agencies as radiation workers (ICRP, 1991).

The last point (i.e., (5)) is particularly illuminating. Aircrews may even receive exposures above recently recommended allowable limits for radiation workers when flying the maximum allowable number of flight hours. Reviews (Hammer et al., 2009) and meta-analyses (Buja et al., 2005; 2006) of cancer studies among flight crew have found excesses of breast cancer and melanoma and possible excesses for other sites. Increasing risk trends with radiation exposure have not been consistently identified, likely due to the limitations of the exposure metrics used (Waters et al., 2000). There is further concern for reproductive health from exposure to high altitude and/or high-latitude flights, as the US National Institute for Occupational Safety and Health (NIOSH) continues to study adverse pregnancy outcomes among commercial flight attendants (Waters et al., 2000). US pilots have been exposed to increasing levels of ionizing radiation since the 1990s (Grajewski et al., 2011). Frequent-flyer business passengers are likely exposed to even higher doses than aircrew, since flight hours are not restricted for airline passengers. In addition, if a large SEP occurs during flight, both passengers and pregnant crew may greatly exceed allowable limits (Barish, 2004).

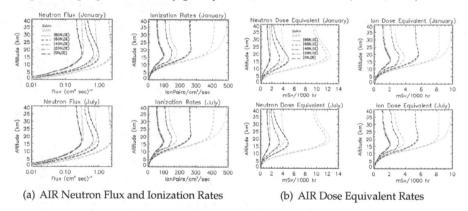

(a) AIR Neutron Flux and Ionization Rates          (b) AIR Dose Equivalent Rates

Fig. 1. (a) Neutron flux and ionization rate profiles for various latitudes, solar activity levels, and seasons. The blue lines represent solar maximum conditions. The green lines represent solar minimum conditions. The top row corresponds to January atmospheric conditions while the bottom row corresponds to July. (b) Neutron and ion dose equivalent rates profiles for the same latitudes, solar activity levels, and seasons indicated in (a).

The original LaRC study (1965 to 1971) commissioned over 300 flights over most of the duration of solar cycle 20 on high-altitude aircraft and balloons to study both the background radiation levels over the solar cycle and to make measurements during SEP events. The LaRC flight package consisted of a 1-10 MeV neutron spectrometer, tissue equivalent ion chamber, and nuclear emulsion for nuclear reaction rates in tissue. Monte Carlo calculations (Lambiotte et al., 1971; Wilson et al., 1970) for incident GCR protons were used to extend the neutron spectrum to high energies. The measured data were combined with the theoretical calculations and integrated into the parametric AIR model, parameterized by neutron monitor count rate, vertical geomagnetic cutoff rigidity, and atmospheric depth. Solar cycle modulation of the GCR spectrum is parameterized by the ground-level neutron monitor count rates. Geomagnetic momentum shielding and overhead atmospheric shielding are parameterized by the vertical geomagnetic cutoff rigidity and atmospheric depth, respectively. The neutron flux ($cm^{-2}$ $sec^{-1}$) component to the atmospheric radiations is converted to dose equivalent rate and total dose rate using 3.14 Sv $cm^2$ sec $hr^{-1}$ and 0.5 Gy $cm^2$ sec $hr^{-1}$, respectively. The charged particle component of the atmospheric radiations is obtained from data taken by Neher (1961; 1967; 1971) and Neher & Anderson (1962) as compiled S. B. Curtis (Boeing 1969) and utilized by Wallance & Sondhaus (1978). The charge particle atmospheric ionization rates are directly converted to dose equivalent rate and total dose rate using measurement data from the tissue equivalent ion chamber. Nuclear stars in tissue are estimated from the nuclear emulsion measurement data after subtraction of the neutron-induced stars (Wilson et al., 1991). Recent updates to the parametric AIR model are described by Mertens et al. (2007a).

Figure 1a shows altitude profiles of neutron flux and ionization rates computed from the AIR model for summer and winter seasons at various latitudes in the northern hemisphere, for both solar maximum and solar minimum conditions. Figure 1b shows the corresponding profiles of dose equivalent rates. In this context, dose rate refers to the rate at which radiation energy is absorbed in human tissue per unit mass. The unit of dose is Gray (1 Gy = $6.24 \times 10^{12}$ MeV $kg^{-1}$). Dose equivalent rate is the sum of the dose (Gy) from each radiation particle, which each particle dose weighted by a factor related to the potential to inflict biological damage (Mertens et al., 2007a). The unit of dose equivalent is Sievert (Sv). Dose equivalent is closely related to biological risk due to radiation exposure (Wilson, 2000). The dosimetric quantities are discussed in more detail in section 3.6.

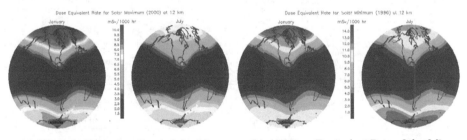

(a) AIR Dose Equivalent Rates: Solar Max          (b) AIR Dose Equivalent Rates: Solar Min

Fig. 2. (a) Global distribution of dose equivalent rate (mSv/1000 hr) predicted by the parametric AIR model at 12 km for solar maximum conditions (year 2000) of cycle 23. (b) The same as in (a) but for solar minimum conditions (year 1996) of cycle 22.

The two most noticeable features in Figure 1 are: (1) the significant increase in flux, ionization rates, and dose equivalent rates at high-latitudes, and (2) the peak in these quantities occur near the typical cruising altitudes of commercial aircraft flying international routes ($\sim$ 10-12 km). The low altitude results are less reliable because of the limited altitude range of the balloon and flight measurements used to develop the AIR parameterizations. The first point is underscored in Figure 2, which shows global maps of the dose equivalent rates computed from the parametric AIR model for summer and winter northern hemisphere seasons at 12 km for both solar minimum and solar maximum conditions. The solar cycle variation in the dose equivalent rates is governed by modulation of the local interstellar spectrum of GCR particles by the heliospheric solar wind and interplanetary magnetic field (Mertens et al.,

2008; 2007a; Wilson et al., 2003; 1991). The latitudinal variation in the dose equivalent rates is determined by the low-momentum shielding provided by the Earth's magnetic field (Mertens et al., 2010a; 2009; 2008; 2007a; Wilson et al., 2003; 1991). The seasonal variations in the dose equivalent rates are determined by the seasonal variations in the overhead atmospheric mass at a fixed altitude (Mertens et al., 2008; 2007a; Wilson et al., 2003; 1991). These space weather and meteorological influences on atmospheric radiation exposure are discussed more fully in sections 3 and 4.

An important point to note is that the North Atlantic corridor region is one of the busiest in the world and comprises the most highly exposed routes in airline operations. This is clearly seen in Figure 2. High-latitude flights, polar flights, and flights over Canada are among the most highly exposed. The maximum GCR radiation exposure occurs during solar minimum conditions. The annual level of GCR exposure to pilots flying high-latitude routes and logging maximum flight hours (1000-hours) is sufficient to trigger individual monitoring and medical surveillance in the European Union (EU) states. Low latitude flights, on the other hand, are minimally exposed to cosmic ray radiation.

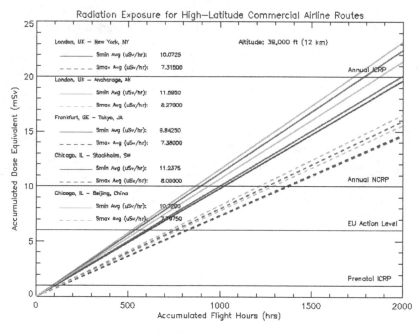

Fig. 3. Accumulated dose equivalent (mSv) predicted by the parametric AIR model as a function of accumulated flight time (hrs) at 12 km for representative high-latitude commercial routes. The solid lines denote solar minimum conditions (year 1996) while the dashed lines denote solar maximum (year 2000) conditions. The top two horizontal lines indicate the recommended ICRP and NCRP annual exposure limits for radiation workers, respectively. The bottom horizontal line shows the ICRP recommended prenatal and annual public exposure limits. The "EU Action Level" horizontal line is described in section 2.3

## 2.3 The NAIRAS decision support system

To place biological risk to high-latitude commercial aircrew and passengers from high-LET radiation exposure into context, consider the accumulated dose equivalent shown in Figure 3 for representative high-latitude commercial routes. The accumulated dose equivalent in Figure 3 is from GCR radiation exposure predicted by the parametric AIR model for both solar maximum and solar minimum conditions. The ICRP recommends 20 mSv as the annual occupational radiation worker limit (ICRP, 2008; 1991; Wilson, 2000; Wilson et al., 2003). The National Committee on Radiological Protection (NCRP) recommends 10 mSv as the annual occupational radiation worker limit (a conservative reconciliation of the NCRP 10*age occupational lifetime limit with its annual occupational limit of 50mSv/yr; see Wilson (2000); Wilson et al. (2003).). The recommended ICRP annual public limit and prenatal exposure limit is 1 mSv. These limits are shown as horizontal black lines in Figure 3.

The Council of the EU adopted Directive 96/20/EURATOM on 13 May 1996. Article 42 of the EU Directive imposes requirements relating to the assessment and limitation of aircrew cosmic ray radiation exposure (AMS, 2007). EU Member States were required to implement the Directive by 13 May 2000 through national legislation. Consistent with "best practice" radiation protection procedures, which is to keep all radiation exposures as low as reasonably achievable (i.e., the ALARA principle), the EU has also adopted an "action level" of 6 mSv/yr. The EU action level is also shown as a horizontal black line in Figure 3. For those likely to exceed 6 mSv/yr, individual record keeping and medical surveillance is required of the aircraft operators (Dyer & Lei, 2001). For exposures less than 6 mSv/yr, monitoring is only recommended and the actual implementation of these recommendation varies among the EU Member States (Meier et al., 2009). The EU Directive recommendation is that individual exposure is assessed by the aircraft operators if one is likely to exceed 1 mSv/yr, and that workers are educated on radiation health risks and work schedules are adjusted to ensure that the 6 mSv/yr level is not exceeded (EURADOS, 1996).

The results from the parametric AIR model in Figure 3 indicate that high-latitude flying commercial aircrew will trigger the "EU action level" for annual flight hours more than ~ 500-600 hours during solar minimum and more than ~ 800-900 hours during solar maximum conditions. The recommended ICRP annual public and prenatal limit is quite low. From Figure 3, the public/prenatal limit can be exceeded in 100 hours of flight time. For high-latitude and polar flights, the dose equivalent rate is on the order of ~ 10 uSv/hr (Copeland et al., 2008; Dyer et al., 2009; Mertens et al., 2008; 2007a). An accumulated 100 hours of flight time can be accrued from 10-20 international flights with roughly 5-10 hours of flight time per flight. Thus, at an average high-latitude GCR exposure rate of 10 uSv/hr, the ICRP annual public and prenatal limit can be exceeded in 10-20 flights.

Recognizing the potential impact on present day passenger and crew exposures - due to the changes since the original work of Foelsche, as described in section 2.2, combined with the results shown in Figure 3 - further studies were initiated at LaRC. A new radiation measurement flight campaign was formulated and executed, consisting of a collaboration of fourteen institutions in five countries with a contribution of eighteen instruments to the flight program. New measurements and advances in theoretical modeling followed (Clem et al., 1996; Wilson et al., 2003), which culminated in the AIR workshop (Wilson et al., 2003).

Following the recent LaRC-sponsored AIR workshop, a number of recommendations for future work were put forth (Wilson et al., 2003). The recommendations relevant to the NAIRAS model development are: (1) utilize satellite input data to provide real-time mapping of GCR and SEP radiation levels to provide guidance in exposure avoidance; and (2) utilize state-of-the-art transport codes and nuclear databases to generate input data to the AIR model. The NAIRAS model addresses these two recommendations, but significantly goes beyond recommendation (2) by using physics-based, state-of-the-art transport code directly in simulating the atmospheric radiation exposure levels. The details of the NAIRAS model are given in section 3.

There are also economic consequences to the issue of aircraft radiation exposure. Over the last decade, airspace over Russian and China has opened up to commercial traffic, allowing for polar routes between North America and Asia (AMS, 2007). These cross-polar routes reduce flight time and operational cost; thus, the number of cross-polar commercial routes has increased exponentially. The typical cost savings from a cross-polar route from the US to China is between $35,000 and $45,000 per flight compared to the previous non-polar route (DOC, 2004). However, the polar region receives the largest quantity of radiation because the shielding provided by Earth's magnetic field rapidly approaches zero near the magnetic pole. On the other hand, the economic loss to an airline from rerouting a polar flight in response to an SEP warning, for example, can be a factor of three greater than the original cost-savings of flying the polar route if fuel stops and layovers are necessary. Thus, the cost to reroute a cross-polar route can be as much as $100,000 per flight (DOC, 2004). Consequently, an aircraft radiation prediction model must also be accurate to minimize radiation risks while simultaneously minimizing significant monetary loss to the commercial aviation industry.

The goal of the NAIRAS model is to provide a new decision support system for the National Oceanic and Atmospheric Administration (NOAA) Space Weather Prediction Center (SWPC) that currently does not exist, but is essential for providing the commercial aviation industry with data products that will enable airlines to achieve the right balance between minimizing flight cost while at the same time minimizing radiation risk. NOAA/SWPC is the main provider of real-time space weather nowcasts and forecasts, both nationally and internationally. However, there are no existing data or models, within NOAA/SWPC or outside NOAA/SWPC, that provide a comprehensive (i.e., comprehensive in terms of input observation data included and comprehensive in terms of the transport physics included in the real-time calculations), global, real-time assessment of the radiation fields that affect human health and safety. NAIRAS is the first-ever, global, real-time, data-driven model that predicts aircraft ionizing radiation exposure, including both GCR and SEP exposures. Thus, the NAIRAS model addresses an important national and international need with broad societal, public health and economic benefits. A detailed description of the NAIRAS model is given in the next section.

## 3. Description of NAIRAS model components

NAIRAS is an operational model for predicting aircraft radiation exposure from galactic and solar cosmic rays (Mertens et al., 2010a; 2009; 2008; 2007a). The NAIRAS prototype model development was funded by the NASA Applied Sciences / Aviation Weather Program. Real-time exposure rate graphical and tabular data products from the operational prototype are streaming live from the NAIRAS public web site at http://sol.spacenvironment.net/~nairas/ (or, just google NAIRAS to locate web site). A subset of the NAIRAS real-time graphical products are available on the SpaceWx smartphone app for iPhone, IPad, and Android.

NAIRAS provides data-driven, global, real-time predictions of atmospheric ionizing radiation exposure rates on a geographic 1x1 degree latitude and longitude grid from the surface of the Earth to 100 km with a vertical resolution of 1 km. The real-time, global predictions are updated every hour. NAIRAS has adopted, as far as possible, the meteorological weather forecasting paradigm of combining physics-based forecast models with data assimilation techniques. Physics-based models are utilized within NAIRAS to transport cosmic rays through three distinct material media: the heliosphere, Earth's magnetosphere, and the neutral atmosphere. While the quantity of observations relevant to radiation exposure predictions is currently too sparse to apply data assimilation techniques per se, nevertheless, as much real-time measurement data as possible are utilized. The real-time measurement data are used to: (1) specify the ionizing radiation field at the boundaries of the aforementioned material media, and (2) characterize the internal properties of the aforementioned material media. The real-time measurements provide necessary observational constraints on the physics-based models that improve simulations of the transport and transmutations of cosmic ray radiation through the heliosphere, magnetosphere, and atmosphere.

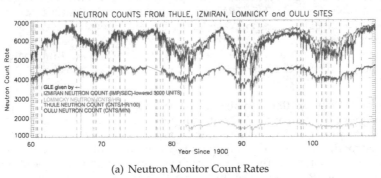

(a) Neutron Monitor Count Rates

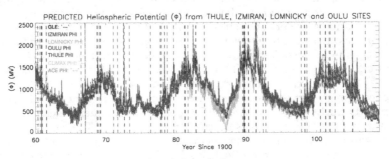

(b) Solar Modulation Parameter

Fig. 4. (a) Monthly-mean neutron monitor count rates from Thule, Izmiran (or Moscow), Lomnicky, and Oulu sites. The vertical dashed lines show ground level enhancement (GLE) events. The time period is 1960-2005. (b) Monthly-mean solar modulation parameter ($\Phi(t)$) predicted by the monthly-mean neutron monitor count rates shown in (a). The reference solar modulation parameters $\Phi_{CLIMAX}$ and $\Phi_{ACE}$ are also shown.

There are a number of models currently in use for calculating GCR radiation exposure at aircraft altitudes. The CARI-6 model utilizes a database of transport calculations generated by the deterministic LUIN code for a wide variety of geographic locations, altitudes, and solar activity levels (O'Brien et al., 1998; 2003). The EPCARD model is based on a similar approach, but uses the Monte Carlo FLUKA code for the transport calculations (Schraube et al., 1999). PC-AIRE is a semi-empirical model based on fits to measurement data (Lewis et al., 2002). Other aircraft radiation exposure models are described in the recent European Radiation Dosimetry Group report (Lindborg et al., 2004). Currently, the above models calculate SEP atmospheric radiation exposure post-storm on a case-by-case basis, although PC-AIRE incorporated low-earth orbit measurements to develop a simple extrapolation to SEP events (Lewis et al., 2002). Recently, Copeland et al. (2008) calculated adult and conceptus aircraft exposure rates for 170 SEP events for years 1986-2008 using the Monte Carlo MCNPX transport code.

The main differences that distinguish the NAIRAS model from the models discussed above are the following. Dynamical solar wind-magnetospheric interactions and the accompanying geomagnetic effects that govern the transport of cosmic rays through the magnetosphere are included in real-time in the NAIRAS radiation exposure calculations (Kress et al., 2004; 2010; Mertens et al., 2010a). Furthermore, the physics-based deterministic High Charge (Z) and Energy TRaNsport code (HZETRN) is used in transporting cosmic rays through the atmosphere. The HZETRN transport calculations are continuously updated using real-time measurements of boundary condition specifications of the space radiation environment and of atmospheric density versus altitude (Mertens et al., 2008; 2007a). And finally, both GCR and SEP atmospheric radiation exposure predictions are included in real-time (Mertens et al., 2010a; 2009).

The remainder of this section contains a detailed description of the salient features of the NAIRAS physics-based modules that transport cosmic rays through the heliosphere, Earth's magnetosphere, and the neutral atmosphere. The input measurement data to the physics-based modules, used to specify boundary conditions of the material medium or to characterize the internal properties of material medium through which the cosmic rays are transported, are also described. The first material medium considered is the heliosphere. GCR are transported from outside the heliosphere to 1 AU using real-time measurements of ground-based neutron monitor count rates. This NAIRAS module is described in section 3.1. SEP are not transported per se but determined in the geospace environment in-situ using a combination of NOAA Geostationary Operational Environmental Satellite (GOES) and NASA Advanced Composition Explorer (ACE) ion flux measurements. This module is described in section 3.2. Both sources of cosmic rays, GCR and SEP, are transported through Earth's magnetosphere using a semi-physics-based geomagnetic shielding model. The geomagnetic shielding model utilizes real-time NASA/ACE solar wind and interplanetary magnetic field (IMF) measurements. This module is described in section 3.3. The cosmic rays that have passed through the heliosphere and magnetosphere are subsequently transported through the neutral atmosphere using the NASA LaRC HZETRN deterministic transport code. The internal properties of the atmosphere important to cosmic ray transport are provided in real-time by the global atmospheric mass density distribution obtained from the NOAA Global Forecasting System (GFS). These two NAIRAS modules are described in sections 3.4 and 3.5, respectively. Finally, the method of quantifying human exposure and the associated biological risk from the atmospheric ionizing radiation field is summarized in section 3.6.

### 3.1 Heliospheric GCR transport

GCR are transported through the heliosphere to the geospace environment using an expanded version of the Badhwar and O'Neill model (Badhwar & O'Neill, 1991; 1992; 1993; 1994; Badhwar & ONeill, 1996). The Badhwar and O'Neill model, which is simply referenced to as the GCR model, has been updated recently by O'Neill (2006) to use ground-based neutron monitor count rate measurements from the Climate neutron monitor site, in order to provide a measurement constraint on the simulated solar cycle modulation of the GCR spectrum at 1 AU. Comparisons between the GCR model and NASA/ACE measurements of the GCR spectra have shown that this step has enabled accurate predictions of GCR spectra in the geospace environment, at least on monthly to seasonal time scales (O'Neill, 2006). The NAIRAS team has extended the work of O'Neill (2006) by incorporating four high-latitude neutron monitor count rate measurements into the GCR model predictions at 1 AU. These additional high-latitude neutron monitor stations are Thule, Oulu, Izmiran (or Moscow), and Lomnicky. The reasons for utilizing these neutron monitor data are two-fold: (1) high-latitude locations are sensitive to the GCR spectral region most influenced by solar cycle variability, and (2) the data from these stations are available in real-time or near real-time.

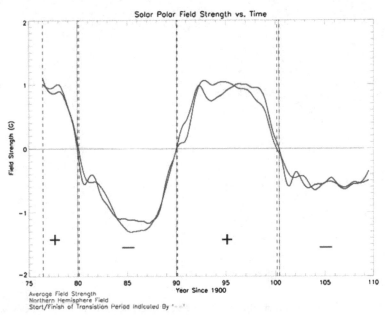

Fig. 5. Solar polar magnetic field data taken by measurements made at the Wilcox Solar Observatory. The blue line is the mean field strength of the northern and southern solar hemispheres. The red line is the northern hemisphere polar magnetic field strength. The '+' and '-' symbols between the vertical dashed lines indicate the time periods of solar positive and negative polarity, respectively. The transition region is the gap between the vertical dashed lines.

The GCR model propagates the local interstellar spectrum (LIS) of each element of the GCR composition to 1 AU by solving a steady-state, spherically symmetric Fokker-Planck transport equation, which accounts for diffusion, convection, and adiabatic deceleration of cosmic rays entering the heliosphere (Parker, 1965). The transport physics described above enables the temporal and spatial dependence of the GCR transport to be absorbed into a ratio of the diffusion coefficient to the bulk solar wind speed. The functional form of this ratio is given by

$$\tilde{k}(r,t) \equiv k(r,t)/V_{SW}(r,t) = (k_0/V_{SW})\beta R \left[1 + (r/r_0)^2\right]/\Phi(t) \tag{1}$$

where $V_{SW}$ is the bulk solar wind speed (nominally set to 400 km/s for all time t), $r$ is the distance from the sun in AU, $t$ is time in years, $k_0$ and $r_0$ are constants, $\beta$ is the particle's speed relative to the speed of light, $R$ is the particle's magnetic rigidity in MV, and $\Phi$ is the so-called solar modulation parameter. Thus, the time-dependent behavior of the GCR spectral flux, due to the level of solar activity, is completely embedded in the solar modulation parameter. The solar modulation parameter is physically related to the energy that interstellar nuclei must have in order to overcome the heliospheric potential field, established by the large-scale structure of the IMF, and propagate through the heliosphere to the radius in question. The solar modulation parameter is determined by fitting the solution of the Fokker-Planck equation for a specified GCR nuclei to corresponding spectral flux measurements throughout the solar cycle, as described in the paragraph below (O'Neill, 2006).

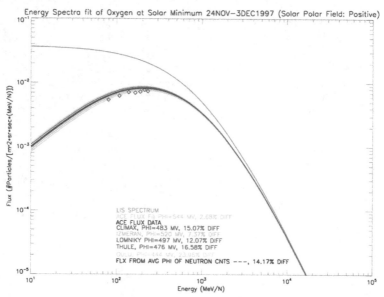

Fig. 6. NAIRAS GCR model spectral oxygen nuclei flux comparisons with ACE/CRIS measurements for solar minimum conditions. The GCR model flux was computed for neutron monitor count rates measured from the four high-latitude sites shown in Figure 4. Also shown are the oxygen spectral flux predicted by the reference Climax-based solar modulations parameter and the GCR model oxygen spectral flux computed using the average solar modulation parameter predicted by each neutron monitor site.

For a fixed parameterization of the LIS, the solar modulation parameter in (1) (i.e., $\Phi(t)$) was determined by fitting the solution of the steady-state Fokker-Planck equation for oxygen nuclei to measurements of the corresponding spectral flux. For energies below roughly 1 GeV (i.e. $\sim$ 50-500 MeV/nucleon), the measurement data were obtained from the Cosmic Ray Isotope Spectrometer (CRIS) instrument on the NASA/ACE satellite. For higher energies (1-35 GeV), the model was fit to data from the C2 instrument on the NASA High Energy Astrophysical Observatory (HEAO-3) satellite (Engelmann et al., 1990).

It is difficult to distinguish the GCR and solar components for protons and alpha spectra observed by CRIS. Fortunately, Lopate (2004) provided an extensive database of quiescent proton and alpha specta from IMP-8 measurements. Thus, the proton and alpha spectra in the

GCR model were fit to IMP-8 data. The high energy proton and alpha spectra were fit to the balloon-borne Isotope Matter-Antimatter Experiment (IMAX) measurements (Menn & et al., 2000).

Once the solar modulation parameter was derived based on the ACE/CRIS oxygen spectra, as described in the above paragraph, the LIS for the remaining elements (i.e., lithium (Z=3) through nickel (Z=28)) were similarly determined by fitting the solutions of the Fokker-Planck equation to the CRIS spectral flux measurements. A simple power law form of the differential LIS was assumed,

$$j_{LIS}(E) = j_0 \beta^\delta (E + E_0)^{-\gamma} \tag{2}$$

where $E$ is the particle kinetic energy per nucleon and $E_0$ is the rest mass energy per nucleon (938 MeV/n). The free parameters ($\gamma, \delta$, and $j_0$) were determined from the fit of the GCR model to the CRIS measurements.

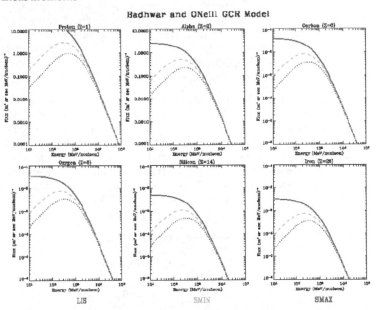

Fig. 7. GCR spectral flux for various nuclei predicted by the Badhwar and O'Neill model for solar cycle 23. The local interstellar spectrum (LIS) is denoted by the red lines. Solar minimum spectra are represented by June 1996 conditions, and are denoted by green lines. Solar maximum spectra are represented by June 2000 conditions, and are denoted by blue lines.

The GCR model was extended beyond the time period of the ACE/CRIS measurements in the following way. First, the solar modulation parameter was alternatively derived from the IMP-8 channel 7 (Z >8, high energy) measurements over three solar cycles from 1973 through 2001, and was calibrated against the solar modulation parameter derived from ACE/CRIS for the period of data overlap (1997.6 to 2001.8). GCR flux comparisons using both sets of solar modulation parameters correlated to within 98.9%. Next, linear fit coefficients were derived between the IMP-8 solar modulation parameter and Climax neutron monitor count rates from 1973-2001. The solar modulation parameter computed using the Climax neutron count rates correlated with the solar modulation parameter derived from IMP-8 data within 97%. Linear fits were derived for the three polarity states of the solar polar magnetic field: (1) positive solar cycle (outward field), (2) negative solar cycle (inward field), and (3) transition state (intermediate between positive and negative polarities with a high degree of modulation). The solar modulation parameter derived from Climax neutron count rates has been recently extended from 1958-2009. This extended Climax-based solar modulation parameter provides the reference solar modulation parameter from which to derive a real-time GCR model suitable for integration into the NAIRAS model.

Four neutron monitor sites - Thule, Oulu, Izmiran, and Lomnicky - were chosen to develop the NAIRAS GCR model. These high-latitude sites were chosen to maximize the solar cycle information content contained in the GCR spectrum and embedded in the ground-based neutron count rates. The neutrons detected on the ground are secondary particles produced by nuclear fragmentation reactions between the incoming GCR particles and the atmospheric constituents (Wilson et al., 1991). At high-latitudes the geomagnetic shielding of the incoming GCR particles is low. Thus, the information contained in the ground-level neutron counts on the low- and medium energy region of the GCR spectrum is high. This is highly desirable since this energy range of the GCR spectrum is most modulated by the solar wind and IMF, and thus closely related to the solar activity cycle (Mertens et al., 2008).

The real-time NAIRAS GCR model was developed by cross-correlating the Climax-based solar modulation parameter (denoted $\Phi_{CLIMAX}$) with the neutron count rates measured at the four high-laitude sites mentioned above. Thus, linear fit coefficients were derived between $\Phi_{CLIMAX}$ and the neutron data at the four high-latitude sites. Monthly-mean neutron count rates at the four high-latitude sites are shown in Figure 4a. Figure 4b shows the solar modulation parameter predicted by the monthly-mean neutron count rates shown in Figure 4a. For comparison, the reference Climax-based solar modulation parameter ($\Phi_{CLIMAX}$) is shown, along with the solar modulation parameter derived from NASA/ACE measurements (denoted $\Phi_{ACE}$).

The heliospheric GCR diffusion coefficient depends on the large-scale structure of the IMF (Parker, 1965). As a result, the diffusion coefficient will depend on the polarity of the Sun's polar magnetic field (O'Neill, 2006). Consequently, to improve the accuracy of the neutron count rate fits to the reference $\Phi_{CLIMAX}$, the data were sorted according to the polarity of the Sun's polar magnetic field, and three sets of fit coefficients were derived: (1) positive solar cycle (outward field), (2) negative solar cycle (inward field), and (3) transition state (intermediate between positive and negative polarities with a high degree of modulation), as described previously for the $\Phi_{CLIMAX}$ reference solar modulation parameter. Figure 5 shows the solar polar magnetic field data since 1978, which are obtained from measurements taken at the Wilcox Solar Observatory (WSO) located at Stanford University. The solar polar magnetic field data from WSO have been added to the NAIRAS input data stream.

Figure 6 shows a comparison between the NAIRAS GCR model spectral oxygen nuclei flux and measurements taken by ACE/CRIS for solar minimum conditions. Also shown is the oxygen spectral flux predicted by the reference solar modulation parameter ($\Phi_{CLIMAX}$). Nominal NAIRAS operations uses the average solar modulation parameter determined from the available high-latitude neutron monitor count rate data. Figure 6 shows the nominal NAIRAS GCR oxygen spectral flux determined from all four high-latitude neutron sites (see Figure 4). In this case, the error in the nominal NAIRAS prediction of the incident GCR oxygen flux is comparable to the error of the predicted flux based on the reference solar modulation parameter ($\sim 15\%$).

A comprehensive set of error statics on the real-time NAIRAS GCR model has been compiled. Generally, the error in using the count rates from the high-latitude neutron monitor stations is comparable to the error in the Badhwar and O'Neill GCR model based on the reference solar modulation parameter. All errors are much larger for solar maximum conditions compared to solar minimum conditions. The main reason for this is that the ACE data are more suspect during solar maximum due to the combination of higher noise levels (GCR flux is minimum during solar maximum) and contamination by particles of solar origin.

Figure 7 shows the solar cycle variation in the GCR spectrum for several nuclei. The spectra were computed by the Badhwar and O'Neill model (Mertens et al., 2008; 2007a). The figure shows the LIS spectra and the spectra at solar maximum and solar minimum conditions. The solar cycle modulation of LIS as the GCR nuclei are transported through the heliosphere to 1 AU is clearly evident.

## 3.2 Geospace SEP fluence rate specification

The solar cosmic rays from SEP events are not transported from the Sun to the geospace environment using a model. Rather, in-situ satellite measurements of ion flux are used to constrain analytical representations of the SEP fluence rate spectrum. However, the analytical functions are guided by the state-of-the-art understanding of the origin, acceleration, and transport of energetic particles from the solar atmosphere to the geospace environment via the interplanetary medium.

The current understanding of SEP processes is that the energy spectrum is a result of injected particle seed populations that are stochastically accelerated in a turbulent magnetic field associated with a CME-driven interplanetary shock (Tylka & Lee, 2006). An analytical expression that represents the differential energy spectrum for this shock acceleration mechanism was given by Ellison & Ramaty (1985), where the spectrum has the form

$$\frac{d^2 J}{dEd\Omega} = C_a E^{-\gamma_a} \exp\left(-E/E_0\right). \tag{3}$$

The differential energy spectrum on the left hand side of (3) has units of $(cm^2\text{-sr-hr-MeV}/n)^{-1}$, and the energy ($E$) has units of MeV/n (i.e., MeV/nucleon). The constant $C_a$ is related to the injected seed population far upstream of the shock. The power-law energy dependence of the spectrum is due to shock acceleration of the seed population by random first-order Fermi acceleration (scattering) events in a turbulent magnetic field, with the power index ($\gamma_a$) related to the shock compression ratio. The exponential turnover in (3) represents high-energy limits to the the acceleration mechanism, such as escape from the shock region. Using the above analytical form, the three parameters ($C_a$, $\gamma_a$, and $E_0$) can be determined by fitting (3) to ion flux measurements.

Recently, Mewaldt et al. (2005) found that the Ellison-Ramaty spectral form failed to fit NOAA-GOES ion flux measurements at the highest energy channels during the Halloween 2003 SEP events. To circumvent this deficiency, Mewaldt et al. proposed using a double power-law spectrum. The low-energy spectrum is assumed to follow the Ellison-Ramaty form. The high-energy spectrum is assumed to have a power-law energy dependence with a different power index, such that

$$\frac{d^2 J}{dEd\Omega} = C_b E^{-\gamma_b}. \tag{4}$$

The power-law expressions in (3)-(4) can be merged into one continuous spectrum by requiring that the differential energy spectra in (3)-(4) and their first derivatives are continuous at the merge energy. The result is given by the expression below:

$$\frac{d^2 J}{dEd\Omega} = C E^{-\gamma_a} \exp\left(-E/E_0\right), \quad E \leq (\gamma_b - \gamma_a)E_0 \tag{5}$$

$$= C E^{-\gamma_b} \left\{ \left[(\gamma_b - \gamma_a)E_0\right]^{(\gamma_b - \gamma_a)} \exp\left(\gamma_b - \gamma_a\right) \right\}, \quad E > (\gamma_b - \gamma_a)E_0. \tag{6}$$

Physically, the double power-law spectrum in (5)-(6) represents SEP sources from two different injected seed populations. For example, the low-energy spectrum, with $\gamma_a$ power index and the e-folding energy $E_0$, is likely associated with solar corona (solar wind) seed populations while the high-energy spectrum, with $\gamma_b$ power index, is likely associated with flare suprathermal seed populations (Tylka et al., 2005).

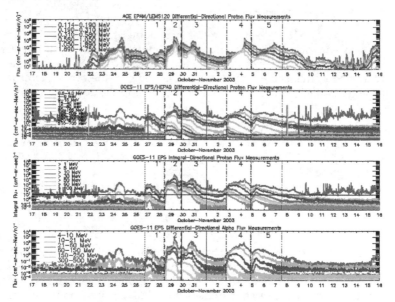

Fig. 8. Proton and alpha flux measurements used to derive the SEP fluence spectra. Row 1: ACE EPAM/LEMS120 differential-directional proton flux measurements. Row 2: GOES-11 EPS and HEPAD differential-directional proton flux measurements. Row 3: GOES-11 EPS integral-directional proton flux measurements. Row 4: GOES-11 EPS differential-directional alpha flux measurements. The different styled vertical lines bound the five SEP events during the Halloween 2003 solar-geomagnetic storm period, which are numbered in all panels. The horizontal line in Row 3 indicates the SEP threshold for the > 10 MeV integral proton flux channel.

Another widely used analytical representation of a SEP energy spectrum is the so-called Weibull distribution (Townsend et al., 2006; 2003). The Weibull distribution has been successful at fitting satellite ion flux measurements, and the differential energy spectrum is given by

$$\frac{d^2 J}{dEd\Omega} = C k \alpha E^{\alpha - 1} \exp\left(-kE^{\alpha}\right). \tag{7}$$

Notice that the Weibull distribution has an analytical form similar to the Ellison-Ramaty distribution in (3). The exponential energy dependence in the Weibull differential energy spectrum could be due to dissipation of the high-energy SEP ions by scattering from self-generated waves. However, further investigation is required to determine if this type of physical process is related to the Weibull distribution (Xapsos et al., 2000).

The NAIRAS model fits four analytical SEP spectral fluence rate functions to the satellite ion flux measurements. The analytical forms that are fit to the measurements are: (1) single power-law in (4), (2) Ellison-Ramaty in (3), (3) double power-law in (5)-(6), and (4) Weibull in (7). The free parameters for each analytical differential energy distribution are derived by a non-linear least-square fit to differential-directional ion flux measurements. The spectral fitting algorithm uses a Marquardt-Levenberg iteration technique (Brandt, 1999). The analytical form that yields the minimum chi-square residual in the fit to the ion flux measurements is the SEP spectral fluence rate distribution used in all subsequent model simulations.

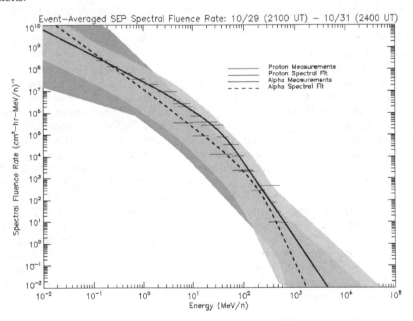

Fig. 9. Event-Averaged SEP spectral fluence rates for Halloween 2003 SEP event 3 [10/29/2003 (2100 UT) - 10/31/2003 (2400 UT)]. The shaded regions show the range of 1-hour averaged ion flux measurements and fitted spectra in the time interval of event 3. The peach shaded regions corresponds to proton flux measurements. The blue shaded regions corresponds to the alpha flux measurements.

NAIRAS utilizes available real-time measurements of proton and alpha differential-directional particle flux $(cm^2\text{-sr-sec-MeV/n})^{-1}$ for the SEP spectral fitting described above. SEP spectral fluence rates $(cm^2\text{-hr-MeV/n})$ incident on Earth's magnetosphere are obtained by time-averaging the particle flux measurements in 1-hr time bins and projecting the incident flux onto the vertical direction assuming an isotropic angular distribution for the solar ions. Low-energy proton data are obtained from the

Electron, Proton, and Alpha Monitor (EPAM) instrument onboard the NASA/ACE satellite (Gold & et al., 1998). EPAM is composed of five telescopes and the LEMS120 (Low-Energy Magnetic Spectrometer) detector in used in the SEP spectral ion fit, which measures ions at 120 degrees from the spacecraft axis. LEMS120 is the EPAM low-energy ion data available in real-time, for reasons described by Haggerty & Gold (2006). The other proton channels used in the SEP spectral fitting algorithm are obtained from NOAA/GOES Space Environment Monitor (SEM) measurements. The Energetic Particle Sensor (EPS) and the High Energy Proton and Alpha Detector (HEPAD) sensors on GOES/SEM measure differential-directional proton flux (Onsager & et al., 1996). Additional differential-directional proton flux measurement channels are generated by taking differences between the EPS integral proton flux channels. The channels used to derive SEP alpha spectral fluence rates are also obtained from EPS measurements. Five-minute averaged ACE and GOES data are used to derive the incident SEP spectral fluence rates.

Figure 8 shows the time variation of the proton and alpha flux measurements used to derive incident SEP spectral fluence rates for the Halloween 2003 storm period. The top panel displays the ACE low-energy proton flux measurements. The next two panels show the GOES-11 EPS and HEPAD proton flux spectra, and the integral proton flux measurements, respectively. By definition, a SEP event occurs when the >10 MeV integral proton flux exceeds 10 proton flux units (pfu $\equiv$ cm$^{-2}$ sr$^{-1}$ sec$^{-1}$) in three consecutive 5-minute periods (NOAA, 2009). The SEP event threshold is denoted by the horizontal line on the integral proton flux panel. There are a total of five SEP events during the Halloween 2003 storm period, which are denoted by the vertical lines in all panels in Figure 8. These events were associated with many simultaneous, complex phenomena such as solar flares, coronal mass ejections (CME), interplanetary shocks, and solar cosmic ray ground level events (GLE) (Gopalswamy et al., 2005). Different line styles are used to bound each of the five events, and the event number is shown between the vertical lines. Note that the onset of event 3 doesn't follow the conventional SEP threshold definition. It is clear from the integral proton flux that two events overlap: event 3 arrives before event 2 decreases below the SEP threshold level. However, there is an important distinguishing feature between the two events. That is, the beginning of our definition of event 3 is accompanied by a sudden increase in high-energy protons associated with the arriving SEP event, as noted by the sudden increase in the 510-700 MeV differential-directional proton flux measurements in Figure 8. Partitioning the simultaneous SEP events 2 and 3 into separate events is useful for the aircraft radiation exposure analysis discussed in section 4, since the high-energy portion of the differential-directional proton flux distribution penetrates deeper in the atmosphere.

In section 4, atmospheric ionizing radiation exposure during SEP event 3 [10/29 (2100 UT) - 10/31 (2400)] will be analyzed since the associated interaction between the arriving CME-driven interplanetary shock and Earth's magnetosphere caused the largest geomagnetic effects during the Halloween 2003 storm period, which is the focus of the case study in section 4.3. In order to isolate the geomagnetic effects, the event-averaged SEP spectral fluence rates is derived and shown in Figure 9. The horizontal lines in Figure 9 are the event-averaged differential ion fluence rate measurements. The width of the horizontal lines correspond to the energy width of the measurement channels. The black lines are the proton and alpha spectral fluence rates derived using the double power-law spectrum and fitting technique describe above. The shaded regions show the range of 1-hour averaged ion spectral fluence rates in the time interval of event 3. The peach colored region corresponds to the range of proton energy spectra and the blue colored region corresponds to the range of alpha energy spectra.

## 3.3 Magnetospheric cosmic ray transport

Lower energy cosmic rays are effectively attenuated by the geomagnetic field (internal field plus magnetospheric contributions) as these charges particles are transported through the magnetosphere and into the neutral atmosphere. The geomagnetic field provides a form of momentum shielding, or attenuation, by deflecting the lower-energy charged particles back out to space via the Lorentz force. This spectral filtering effect is quantified by a canonical variable, in the mathematical sense, called the geomagnetic cutoff rigidity. Once the cutoff rigidity is known, the minimum access energy to the neutral atmosphere is determined for each incident charged particle through the relativistic energy equation (Mertens et al., 2010a).

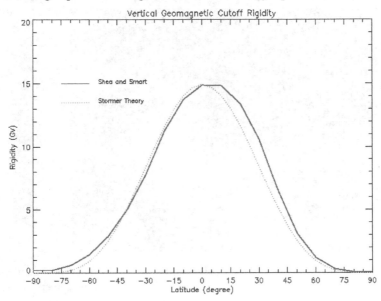

Fig. 10. Zonal-averaged vertical geomagnetic cutoff rigidity. The solid red line corresponds to the quiescent cutoff rigidities computed from particle trajectories and the IGRF model. The green dotted line is the analytical solution for the vertical cutoff rigidity using Störmer Theory.

The utility of the geomagnetic cutoff rigidity quantity is motivated by considering the motion of a charged particle in a magnetic field. The particle motion is determined by solving Newton's equation of motion for a charged particle subject to the Lorentz force. For a positively charged particle, the equation of motion is

$$\frac{d\mathbf{p}}{dt} = \frac{Ze}{c}\mathbf{v}\times\mathbf{B} \tag{8}$$

in cgs units. The bold-faced quantities are vectors and x designates the vector cross product. The charged particle momentum and velocity are $\mathbf{p}$ and $\mathbf{v}$, respectively, and $\mathbf{B}$ is the magnetic field strength. The magnitude of the charge of an electron is denoted $e$ and $Z$ is the number of electron charge units. The equation of motion in (8) can be written, equivalently, as

$$\frac{R}{B}\frac{d\hat{\mathbf{v}}}{dt} = \hat{\mathbf{v}}\times\hat{\mathbf{B}} \tag{9}$$

where the ^ symbol denotes units vectors and

$$R \equiv \frac{pc}{Ze} \tag{10}$$

is defined as the rigidity. The canonical aspect of the rigidity is evident in the above equation. For a given magnetic field strength(B), charged particles with the same rigidity follow identical trajectories.

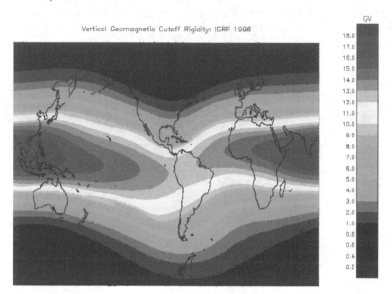

Fig. 11. Global grid of quiescent vertical geomagnetic cutoff rigidities (GV) calculated from charged particle trajectory simulations using the IGRF model for the 1996 epoch (solar cycle 23 minimum).

Motions of charged particles in a pure magnetic dipole field were examined by Störmer (1965). Because of the azimuthal symmetry in a pure dipole field, the azimuthal angular momentum is a conserved quantity. A main feature of Störmer theory is that regions of bounded and unbounded motion can be derived analytically from the integral of motion found from the conservation of azimuthal angular momentum (Störmer, 1965; VanAllen, 1968). It can be shown that the minimum rigidity that a vertically arriving particle must have in order to reach an altitude $z$ above the Earth's surface is

$$R_{vc} = \frac{\mathcal{M}}{(R_e + z)^2} \cos^4 \lambda_m \approx 15 \cos^4 \lambda_m \text{ (GV)}. \tag{11}$$

In the above equation, $R_{vc}$ designates the vertical geomagnetic cutoff rigidity, $\mathcal{M}$ is the Earth's magnetic dipole moment, $R_e$ is the average radius of the Earth, and $\lambda_m$ denotes magnetic latitude. Therefore, vertically arriving charged particles with energies ($E$) less than the cutoff energy ($E_{vc}$) will be deflected by the Lorentz force and not reach altitude $z$. The cutoff energy for each charged particle of charge $Z$ and mass number $A$ is determined from the canonical cutoff rigidity through the relativistic energy equation, such that

$$E_{vc} = \left[ \sqrt{R_{vc}^2 \left( Z/A \cdot amu \cdot c^2 \right)^2 + 1} - 1 \right] \cdot amu \cdot c^2, \qquad (12)$$

where $E$ is kinetic energy per nucleon (MeV/n), $R_{vc}$ is vertical geomagnetic cutoff rigidity (MV), $c$ is the speed of light in vacuum, and amu = 931.5 MeV/c$^2$ (atomic mass unit). Thus, the geomagnetic field has the effect of filtering out lower-energy charged particles as they are transported through the magnetosphere and into the neutral atmosphere.

The vertical cutoff rigidity at the Earth's surface derived from Störmer theory is plotted versus geographic latitude in Figure 10. The maximum cutoff rigidity is at the equator since a vertically arriving charged particle is perpendicular to the dipole magnetic field lines at the equator. The affect of the vector cross product in the Lorentz force in (8) is that charged particle motions perpendicular to magnetic field lines will experience the maximum deflection while particle motions parallel to the magnetic field will experience no deflecting force whatsoever. Figure 10 illustrates that a vertically arriving proton at the equator must have a kinetic energy of $\sim$ 15 GeV to arrive near the surface of the Earth. In the polar regions, vertically arriving charged particles are parallel to the magnetic field lines. Therefore, the cutoff rigidity is zero; particles of all energies can arrive at the Earth's surface in this case.

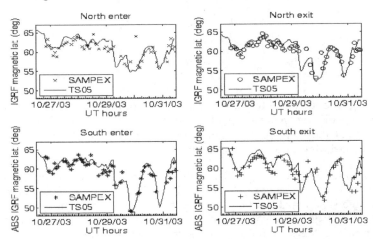

Fig. 12. The four panels show comparisons between cutoff latitudes determined in the TS05 geomagnetic field model and cutoff latitudes extracted from SAMPEX/PET energetic particle data during 28-31 October (Halloween) 2003 storms. Cutoff latitudes as SAMPEX enters and exists the north and south polar cap regions are shown separately.

The Earth's geomagnetic field is not a pure dipole field. On the contrary, the internal geomagnetic field is comprised of dipolar and non-dipolar contributions (Langlais & Mandea, 2000). The dipole moment is off-center and tilted with respect to the rotational axis. Furthermore, the geomagnetic field is distorted at large radial distances ($r \geq 4R_e$) by its interaction with the solar wind. A balance between the solar wind dynamic pressure and the magnetic field pressure, from the internal geomagnetic field, is established by inducing five magnetospheric current systems (Tsyganenko, 1989; 2002). These current systems generate their own magnetic fields which add vectorally to the internal geomagnetic field.

The complexities of the actual internal geomagnetic field, with dipolar and non-dipolar contributions, and the magnetospheric magnetic field contributions prohibit an analytical solution for the vertical geomagnetic cutoff rigidity. Numerical methods must be employed, as described below.

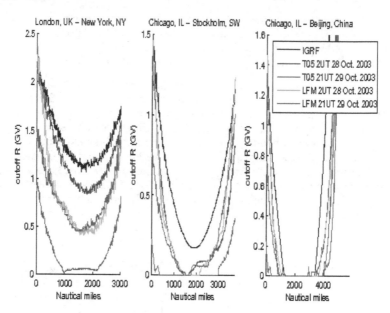

Fig. 13. Geomagnetic cutoff rigidities along three high-latitude commercial flight routes computed in three different magnetic field models: IGRF, TS05, and LFM MHD. The October 28 (0200 UT) calculations represent a magnetically quiet time, while the October 29 (2100 UT) calculations occur during a geomagnetic storm.

The internal geomagnetic field is specified by the International Geomagnetic Reference Field (IGRF) model (Langlais & Mandea, 2000). In the IGRF, the internal field is represented in terms of a magnetic potential function ($\Phi_M(r, \theta, \phi, t)$). Outside of the internal source region, i.e, for $r \geq R_e$, the magnetic potential function must be a solution of the Laplace equation. In spherical coordinates, this solution is expressed in the following form:

$$\Phi_M(r, \theta, \phi, t) = R_e \sum_{n=1}^{N} \sum_{m=0}^{n} \left( \frac{R_e}{r} \right)^{n+1} [g_n^m(t) \cos m\phi + h_n^m(t) \sin m\phi] \, P_n^m(\cos \theta). \quad (13)$$

In the above equation, $r$ denotes the radial distance from the center of the Earth, $\theta$ and $\phi$ denote the geocentric colatitude and longitude at a given location. Schmidt-normalized associated Legendre functions of degree $n$ and order $m$ are denoted by $P_n^m(\cos \theta)$, and $g_n^m$ and $h_n^m$ denote the Gauss coefficients. The magnetic field components are given by the gradient of the potential function, such that

$$B_r = \frac{\partial \Phi_M}{\partial r}, \; B_\theta = \frac{1}{r} \frac{\partial \Phi_M}{\partial \theta}, \; B_\phi = \frac{1}{r \sin \theta} \frac{\partial \Phi_M}{\partial \phi}. \quad (14)$$

The Gauss coefficients are derived from a global set of magnetic field measurements, using the method of least-squares, and updated every five years (Langlais & Mandea, 2000). Secular variations in the Gauss coefficients are also derived from magnetic field measurements so that derivatives of the Gauss coefficients can be computed. In this way, the temporal dependence of the internal geomagnetic field is represented by

$$g_n^m(t) = g_n^m(T_0) + \dot{g}_n^m(t - T_0) \tag{15}$$

$$h_n^m(t) = h_n^m(T_0) + \dot{h}_n^m(t - T_0) \tag{16}$$

where $\dot{g}_n^m$ and $\dot{h}_n^m$ are first-order derivatives of the Gauss coefficients. The epoch of the IGRF model is denoted by $T_0$ and $t$ is such that $T_0 \leq t \leq T_0 + 5$, where time is expressed in decimal years.

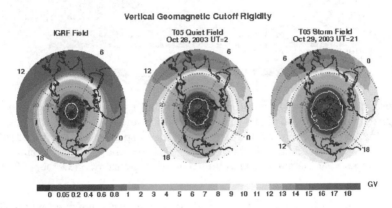

Fig. 14. Simulated vertical geomagnetic cutoff rigidity shown over the northern hemisphere in October 2003. The cutoff rigidities in the left column were calculated using the IGRF model. The cutoffs in the middle column were calculated using the TS05 model during a geomagnetically quiet period. Cutoff rigidities in the right column were calculated using the T05 model during the largest geomagnetically disturbed period of SEP event 3. Also shown are the magnetic latitude circles and the meridians at 0, 6, 12, and 18 magnetic local time.

The vertical cutoff rigidities in a realistic geomagnetic field are determined by numerical solutions of charged particle trajectories in the magnetic field using the techniques advanced by Smart & Shea (1994; 2005). Figure 11 shows the vertical cutoff rigidities at 20 km in the internal IGRF field. The longitudinal variations in the cutoff rigidity are due to a combination of geocentric offset and relative tilt of the magnetic dipole, with respect to the rotational axis, and the non-dipolar contributions to the internal geomagnetic field. A zonal-average of the IGRF cutoff rigidities in Figure 11 are compared to the analytical Störmer theory in Figure 10. The simple Störmer theory represents the latitudinal behavior of the vertical cutoff rigidity quite well. The displacement of the Störmer theory cutoffs in Figure 10 relative to the numerical solutions of the cutoffs in the IGRF field is due to the fact that the true dipole contribution to the internal geomagnetic field is off-centered and tilted with respect to the rotational axis.

NAIRAS real-time geomagnetic cutoff rigidities are computed from numerical solutions of charged particle trajectories in a dynamically varying geomagnetic field that includes both the internal magnetic field and the magnetospheric magnetic field contributions (Kress et al.,

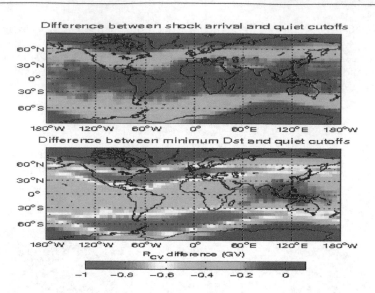

Fig. 15. Storm-quiet geomagnetic cutoff rigidities differences calculated using the TS05 magnetospheric magnetic field model. The magnetically quiet time cutoffs were calculated on October 28 (0200 UT), 2003. The storm cutoffs in the top panel were computed when the interplanetary shock arrived on October 29 (0612 UT), 2003. The storm cutoffs in the bottom panel were computed at the maximum build-up of the ring current on October 29 (2100 UT), 2003.

2010; Mertens et al., 2010a; 2009). The cutoff rigidity code was developed by the Center for Integrated Space Weather Modeling (CISM) at Dartmouth College. The CISM-Dartmouth geomagnetic cutoff model can be run using several different empirical, semi-physics-based, and physics-based models (Kress et al., 2004). In particular, the specification of the geomagnetic field due to Earth's internal field source is provided by the IGRF model [Langlais and Mandea, 2000], as discussed above. The real-time dynamical response of the magnetospheric magnetic field to solar wind conditions and IMF can be provided by the semi-physics-based TS05 model (Tsyganenko & Sitnov, 2005), or by the Lyon-Feder-Mobarry (LFM) global MHD (magnetohydrodynamic) simulation code (Fedder & Mobarry, 2004). Routines were developed and tested to couple the geomagnetic cutoff model with the different magnetic field models.

The LFM MHD code may be run as a stand alone model or coupled with other geospace models currently under development within CISM. For example, the LFM magnetospheric magnetic fields may be coupled with the Thermosphere-Ionosphere Nested Grid (TING) model (Wang et al., 2004) and/or with the Rice Convection Model (RCM) (Toffoletto et al., 2004), which models the ring current. The semi-physics-based TS05 model provides more accurate cutoff rigidities than the stand alone LFM MHD model, as determined by comparisons with satellite observations during a Halloween 2003 geomagnetic storm. Figure 12 shows comparisons between cutoff latitudes for $\sim$ 20 MeV protons computed using the TS05 model and measured by the Proton Electron Telescope (PET) instrument on the Solar Anomalous and Magnetospheric Particle Explorer (SAMPEX) satellite (Baker et al., 1993).

The agreement between SAMPEX/PET measurements and the TS05 model in Figure 12 are quite good. The reason the physics-based LFM MHD model doesn't calculate cutoffs as accurate as the semi-physics-based TS05 model is mainly due to the lack of a full kinetic description of the ring current in the MHD model, which typically causes the LFM fields to be too high. This is evident in Figure 13, which compares geomagnetic cutoff rigidities calculated along three representative high-latitude commercial flight routes from the IGRF, TS05, and LFM MHD magnetic field models. The small differences in cutoffs using the LFM MHD model between quiet and geomagnetic storm conditions is indicative of an inadequate modeling of the ring current build-up during the geomagnetic storm.

It is anticipated that the fully coupled LFM-RCM-TING model currently under development will significantly improve the simulations of cutoff rigidities compared to the stand alone LFM MHD model. Furthermore, the physics-based LFM-RCM-TING model will be able to incorporate short time-scale dynamics not included in semi-physics-based (empirical) magnetospheric magnetic field models. When the code development within CISM reaches sufficient maturity, the influence of short time-scale magnetospheric dynamics on the atmospheric ionizing radiation field using the fully coupled LFM-RCM-TING model will be assessed. For the present work, the simulated real-time geomagnetic cutoff rigidities are calculated using the TS05 model, and using the IGRF model for comparison.

Figure 14 shows the vertical cutoff rigidity over the northern hemisphere for three different models of the geomagnetic field during the Halloween 2003 storm period. The left column is cutoff rigidity computed using the IGRF field. Since total flight-path exposure at aviation altitudes do not change significantly ($< \sim 1\%$) for cutoffs less than 0.05 GV, the cutoffs are set to zero at geographic locations poleward of the 0.05 GV contour (see the bold-white 0.05 GV color contour in Figure 14). The middle column in Figure 14 shows the cutoff rigidities computed using the TS05 field under geomagnetically quiet conditions, October 28 (0200 UT), prior to the onset of the Halloween 2003 SEP event 3 (see section 3.2). One can see that even during magnetically quiet conditions, the cutoff rigidities predicted from the TS05 field are lower than predicted from the IGRF field, and the polar cap region (i.e., inside the bold-white 0.05 GV contour in Figure 14) is expanded to lower latitudes. A weaker field predicted by the TS05 model, compared to IGRF, is due in part to the diamagnetic effect of the magnetospheric ring current included in the TS05 model. Lower cutoff rigidities correspond to less momentum shielding and higher radiation exposure levels. The right column in Figure 14 shows the cutoff rigidities during peak geomagnetic storm conditions, October 29 (2100 UT), during SEP event 3. The cutoffs are lower at all latitudes compared to the two previous simulations, and the polar cap region has expanded to much lower latitudes than during the magnetically quiet period. These geomagnetic effects are discussed in more detail in section 4.3.

The difference in cutoffs between storm and quiet conditions is shown in Figure 15. The cutoffs were calculated using the TS05 model. The magnetically quiet period is the same as above, October 28 (0200 UT), 2003. The top panel shows the cutoff difference between the magnetically quiet time and the arrival of an interplanetary shock at the magnetosphere on October 29 (0612 UT), 2003. The bottom panel shows the storm-quiet cutoff difference when the Disturbed Storm Time (Dst) index is near its minimum on October 29 (2100 UT), 2003. The cutoffs are most suppressed at mid-latitudes during the night. The storm-quiet cutoff difference can be as much as $\sim 1$ GV, which has a significant effect on radiation exposure.

An important aspect of these model studies is our assessment of the impact of the changes in cutoff rigidity due to the magnetospheric field effect on atmospheric radiation exposure, and the identification of the need for accurate and computationally efficient geomagnetic cutoff rigidity models with solar wind-magnetospheric dynamical responses included. The $\sim 1$ GV suppression in cutoff at mid-latitudes during a geomagnetic storm means that high-level SEP radiation exposure normally confined to the polar cap region will be extended to mid-latitudes. More details of these findings are included in section 4.3.

## 3.4 Atmospheric cosmic ray transport

The transport of cosmic rays through the neutral atmosphere is described by a coupled system of linear, steady-state Boltzmann transport equations, which can be derived on the basis of conservations principles (Wilson et al., 1991). The transport equation for the directional fluence $\phi_j(\mathbf{x}, \Omega, E)$ of particle type $j$ is given by (Mertens et al., 2008; 2007a)

$$\Omega \bullet \nabla \Phi_j(\mathbf{x}, \Omega, E) = \sum_k \int \int \sigma_{jk}(\Omega, \Omega', E, E') \Phi_k(\mathbf{x}, \Omega', E') d\Omega' dE' - \sigma_j(E) \Phi_j(\mathbf{x}, \Omega, E) \quad (17)$$

where $\sigma_j(E)$ and $\sigma_{jk}(\Omega, \Omega', E, E')$ are the projectile-target macroscopic interaction cross sections. The $\sigma_{jk}(\Omega, \Omega', E, E')$ are double-differential particle production cross sections that represent all processes by which type $k$ particles moving in direction $\Omega'$ with energy $E'$ produce a particle of type $j$ moving in direction $\Omega$ with energy $E$, including radioactive decay processes. The total interaction cross section $\sigma_j(E)$ for each incident particle type $j$ is

$$\sigma_j(E) = \sigma_{j,at}(E) + \sigma_{j,el}(E) + \sigma_{j,r}(E), \quad (18)$$

where the first term refers to projectile collisions with atomic electrons of the target medium, the second term refers to elastic ion-nucleus scattering, and the third term contains all relevant nuclear reactions. The corresponding differential cross sections are similarly ordered.

Consider the transport of cosmic ray ions through the atmosphere. In this case, the second term in (18) represents elastic ion-nucleus Coulomb scattering between the incident ions and the atoms that comprise the neutral atmosphere. Figure 16 shows the characteristic elastic scattering length versus kinetic energy of various ions colliding with the neutral atmosphere (Mertens et al., 2008). Ion-nucleus scattering becomes important in the atmosphere only at low energies. For example, the length of the Earth's atmosphere in units of areal density is $\sim 1000$ g/cm$^2$. Thus, Figure 16 shows that cosmic ray ions will not elastically scatter off an atmospheric nucleus before reaching the surface unless the ion kinetic energy is well below 1 MeV/amu. However, ions with kinetic energy less than 1 MeV/amu are stopped via ionization and/or atomic excitation energy loss processes at high altitudes before a scattering event can take place (see Figure 18). Multiple Coulomb scattering and coupling with ionization energy loss become important factors in the transport of ions within living tissue (Mertens et al., 2010b; 2007b), which are related to the degree of biological damage inflicted on sensitive components within the living cell. However, for cosmic ray transport through the atmosphere, the ion-nucleus scattering term in (18) can be neglected to a good approximation.

The principle mechanism for atomic interactions between the cosmic ray ions and the target medium is ionization and/or atomic excitation. This process is represented by the first term in (18). The result of this interaction is the transfer of energy from the projectile ions to the atomic electrons of the target medium via the Coulomb impulse force. Since the projectile ion mass is much greater than the electron mass, the ion travels essentially in a straight line as it looses energy through ionization of the target medium. The ionization and atomic excitation

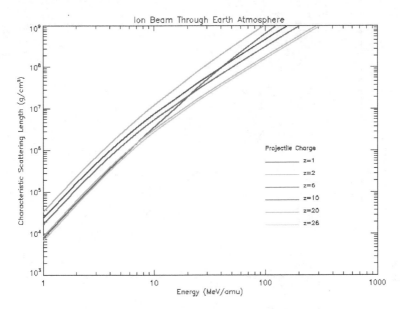

Fig. 16. Characteristic scattering length for ion beam transport through the Earth's atmosphere. The charge number (z) of the ion is specified in the legend.

energies, as well as the energies of ejected orbital electrons, are usually small in comparison to the incident ion kinetic energy. As a result, the ionization energy loss processes by which the projectile ions transfer energy to the target bound and/or ejected orbital electrons can be considered continuous. Because of this so-called continuous slowing down approximation (CSDA), the energy $dE$ which is lost by the incident ion and transferred to the orbital electrons of the target medium by ionization and/or atomic excitation within an element of path $dx$ is given by the stopping power, $S$, (Tai et al., 1997), i.e.,

$$S = -\frac{dE}{dx} = \frac{4\pi Z_P^2 Z_T e^4}{mv^2} N \left\{ B_0 - \frac{C(\beta)}{Z_T} + Z_P L_P(\beta) + Z_P^2 L_T(\beta) + \frac{1}{2}[G(M_P, \beta) - \delta(\beta)] \right\} \quad (19)$$

where

$$B_0 = \ln\left(\frac{2mc^2\beta^2}{I(1-\beta^2)}\right) - \beta^2. \quad (20)$$

In the above equations, $Z_P$ and $Z_T$ are the projectile ion charge and the number of electrons per target atom, respectively, $v$ is the projectile velocity, $c$ is the speed of light, $\beta = v/c$, $N$ is the density of atoms in the target medium, and $I$ is the mean ionization potential of the target medium. The electron charge and mass are denoted, respectively, by $e$ and $m$.

The various terms in (19) have the following interpretation (Tai et al., 1997; Wilson et al., 1991). The $B_0$ term is the high-energy asymptotic limit of the stopping power assuming that the orbital electrons of the target atoms can be treated as essentially free electrons. This requires that the projectile's velocity be much greater than the orbital velocities of the bound atomic electrons, which is an inadequate approximation for inner shell electrons of heavy element target media. The $C(\beta)/Z_T$ term provides a correction for inner shell electrons. The $L_P(\beta)$ term arises from polarization of the target electrons by the incident ion, and is referred to as

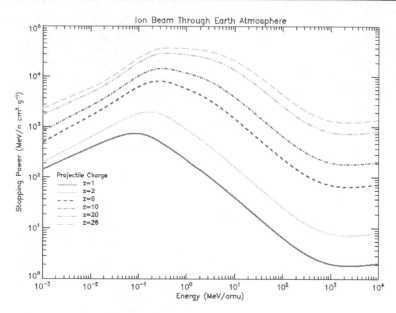

Fig. 17. Stopping power versus kinetic energy of incident ions on Earth's atmosphere. The charge number (z) of the ion is specified in the legend.

the Barkas effect. The $L_T(\beta)$ is the Bloch term which provides a correction to the assumption that the ejected orbital electrons in an ionization event can be represented as a plane wave for close collisions with the incident ion. The Mott term is denoted by $G(M_P, \beta)$, where $M_P$ is the mass of the projectile ion, which includes a kinetic correction for the recoil of the target nucleus. The $\delta(\beta)$ term is a density correction that originates from the dielectric response of a solid target material to the electric field generated by the projectile ion. Finally, at low energy charge exchange processes begin to dominate, which leads to electron capture by the projectile ion and reduces the atomic excitation and/or ionization energy loss. This effect is included by introducing an effective charge for the projectile ion (Tai et al., 1997).

Figure 17 shows the stopping power for various cosmic ray ions incident on Earth's atmosphere. The stopping power decreases inversely with projectile energy between $\sim 100$ keV/amu and 2 GeV/amu. The stopping power begins to increase with increasing projectile energy above 2 GeV/amu due to the relativistic corrections in the $B_0$ term in (19). The stopping power decreases for projectile energies less than $\sim 100$ keV/amu due to electron capture by the projectile ion and the other correction terms in (19). The dependence of the stopping power on the projectile kinetic energy plays a major role in determining the spectral shape of the cosmic ray fluence rates in the atmosphere, as indicated by Figures 20 and 21.

The range of an ion is the mean path length traveled in the target medium before coming to rest after losing its initial kinetic energy through ionization and/or atomic excitation energy loss. In the CSDA, the range is defined by

$$R_j(E) = A_j \int_0^E \frac{dE'}{S_j(E')} \tag{21}$$

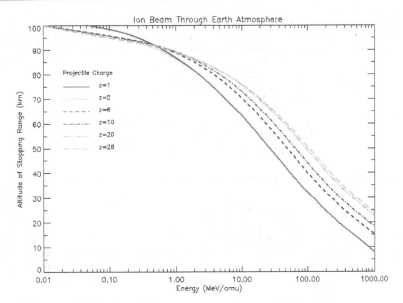

Fig. 18. Average range versus kinetic energy of an ion beam incident at the top of Earth's atmosphere. The charge number (z) of the ion is specified in the legend.

where $A_j$ is the atomic mass number of ion particle type $j$. The above equation is referred to as the range-energy relation. Figure 18 shows the range of various cosmic ray ions incident on the neutral atmosphere as a function of ion kinetic energy. At 1 GeV/amu, only protons and alpha particles can reach the typical cruising altitudes of $\sim$ 10-12 km, or $\sim$ 200 g/cm$^2$, for commercial aircraft before coming to rest due to ionization energy loss. Below 1 GeV/amu, all cosmic ray ions loose their kinetic energy before reaching commercial aircraft cruising altitudes. For energies greater than 1 GeV/amu, the cosmic ray particle flux densities decrease with a power-law dependence on energy (see Figure 7). Although secondary charged particles can be produced by nuclear fragmentation reactions, which is represented by the third term in (18), the stopping range in Figure 18 explains why the heavy-ion fluence rates in Figure 20 are significantly less than the proton fluence rate. The same is true for the light-ions in Figure 21.

Two approximations have been made to the total ion-target interaction cross section in (18). First, elastic ion-nucleus scattering has been neglected for cosmic ray transport through the atmosphere. Second, the CSDA has been invoked in the representation of atomic ion-electron energy transfer collisions. As a consequence of these two approximations, the coupled Boltzmann transport equations in (17) can be expressed, alternatively, as

$$\overline{B}[\Phi_j(\mathbf{x}, \mathbf{\Omega}, E)] = \sum_k \int \int \sigma_{jk,r}(\mathbf{\Omega}, \mathbf{\Omega}', E, E') \Phi_k(\mathbf{x}, \mathbf{\Omega}', E') d\mathbf{\Omega}' dE' \qquad (22)$$

where

$$\overline{B}[\Phi_j(\mathbf{x}, \mathbf{\Omega}, E)] \equiv \left[ \mathbf{\Omega} \bullet \nabla - \frac{1}{A_j} \frac{\partial}{\partial E} S_j(E) + \sigma_{j,r}(E) \right] \Phi_j(\mathbf{x}, \mathbf{\Omega}, E). \qquad (23)$$

The $\overline{B}[\Phi_j]$ in the above equations denote a differential operator acting on the directional fluence.

The differential operator in (22) can be inverted using the method of characteristics in order to transform the integro-differential equation into a Volterra-type integral equation (Wilson, 1977). As a result, the integral equation for cosmic ray transport is given by

$$\Phi_j(x, \Omega, E) = \frac{S_j(E_\gamma)P_j(E_\gamma)}{S_j(E)P_j(E)}\Phi_j(\Gamma_{\Omega,x}, \Omega, E_\gamma)$$

$$+ \sum_k \int_E^{E_\gamma} \frac{A_j P_j(E')}{S_j(E)P_j(E)} dE' \int_{E'}^{\infty} dE'' \int d\Omega' \, \sigma_{jk,r}(\Omega, \Omega', E', E'')$$

$$\times \Phi_k[x + (R_j(E) - R_j(E'))\Omega, \Omega', E'']. \tag{24}$$

In the above equation, $\Gamma_{\Omega,x}$ is a position vector of a point on the boundary surface and $E_\gamma$ is given by

$$E_\gamma = R_j^{-1}[R_j + \Omega \bullet (x - \Gamma_{\Omega,x})] \tag{25}$$

The $R_j^{-1}$ operator in (25) is the inverse operation of obtaining the energy given the range using the range-energy relation in (21). The expression for the integral cosmic ray transport equation in (24) was made compact by introducing the total nuclear survival probability, which is defined by

$$P_j(E) \equiv \exp\left[-A_j \int_0^E \frac{\sigma_{j,r}(E')dE'}{S_j(E')}\right]. \tag{26}$$

The first term in (24) describes the attenuation of the directional fluence specified at the boundary as a result of transport through the target medium. For atmospheric cosmic ray transport, an isotropic distribution is assumed for the directional fluence and the boundary specification is defined to be the cosmic ray particle fluence rates that have been transported through the heliosphere and magnetosphere and incident at the top of the neutral atmosphere. These incident cosmic ray ions are attenuated by ionization energy loss ($S(E)$) and nuclear absorption ($P(E)$), as indicated by the first term in (24). The second term in (24) describes the generation of type $j$ particles from projectile-target nuclear fragmentation reactions by type $k$ particles. The second term in (24) includes the production of type $j$ particles from type $k$ particles at all intervening positions between the boundary point and the position of observation, accounting for the attenuation by ionization energy loss and nuclear absorption in between the point of production of a type $j$ particle and the observation point.

The representation of the relevant total nuclear absorption cross sections ($\sigma_{j,r}$) and nuclear fragmentation production cross sections ($\sigma_{jk,r}$) can not be expressed in a simple, compact form such as the stopping power in (19)-(20). Nevertheless, important insight into the influence of nuclear reactions on the atmospheric transport of cosmic rays can be gained by examining the probability of a nuclear reaction as a function of incident ion kinetic energy. The probability of a nuclear reaction is one minus the total nuclear survival probability in (26) (i.e., $1 - P_j(E)$), which is shown in Figure 19. For particles with kinetic energy below 100 MeV/amu, there is a small chance of a nuclear reaction. Recall from Figure 7 that the peak of the incident GCR spectrum is between $\sim$ 200-500 MeV/amu. At these energies, one out of every two particles will undergo some kind of nuclear reaction. For kinetic energies greater than 1 GeV/amu, nearly every particle will be subject to some type of nuclear reaction. Combining this discussion with the discussion of Figure 18, the only primary cosmic ray particles that can survive transport through the atmosphere and reach the cruising altitudes of typical commercial aircraft are protons with kinetic energy on the order of 1 GeV or greater. The

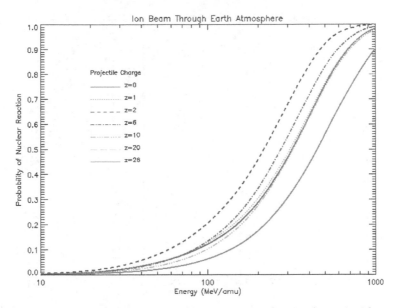

Fig. 19. Nuclear survival probability versus kinetic energy of an ion beam incident at the top of Earth's atmosphere. The charge number (z) of the ion is specified in the legend.

high-LET particles present at 10-12 km with energies less than 1 GeV/amu are secondary particles created at higher altitudes from nuclear fragmentation reactions, most of which are neutrons.

The coupled cosmic ray integral transport equations in (24) are solved in the NAIRAS model using NASA LaRC's deterministic HZETRN code. Details of the early analytical and computation approaches to solving (24) are given by Wilson et al. (1995a; 1997; 1991; 2005a). The stopping power parameterization used in HZETRN is described by Tai et al. (1997). The nuclear cross sections for neutron and proton interactions are described extensively in Wilson et al. (1989). The model for calculating the heavy-ion nuclear fragmentation cross sections are described by Wilson et al. (1995b). HZETRN is used in a wide variety of radiation transport applications: e.g., the calculation of dosimetric quantities for assessing astronaut risk to space radiations on the International Space Station (ISS) and the Space Transportation System (STS) Shuttle, including realistic spacecraft and human geometry (Badavi et al., 2005; 2007a; Slaba et al., 2009; Wilson & et al., 2006). Extensive summaries of HZETRN laboratory and space-flight verification and validation are found in recent reports by Badavi et al. (2007a); Nealy et al. (2007); Wilson et al. (2005c;a).

The computation methods employed in HZETRN to solve the coupled cosmic ray integral transport equations in (24) are summarized below. The details are given in the references. The numerical procedures fall into two categories based on the fundamental physics of cosmic ray projectile-target nuclear interactions. The first category is heavy-ion transport. Ion beam experiments have shown that projectile fragments have an energy and direction very near to that of the incident heavy-ion projectile (Wilson, 1977; Wilson et al., 1995a). The observation of forward directed projectile fragments is the bases of the so-called straight-ahead approximation, where the integral over solid angle in (24) is neglected and

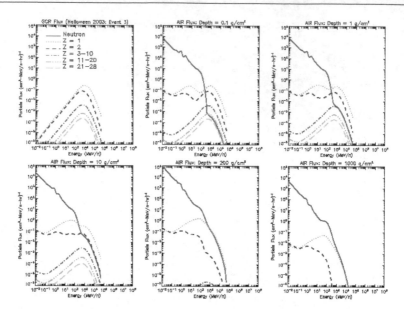

Fig. 20. Event-averaged GCR spectral fluence rates at zero vertical geomagnetic cutoff rigidity during the Halloween 2003 solar-geomagnetic storm [29 October 2003 (2100 UT) to 1 November (0000 UT)]. The panels show the fluence rates at different atmospheric depths. The typical cruising altitudes for commercial aircraft correspond to an atmospheric depth of roughly 200 g/cm$^2$. The fluence rates from different charge groups have been summed together to reduce the number of lines.

the transport is reduced to one-dimension along the direction of the incident heavy-ion beam. Moreover, the observation of equal velocity between the heavy-ion projectile and the projectile fragment suggests a delta function dependence in the projectile fragment production cross section, which effectively eliminates the integral over $dE''$ on the right-hand side of (24).

In addition to the approximations discussed in the previous paragraph, the target fragments are produced at low-energy and distributed nearly isotropically. At low energy, the target fragments do not travel far before coming to rest due to ionization energy loss. These observations justify a decoupling of the target and projectile fragments in the source term on the right-hand side of (24). The advantage of this decoupling is that the target fragments can be neglected in the heavy-ion transport procedure. The absence of the target fragments in the heavy-ion transport solution means that the summation over $k$ type particles in (24) involves only projectiles with masses greater the the mass of the type $j$ particle (i.e., $\sum_{k>j}$). These approximations enable a self-consistent solution of projectile fragment heavy-ion transport using backsubstitution and perturbation theory with rapid convergence. The contribution of the target fragments to the dosimetric quantities is included indirectly using the method described in section 3.6. Significant improvements in the accuracy and computational efficiency of HZETRN's heavy-ion numerical transport procedures have recently been made by Slaba et al. (2010c;d).

The numerical solution of the integral transport equation in (24) for light-particle projectiles does not permit the same approximations as the heavy-ion transport solution. Light-particles are defined in HZETRN as those particles with mass number $A \leq 4$ and charge number

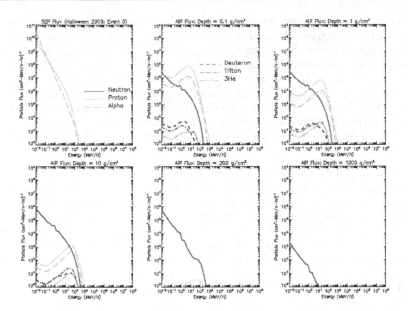

Fig. 21. Event-averaged SEP spectral fluence rates at zero vertical geomagnetic cutoff rigidity during the Halloween 2003 solar-geomagnetic storm [29 October 2003 (2100 UT) to 1 November (0000 UT)]. The panels show the fluence rates at different atmospheric depths. The typical cruising altitudes for commercial aircraft correspond to an atmospheric depth of roughly 200 g/cm$^2$.

$Z \leq 2$. Thus, there are six light-particles: neutrons and five charged particles (protons, deuterons, tritons, helium-3, and helium-4). The equal velocity relationship between projectile and projectile fragments is no longer valid for light-particle transport. This means that the energy integral over $dE''$ on the right-hand side of (24) must be explicitly evaluated, and both the projectile and target fragments are included in the numerical transport procedure. Moreover, the straight-ahead approximation can not be employed in light-particle transport. The integral over solid angle in (24) must be considered, which is especially important for low-energy neutron transport.

Important updates to the numerical solution of light-particle transport in HZETRN have been made recently by Slaba et al. (2010a;c;d;b). These updates have significantly improved accuracy and computation efficiency. The solution approach is to decompose the light-particle fluence into a straight-ahead component and an isotropic component. The transport solution for the light-particle straight-ahead component is described by Slaba et al. (2010c;d). The neutron fluence is further decomposed into semi-isotropic forward and semi-isotropic backward components. The numerical approach for solving the directionally coupled forward-backward neutron transport scheme for the semi-isotropic component is described by Slaba et al. (2010a;b), which also describes the solution of the charged particle isotropic component. The charged particle isotropic transport is approximated by assuming that the source term originates from nuclear fragmentation reactions between the target medium and the low-energy semi-isotropic neutrons, which turns out to be a good approximation (Slaba et al., 2010c;d).

In the NAIRAS model, there are 59 coupled transport equations in the HZETRN description of GCR transport through the atmosphere. This set includes transport equations for neutrons and GCR nuclear isotopes from protons through nickel (Z=28, A=58). Figure 20 shows the GCR spectral fluence at various atmospheric depths during the Halloween 2003 storm period. The top left panel shows the spectral fluence at zero vertical cutoff rigidity incident on the neutral atmosphere. The fluences in this panel are the predictions from NAIRAS GCR model (see section 3.1). There is no neutron fluence in this panel since neutrons are secondary particles created by projectile-target nuclear fragmentation reactions, as discussed above. The remaining panels show the GCR fluences (primaries + secondaries) at different depths within the atmosphere.

Since neutrons do not interact with the target medium via the Coulomb force, there is not an ionization threshold, or an atomic excitation threshold, or a nuclear Coulomb potential barrier to overcome. Thus, the neutrons are not brought to rest as the charged particles are. Neutrons continue to cascade down in energy through neutron-nucleus interactions and the low-energy neutron fluence continues to build, as is evident in Figure 20. The low-energy neutron fluence is quite large, even at a small atmospheric depth of 0.1 g/cm$^2$. The large low-energy neutron fluence at small atmospheric depth is dominated by backscattered neutrons generated at much larger depths, or at much lower altitudes in the atmosphere. At the atmospheric depth of typical cruising altitudes of commercial aircraft, the heavy-ions have largely disappeared, due to a combination of ionization energy loss and nuclear fragmentation reactions into lower energy, lighter particles. Refer to the discussion of Figures 18 and 19.

Solar cosmic rays consist mainly of protons and alpha particles. As a result, only the solution of the six light-particle coupled transport equations defined in HZETRN are required in the description of SEP transport through the atmosphere. Figure 21 shows the light-particle SEP spectral fluences at various atmospheric depths for the same time period during the Halloween 2003 storm as shown in the previous figure. The top left panel shows the spectral fluence at zero vertical cutoff rigidity incident on the neutral atmosphere. These SEP fluence rates were determined using the satellite in-situ ion flux measurements and the spectral fitting algorithm described in section 3.2. The remaining panels show the SEP fluences (primaries + secondaries) at different depths within the atmosphere. Similar to the GCR atmospheric transport properties, the large low-energy neutron fluence at small atmospheric depth for SEP events is due to the large backscattered neutron component, which also originates at larger penetrations depths. Furthermore, only nucleon (protons + neutrons) fluences remain at the typical cruising altitude of commercial aircraft, for the same reasons as previously described.

### 3.5 Meteorological data

This section describes the characterization of the internal properties of the atmosphere that are relevant to cosmic transport. The atmosphere itself provides shielding from incident charged particles. The shielding of the atmosphere at a given altitude depends on the overhead mass. Sub-daily global atmospheric depth is determined from pressure versus geopotential height and pressure versus temperature data derived from the National Center for Environmental Prediction (NCEP) / National Center for Atmospheric Research (NCAR) Reanalysis 1 project (Kalnay & et al., 1996). The NCAR/NCEP Reanalysis 1 project uses a state-of-the-art analysis/forecast system to perform data assimilation using past data from 1948 to the present. The data products are available 4x daily at 0, 6, 12, and 18 UT. The spatial coverage is 17 pressure levels in the vertical from approximately the surface (1000 hPa) to the

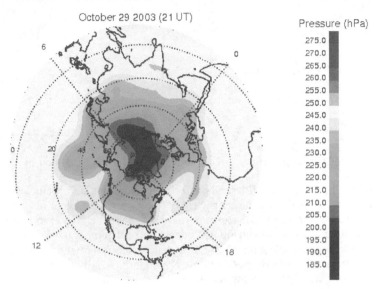

Fig. 22. NCAR/NCEP Reanalysis 1 pressure levels at 11 km corresponding to the date/time of the largest geomagnetically disturbed period of Halloween 2003 SEP event 3 (10/29/2003, 2100 UT) analyzed in section 4. Also shown are the magnetic latitude circles and the meridians at 0, 6, 12, and 18 magnetic local time.

middle stratosphere (10 hPa), while the horizontal grid is 2.5 x 2.5 degrees covering the entire globe.

NCAR/NCEP pressure versus geopotential height data is extended in altitude above 10 hPa using the Naval Research Laboratory Mass Spectrometer and Incoherent Scatter (NRLMSIS) model atmosphere (Picone et al., 2002). NCAR/NCEP and NRLMSIS temperatures are smoothly merged at 10 hPa at each horizontal grid point. NRLMSIS temperatures are produced at 2 km vertical spacing from the altitude of the NCEP/NCAR 10 hPa pressure surface to approximately 100 km. The pressure at these extended altitudes can be determined from the barometric law using the NRLMSIS temperature profile and the known NCAR/NCEP 10 hPa pressure level, which assumes the atmosphere is in hydrostatic equilibrium and obeys the ideal gas law. Finally, the altitudes and temperatures are linearly interpolated in log pressure to a fixed pressure grid from 1000 hPa to 0.001 hPa, with six pressure levels per decade. The result from this step is pressure versus altitude at each horizontal grid point from the surface to approximately 100 km.

Atmospheric depth (g/cm$^2$) at each altitude level and horizontal grid point is computed by vertically integrating the mass density from a given altitude to the top of the atmosphere. The mass density is determined by the ideal gas law using the pressure and temperature at each altitude level. The result from this step produces a 3-D gridded field of atmospheric depth. Atmospheric depth at any specified aircraft altitude is determined by linear interpolation along the vertical grid axis in log atmospheric depth. Figure 22 shows the atmospheric

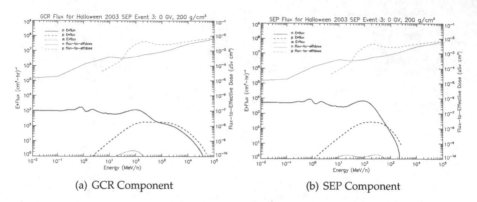

(a) GCR Component  (b) SEP Component

Fig. 23. The green lines show the fluence-to-effective dose conversion coefficients. The blue lines show the product of the kinetic energy (MeV/amu) times the event-averaged spectral fluence rates for the Halloween 2003 storm period [29 October (2100 UT) to 1 November (0000 UT)]. The fluence rates were evaluated at zero vertical geomagnetic cutoff rigidity and at an atmospheric depth of 200 g/cm². The lines denote neutron quantities; dashed lines denote proton quantities; dotted lines denote alpha quantities. All quantities are shown for: (a) GCR component and (b) SEP component.

pressure over the northern hemisphere at 11 km on October 29, 2003 (2100 UT). This is the atmospheric data used in the exposure rate calculations in section 4.3.

### 3.6 Radiation dosimetry

The energy deposited in a target medium by the radiation field of particle $j$ is the dose, which is given by

$$D_j(\mathbf{x}) = K \int_\Omega \int_0^\infty S_j(E)\Phi_j(\mathbf{x}, \Omega, E)d\Omega dE. \tag{27}$$

In the above equation, $S_j(E)$ is the target stopping power for particle $j$ (Mev/g/cm²) and $K$ is a unit conversion factor ($1.602 \times 10^{-10}$) to convert dose to units of Gray (1 Gy = J/kg). The target stopping power is given by (19)-(20), which is shown in Figure 17 for representative cosmic ray ions incident on Earth's atmosphere. Radiation health risk and the probability of biological damage depend not only on the absorbed dose, but also on the particle type and energy of the radiation causing the dose. This is taken into account by weighting the absorbed dose by a factor related to the quality of the radiation. The weighted absorbed dose has been given the name dose equivalent by the ICRP (ICRP, 1991). The unit of dose equivalent is the Sievert (Sv). Dose equivalent in tissue $T$ from particle $j$ ($H_{j,T}(\mathbf{x})$) is defined in terms of the tissue LET dependent quality factor $Q$, such that

$$H_{j,T}(\mathbf{x}) = \int_L Q(L)D_j(\mathbf{x}, L)dL, \tag{28}$$

where $L$ is LET, which can be approximated by the stopping power in units of keV/um; $D_j(\mathbf{x}, L)$ is the spectral dose distribution from particle $j$ in terms of LET, and $Q(L)$ is the tissue LET-dependent quality factor.

The relationship between the probability of biological damage and dose equivalent is found to also depend on the organ or tissue irradiated. A further dosimetric quantity, called the effective dose, is defined to include the relative contributions of each organ or tissue to the total biological detriment caused by radiation exposure. The effective dose ($E(\mathbf{x})$) is the sum of weighted dose equivalents in all the organs and tissues in the human body, such that

$$E(\mathbf{x}) = \sum_T \sum_j w_T H_{j,T}(\mathbf{x}).\tag{29}$$

The organ/tissue weighting factors are given in the ICRP 60 report (ICRP, 1991). A computationally efficient approach is to calculate the effective dose rates directly from the particle spectral fluence rates using pre-computed fluence-to-effective dose conversion coefficients. The NAIRAS model uses neutron and proton conversion coefficients tabulated by Ferrari et al. (1997a;b). The effective dose contributions from the other ions are obtained by scaling the proton fluence-to-effective dose conversion coefficients by $Z_j^2/A_j$, according to stopping power dependence on charge and mass in (19)-(20). All recommended ICRP radiation exposure limits are defined in terms of effective dose.

Figure 23 shows the proton and neutron fluence-to-effective dose conversion coefficients and the event-averaged GCR and SEP spectral fluence rates computed during the Halloween 2003 SEP event 3. The fluence rates were computed at zero vertical geomagnetic cutoff rigidity at an atmospheric depth of 200 g/cm$^2$, which is the depth corresponding to typical cruising altitudes of 10-12 km for commercial aircraft. Above 20 MeV, protons make a larger contribution to effective dose per unit fluence compared to neutrons. Below 20 MeV, the reverse is true. The fluence rates are shown in Figure 23 as a product of the fluence rates times the energy. This is a convenient representation on a log-log scale since the spectral integration with respect to log-energy, which is performed in order to obtain the effective dose rate, is proportional to energy multiplied by the fluence rate. In this representation, neutrons dominate below about 1 GeV. The peak in energy times the proton fluence rate is slightly larger for the SEP component as compared to the GCR component. At 100 MeV and below, the energy times the neutron fluence rate is nearly an order of magnitude greater for the SEP component compared to the GCR component.

The GCR and SEP normalized spectral and accumulated spectral effective dose rates are presented in Figure 24 for the event-averaged Halloween 2003 SEP event 3. Similar to the previous figure, the effective dose rate quantities were computed at zero vertical geomagnetic cutoff rigidity at an atmospheric depth of 200 g/cm$^2$. The spectral effective dose rates are normalized with respect to the peak in the spectrum. The peak in both the proton and neutron spectral effective dose rates occur between 100-200 MeV, which is true for both GCR and SEP contributions. Neutrons make the largest contributions to effective dose at energies below the peak in the spectrum. Protons make the largest contribution to effective dose at energies above the peak in the spectrum. The relative spectral contribution of protons and neutrons to effective dose holds for both the GCR and SEP components. Half of the total effective dose rate comes from spectral contributions at energies less than 100 MeV for the GCR component, and at energies less than roughly 20 MeV for the SEP component. Half of the neutron effective dose comes from spectral contributions at energies less than about 30 MeV for the GCR component, which is similar for the SEP component. Recall that the incident GCR and SEP proton fluence rates spectra are very different. For example, compare Figures 7 and 9. As a result, half of the proton effective dose rates comes from spectral contributions at energies less than 100 MeV for the SEP component, and at energies less than about 500 MeV for the GCR component.

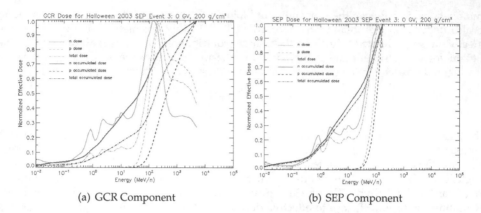

(a) GCR Component                                    (b) SEP Component

Fig. 24. Normalized spectral effective dose rates evaluated during the Halloween 2003 storm period [29 October (2100 UT) to 1 November (0000 UT)] at zero vertical geomagnetic cutoff rigidity and at an atmospheric depth of 200 g/cm². The green lines show the spectral effective dose rate normalized to the peak in the spectrum. The blue lines are normalized to the spectrally integrated effective dose rate, and show the accumulated effective dose rate as the spectral effective dose rate is integrated over energy. The solid lines denote neutron quantities; dashed lines denote proton quantities; dash-dot lines denote total quantities.

## 4. Analysis of atmospheric ionizing radiation exposure

In this section NAIRAS predictions of aircraft radiation exposure are presented and analyzed for three distinct space weather phenomena. The suppression of GCR exposure due to a Forbush decrease is considered briefly in section 4.1. Maximum GCR exposure occurs during solar minimum, as discussed in section 2.1. Results during solar minimum are presented in section 4.2. The Halloween 2003 superstorm is used as a testbed for diagnosing the influence of geomagnetic storm effects on SEP atmospheric radiation exposure. Section 4.3 contains the analysis of the Halloween 2003 SEP exposure and geomagnetic storm effects. When possible, the NAIRAS predictions are compared to other models and/or onboard aircraft radiation measurements.

### 4.1 Case Study 1: GCR during forbush decrease

The Halloween 2003 superstorm was rich in the variety of simultaneous, and often competing, space weather influences (Gopalswamy et al., 2005). Some of the phenomena that occurred during this storm period that exert important influences on atmospheric ionizing radiation exposure are: SEP events, Forbush decreases, geomagnetic storms, cosmic ray anisotropy, and ground level enhancements (GLE). The topic of this section is aircraft radiation exposure during a Forbush decrease that occurred during the Halloween 2003 storm.

A Forbush decrease is a suppression of the GCR exposure due to the interaction of the solar wind with the incident GCR particles. At latitudes with cutoff rigidities greater than 1.0-1.2 GV, the SEP exposure rates are comparable to or less than the GCR exposure rates. Thus, at these latitudes, the total GCR+SEP dose rate can be less than the quiet-time dose rates prior to the SEP event during a Forbush decrease. Getley et al. (2005a;b) observed the apparent influence of a Forbush decrease on dosimetry measurements taken on Qantas Flight 107 from Los Angeles, California to New York, New York on October 29, 2003. These measurements

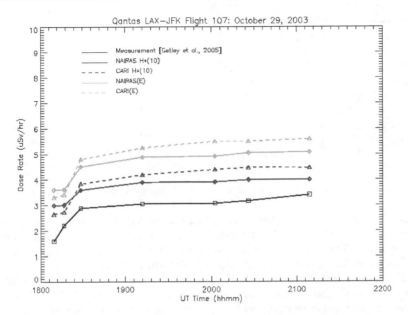

Fig. 25. Comparisons between NAIRAS, FAA/CARI, and TEPC measurements for Qantas Flight 107 from Los Angeles, California to New York, New York on October 29, 2003. The dosimetric quantities compared are effective dose rate (denoted E) and ambient dose equivalent rate (denoted H*(10)).

are well-suited for testing the ability of NAIRAS to model the reduction in the GCR exposure due to Forbush decreases.

Quantitative comparisons were made between NAIRAS GCR exposure rates and aircraft Tissue Equivalent Proportional Counter (TEPC) measurements taken by Getley et al. (2005a;b). The latitudes of the Qantas flight trajectory were too low to observe significant SEP radiation exposure. However, the GCR exposure was suppressed by a Forbush decrease. Figure 25 shows the comparisons between NAIRAS and the TEPC measurements along the Qantas flight trajectory. This figure shows comparisons of effective dose rate and ambient dose equivalent rate, which is a measurement-based proxy for the effective dose rate. Also shown in the figure are the dosimetric quantities computed from the FAA/CARI-6 model. The NAIRAS calculations are closer to the measurement data than the CARI model. The NAIRAS model can account for Forbush decrease effects by virtue of using real-time neutron monitor count rates in specifying the GCR fluence rates incident at the top of the neutral atmosphere, as discussed in section 3.1.

### 4.2 Case Study 2: GCR during solar minimum

In this section NAIRAS model predictions are compared with TEPC measurements during solar minimum conditions where the GCR exposure reaches its maximum during the solar cycle. TEPC measurements are taken on an equatorial flight and a high-latitude flight. The TEPC data were provided courtesy of Dr. Matthias Meier from The German Aerospace Corporation (DLR) in Cologne, Germany. The equatorial flight was from Dusseldorf,

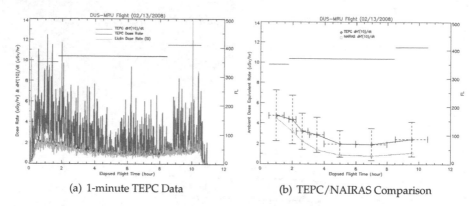

(a) 1-minute TEPC Data                           (b) TEPC/NAIRAS Comparison

Fig. 26. (a) TEPC and Liulin dosimetric measurements on a flight from Dusseldorf to
Mauritius on February 13, 2008. The left ordinate axis indicate dose rates versus elapsed time
of flight (abscissa). The TEPC data shown are: (blue line) absorbed dose rate (uGy/hr) and
(red line) ambient dose equivalent rate (uSv/hr). The Liulin data are (green line) absorbed
dose in silicon (uGy/hr). The right ordinate axis indicates the flight level versus elapsed
flight time (abscissa). The flight levels are the horizontal black lines. (b) TEPC/NAIRAS
comparisons of ambient dose equivalent rate for the same flight. The left ordinate axis
indicates the ambient dose equivalent rate versus elapsed flight time (abscissa). The TEPC
measurements are shown as red diamond symbols. The NAIRAS predictions are shown as
light blue square symbols. The 1-minute TEPC data shown in the left panel have been
averaged over roughly 1-hour periods. The horizontal error bars in the above figure
correspond to the averaging interval. The vertical error bars represent one standard
deviation in TEPC data.

Germany to the island nation of Mauritius on February 13, 2008, and the TEPC 1-minute data
are shown in Figure 26a.

Initial comparison between NAIRAS and TEPC data are focused on the ambient dose
equivalent, since this dosimetric quantity is a fairly reasonable proxy for effective dose - the
quantity directly related to biological risk. Figure 26b shows the NAIRAS/TEPC ambient
dose equivalent rate comparisons. The NAIRAS model predictions are within the statistical
uncertainty of the TEPC measurements. Nevertheless, NAIRAS results are biased low with
respect to the measurements. The largest differences are at the lowest latitudes near the
equator. The equatorial low bias in the NAIRAS results may be due to an underprediction
of electromagnetic cascade processes initiated by pion production. It is known that the
generation and transport of these processes within the HZETRN component of NAIRAS needs
improvement, and these processes are likely to be more important at low latitudes compared
to higher latitudes. Improvements in HZETRN with modeling the transport and dosimetry of
pion initiated electromagnetic cascade effects are currently underway.

The high-latitude dosimetric measurements were taken on a flight from Fairbanks, Alaska to
Frankfurt, Germany on May 23, 2008. The 1-minute TEPC data are shown in Figure 27a.
NAIRAS/TEPC ambient dose equivalent rate comparisons are shown in Figure 27b. The
NAIRAS predictions are at the boundary of one standard deviation in the statistical
uncertainty of the TEPC measurements. For the high-latitude comparison, the NAIRAS
results are biased high with respect to the TEPC measurements. More comparisons between

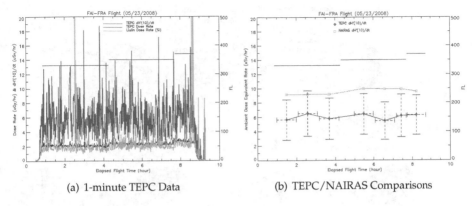

(a) 1-minute TEPC Data                    (b) TEPC/NAIRAS Comparisons

Fig. 27. (a) TEPC and Liulin dosimetric measurements on a flight from Fairbanks to Frankfurt on May 23, 2008. The left ordinate axis indicate dose rates versus elapsed time of flight (abscissa). The TEPC data shown are: (blue line) absorbed dose rate (uGy/hr) and (red line) ambient dose equivalent rate (uSv/hr). The Liulin data are (green line) absorbed dose in silicon (uGy/hr). The right ordinate axis indicates the flight level versus elapsed flight time (abscissa). The flight levels are the horizontal black lines. (b) NAIRAS/TEPC comparisons of ambient dose equivalent rate for the same flight. The left ordinate axis indicates the ambient dose equivalent rate versus elapsed flight time (abscissa). The TEPC measurements are shown as red diamond symbols. The NAIRAS predictions are shown as light blue square symbols. The 1-minute TEPC data shown in the left panel have been averaged over roughly 1-hour periods. The horizontal error bars in the above figure correspond to the averaging interval. The vertical error bars represent one standard deviation in TEPC data.

NAIRAS predictions and dosimetric measurements over a range of latitudes and solar modulation levels are required to fully characterize the uncertainty in the model predictions. Comprehensive flight campaigns are currently underway for extensive model verification and validation.

### 4.3 Case Study 3: A halloween 2003 SEP event

In this section NAIRAS model predictions are presented for SEP effective dose rates and accumulated effective dose along representative high-latitude commercial routes during the Halloween 2003 SEP event 3 [10/29 (2100 UT) - 10/31 (2400 UT)]. The incident SEP spectral fluence rates and meteorological data are fixed in time in the NAIRAS calculations presented in this section, which are given by the event-averaged spectral fluence rates and atmospheric depth-altitude data shown in Figures 9 and 22, respectively.  On the other hand, the geomagnetic cutoff rigidity is allowed to vary in time along the flight trajectories, according to the magnetospheric magnetic field response to the real-time solar wind and IMF conditions (Kress et al., 2010). This enables the geomagnetic influence on SEP radiation exposure to be isolated. The Halloween 2003 superstorm is an ideal event to study geomagnetic effects since this event contained a major magnetic storm which was one of the largest of solar cycle 23. Geomagnetic effects on atmospheric ionizing radiation have not been sufficiently quantified in the past. A unique outcome of this analysis is that it was found that geomagnetic storm effects have a profound effect on SEP atmospheric radiation exposure, especially for flights

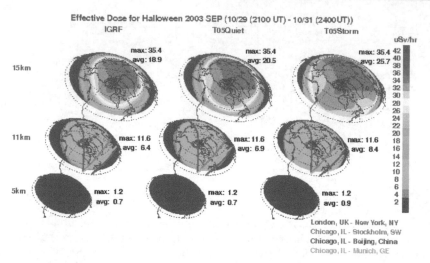

Fig. 28. Effective dose rates computed during Halloween 2003 SEP event 3. The three columns correspond to exposure rates calculated using the geomagnetic cutoff rigidities and magnetic field models shown in Figure 14. The three rows are exposure rates calculated at different altitudes. In each graph, the hemispheric average effective dose rate (uSv/hr) is indicated by the value next to "avg." The maximum exposure rate is indicated by the value next to "max". See text for definition of "avg" and "max."

along the North Atlantic corridor region connecting international flight from the east coast of the US with Europe. The details of this analysis are given in the section below.

### 4.3.1 Global SEP dose distribution

Global SEP atmospheric ionizing radiation exposure are obtained from a pre-computed database. The effective dose rates are calculated on a fixed 2-D grid in atmospheric depth and cutoff rigidity. The atmospheric depth grid extends from zero to 1300 $g/cm^2$, and the cutoff rigidity grid extends from zero to 19 GV. Both grids have non-uniform spacing with the highest number of grid points weighted toward low cutoff rigidities and tropospheric atmospheric depths. The real-time cutoff rigidities are computed on the same 2.5 x 2.5 horizontal grid as the NCEP/NCAR meteorological data. The pre-computed effective dose rates are interpolated to the real-time cutoff rigidity and atmospheric depth specified at each horizontal grid point.

Figure 28 shows global snapshots of atmospheric effective dose rates over the northern hemisphere polar region for the Halloween 2003 SEP event 3. The effective dose rates are shown at three altitudes and for three different magnetic field models used in the cutoff rigidity simulations. The left column shows exposure rates using the IGRF field. The middle column shows exposure rates computed for a geomagnetically quiet time prior to the onset of SEP event 3 using the TS05 field (October 28, 2003, 0002 UT). The right column shows the exposure rates using the TS05 field at the peak of the geomagnetic storm (October 29, 2003, 2100 UT) during SEP event 3. A typical cruising altitude for a commercial high-latitude flight is 11 km. Overlaid on the 11 km effective dose rate altitude surface are great circle routes for three representative high-latitude commercial flights: London, England (LHR) to New York,

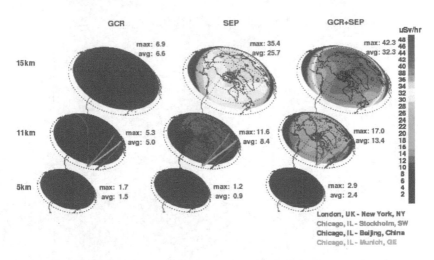

Fig. 29. Effective dose rates computed during Halloween 2003 SEP event 3. The first, middle, and last columns are the GCR, SEP, and total (GCR+SEP) effective dose rates, respectively. The three rows are exposure rates calculated at different altitudes. In each graph, the hemispheric average effective dose rate (uSv/hr) is indicated by the value next to "avg." The maximum exposure rate is indicated by the value next to "max". See text for definition of "avg" and "max."

New York (JFK) (5.75 hour flight time); Chicago, Illinois (ORD) to Stockholm, Sweden (ARN) (8.42 hour flight time), and a combination of two great circle routes from Chicago, Illinois (ORD) to Beijing, China (PEK) (13.5 hour flight time).

There are a number of striking features to be noted from Figure 28. First, the representation of the geomagnetic field has a significant influence on SEP atmospheric ionizing radiation exposure. Comparing the left and middle columns of Figure 28 shows that even during geomagnetically quiet periods, the magnetospheric magnetic field weakens the overall geomagnetic field with a concomitant increase in radiation levels. This is seen as a broadening of the open-closed magnetospheric boundary in the TS05 quiet field compared to the IGRF field. The cutoffs are zero in the region of open geomagnetic field lines. Thus, effective dose rates based on the IGRF field are underestimated even for magnetically quiet times. During strong geomagnetic storms, as shown in the third column of Figure 28, the area of open field lines are broadened further, bringing large exposure rates to much lower latitudes. Effective dose rates predicted using the IGRF model during a large geomagnetic storm can be significantly underestimated. The expansion of the polar region high exposure rates to lower latitudes, due to geomagnetic effects, is quantified by calculating hemispheric average effective dose rates from 40N to the pole. This is denoted by "avg" in Figure 28. At 11 km, there is roughly a 8% increase in the global-average effective dose rate using TS05 quiet-field compared to IGRF. During the geomagnetic storm, there is a $\sim$ 30% increase in the global-average effective dose rate using TS05 storm-field compared to IGRF.

A second important feature to note in Figure 28 is the strong altitude dependence due to atmospheric shielding. The exposure rates are very low at 5 km, independent of geomagnetic field model used. At 15 km, the exposure rates are significantly higher than at 11 km. Figure 28

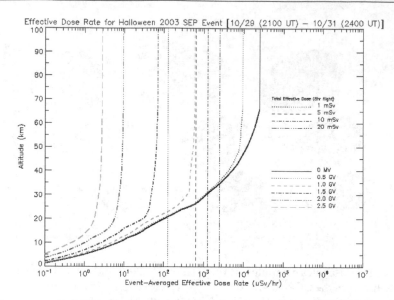

Fig. 30. Event-averaged SEP effective dose rates for Halloween SEP event 3 [(10/29/2003, 2100 UT) - 10/31/2003 (2400 UT)] as a function of altitude for various geomagnetic cutoff rigidities. Different vertical lines indicate constant exposure rates required to reach the corresponding total exposure levels indicated in the legend for a 8-hr flight.

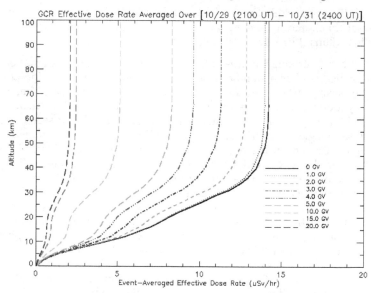

Fig. 31. Event-averaged GCR effective dose rates for Halloween SEP event 3 [(10/29/2003, 2100 UT) - 10/31/2003 (2400 UT)] as a function of altitude for various geomagnetic cutoff rigidities.

shows that the SEP effective dose rates increase (decrease) exponentially with increasing (decreasing) altitude. The SEP exposure rate altitude dependence is a fortunate feature for the aviation community, since radiation exposure can be significantly reduced by descending to lower altitudes. Private business jets will receive more radiation exposure than commercial aircraft if mitigation procedures are not taken, since business jet cruising altitudes are roughly 12-13 km. The altitude dependence of the SEP exposure rates is quantified in Figure 28 by showing the maximum effective dose rate at each altitude, which is the exposure rate at zero cutoff rigidity (i.e., in the polar region of open geomagnetic field lines). The maximum is denoted "max" in Figure 28. The exposure rate increases on average by 160% per km between 5 km and 11 km. Between 11 km and 15 km, the exposure rate increases on average by approximately 75% per km.

Figure 29 shows the event-averaged GCR and SEP effective dose rates separately for the Halloween 2003 SEP event 3. The total (GCR+SEP) effective dose rates are also shown. The SEP exposure rates were computed with the TS05 field at the peak of the geomagnetic storm (29 October 2003, 2100 UT) during SEP event 3, which are identical to the exposure rates from the third column of Figure 28. At 5 km, the GCR and SEP effective dose rates are comparable. The SEP exposure rates in the polar region at 11 km are roughly a factor of two greater than the corresponding GCR exposure rates. At 15 km over the polar region, the SEP exposure rates are greater than the GCR exposure rates by about a factor of five. Due to a substantially larger number of >100 MeV protons during the January 2005 storm event, the peak polar SEP effective dose rates are significantly greater during the January 2005 event compared to the Halloween 2003 storm period (Copeland et al., 2008).

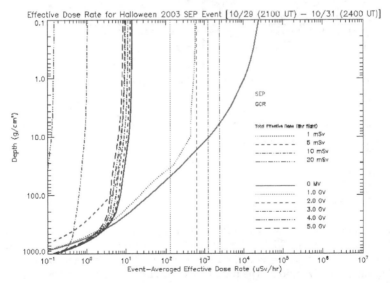

Fig. 32. Event-averaged GCR and SEP effective dose rates for Halloween SEP event 3 [(10/29/2003, 2100 UT) - 10/31/2003 (2400 UT)] as a function of altitude for various geomagnetic cutoff rigidities.

### 4.3.2 Dose on high-latitude flights

Before calculating radiation exposure along specified flight paths, it's constructive to examine a sample of effective dose rate profiles at different cutoff rigidities from our pre-computed database previously described for SEP event 3. Figure 30 shows SEP effective dose rates as a function of altitude for cutoff rigidities from zero to 2.5 GV. This figure clearly shows the exponential dependence of SEP exposure on both cutoff rigidity and altitude. The vertical lines indicate constant exposure rates necessary to receive a total exposure of 1, 5, 10, and 20 mSv on a 8 hour flight. A typical international, high-latitude flight is 8 hours. The rationale for choosing the total exposure identified with the vertical lines is as follows (Wilson et al., 2003): 20 mSv is the ICRP annual occupational radiation worker limit, 10 mSv is the NCRP annual occupational exposure limit, 5 mSv is the NCRP occasional public exposure limit, and 1 mSv is the ICRP annual public and prenatal exposure limit. The cutoff rigidities at high-latitudes are less than 1 GV. The typical commercial airline cruising altitudes correspond to an atmospheric depth between $\sim$ 200-300 g/cm$^2$. Consequently, one can see from Figure 30 that it's not possible for passengers on high-latitude commercial flights during the Halloween 2003 SEP events to approach or exceed the ICRP public and/or prenatal radiation exposure limit. However, the recommended ICRP annual public and prenatal exposure limits were exceeded during the January 2005 storm event (Copeland et al., 2008).

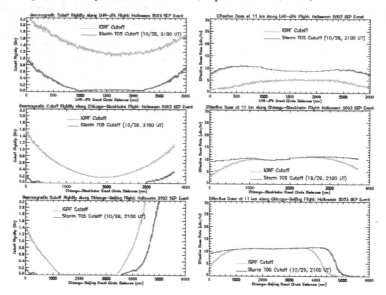

Fig. 33. Geomagnetic cutoff rigidities (left column) and effective dose rates (right column) calculated during Halloween 2003 SEP event 3 along three representative flight paths for a cruising altitude of 11 km. The green line represents cutoff rigidities and exposure rates calculated using the IGRF model. The red lines represent cutoffs and exposure rates computed using the TS05 model during the period of largest geomagnetic activity of event 3. The total flight times are the following: 5.75 hours from JFK-LHR, 8.42 hours for ORD-ARN, and 13.5 hours for ORD-PEK.

For comparison with the previous plot, Figure 31 shows the GCR effective dose rates as function of altitude for cutoff rigidities from zero to 20 GV. In contrast to the SEP exposure rates from Figure 30, the GCR exposure rates do not vary exponentially with respect to altitude or cutoff rigidity. The difference between the GCR and SEP exposure rate dependence on cutoff rigidity and altitude is elucidated further by placing both GCR and SEP effective dose rate profiles computed at the same cutoff rigidities on same figure, which is shown in Figure 32. The geomagnetic field is very effective at shielding SEP radiation. For example, at cutoff rigidities greater than 2 GV, which for North American corresponds to latitudes south of Canada and the upper regions of the United States, the SEP effective dose rates are already below the GCR exposure rates. The significance of geomagnetic storm effects is that the expansion of the polar cap region, or magnetosphere open-closed boundary, can bring the larger SEP exposure rates, normally confined within the polar cap region, to much lower latitude, even mid-latitudes for a large enough geomagnetic storm.

Figure 33 shows the cutoff rigidities and effective dose rates for the three representative high-latitude flights mentioned in the previous subsection, which were calculated along great circle routes. The left column is the cutoff rigidities along the flight paths and the right column is the corresponding effective dose rates along the flight paths. The cutoff rigidities include both latitude and time-dependent variations along the flight paths. The variations of the exposure rates along the flight paths include latitudinal variations in both atmospheric depth and cutoff rigidity. The temporal variations in cutoff rigidity also map into the variations of the exposure rates along the flight path. The top row shows results for the LHR-JFK flight, while the middle and bottom rows show results for the ORD-ARN and ORD-PEK flights, respectively. Each panel in Figure 33 shows cutoff rigidities and corresponding effective dose rates using the IGRF field (green lines) and the TS05 storm-field (red lines) in the cutoff calculations. The largest differences in flight-path cutoff rigidities between IGRF and TS05 storm-field models are for the LHR-JFK flight. The entire LHR-JFK flight path is near the magnetosphere open-closed boundary and is most sensitive to perturbations in cutoff rigidity due to geomagnetic effects. Consequently, the exposure rates along the LHR-JFK flight are most sensitive to geomagnetic effects. The ORD-PEK polar route is the least sensitive to geomagnetic suppression of the cutoff rigidity, since most of the flight path is across the polar cap region with open geomagnetic field lines. The influence of geomagnetic storm effects on the ORD-ARN flight is intermediate between a typical polar route and a flight along the north Atlantic corridor between the US and Europe.

The effective dose rates for the representative high-latitude flights in Figure 33 are within the range of exposure rates measured during other storm periods for similar flight paths. For example, Clucas et al. (2005) reported peak SEP dose rates in the 3.5-4.0 uSv/hr range for a LHR-JFK flight on July 14, 2000, a SEP event without a concomitant geomagnetic storm. From Figure 33, the peak dose rate for the LHR-JFK flight computed using the IGRF field is ~ 4 uSv/hr. Clucas et al. also showed peak measured SEP dose rates on the order of 10-12 uSv/hr for a LHR-JFK flight during an April 2001 SEP event, which was accompanied by a geomagnetic storm. The average LHR-JFK effective dose rate in Figure 33 computed using the TS05 storm-field is 9.4 uSv/hr. Dyer et al. (2005) reported measured peak SEP dose rates on the order of 9.5 uSv/hr for a flight from Prague to New York during the April 2001 SEP event. Thus, our computed effective dose rates are in qualitative agreement with measured dose rates for similar flight paths during other storm periods.

The high sensitivity of SEP atmospheric dose rates to geomagnetic conditions near the open-closed magnetospheric boundary is also responsible for the high sensitivity of SEP dose rates to the exact flight path in the north Atlantic corridor region. For example, Dyer et al. (2007) found that the differences in peak dose rates between great circle and actual flight paths for LHR-JFK Concorde flights were a factor of five during the September 1989 SEP event and a factor of 2.5 during the October 1989 event, which was geomagnetically quiet. Furthermore, the difference in peak SEP dose rates between great circle and actual flight paths for the commercial flight from Prague to New York during the April 2001 event was a factor of two.

The actual flight paths for the Chicago to Munich flights reported by Beck et al. (2005) may have been equatorward of a great circle route, which could explain their low dose rates compared to our calculated LHR-JFK great circle route dose rates during the Halloween 2003 superstorm. During the Halloween 2003 SEP event 3, the Chicago-Munich flight measured a mean SEP dose rate of 3.6 uSv/hr and an accumulated dose of 0.032 mSv for the 8.75 hour flight. These results are comparable to our 5.75 hour JFK-LHR for geomagnetically quiet conditions presented in Figure 33. In other words, the Chicago-Munich flight didn't seem to experience the dose rate enhancement due to the large geomagnetic storm. By comparing the LHR-JFK dose rates in Figure 33 computed using the IGRF field and the TS05 storm-field with the dose rates computed for the ORD-ARN and ORD-PEK flights, it is clear that geomagnetic effects enable flights along the North Atlantic corridor, or near the magnetosphere open-closed boundary, to experience the same dose rates that are confined to the polar region under geomagnetically quiet conditions.

Another possible explanation for the relative difference in the Chicago-Munich measured dose rates reported by Beck et al. (2005) and our computed dose rates for the LHR-JFK flight is our use of the event-averaged incident SEP spectral fluence rate, which we employed for the purpose of isolating geomagnetic effects in our case study. It is clear from the shaded regions in Figure 9 that our event-averaged SEP spectra fluence rate is weighted more toward the peak ion flux measurements observed during SEP event 3. The measurements in Figure 8 show that the SEP ion flux rapidly decreased in time from the peak values present at the beginning of event 3. Thus, our use of an event-averaged incident SEP spectral fluence rate will tend to overestimate the accumulated effective dose over the flight-paths in Figure 33. However, employing this constraint on the incident SEP spectral fluence rate is necessary to unambiguously isolate the geomagnetic effects. Despite these caveats, the results discussed in this paper are within the current factor of two uncertainty in SEP atmospheric dose rates (Clucas et al., 2005).

## 5. Conclusions

The NAIRAS (Nowcast of Atmospheric Ionizing Radiation for Aviation Safety) prototype operational model is currently streaming live from the project public website (google NAIRAS). NAIRAS predicts biologically hazardous radiation exposure globally from the surface to 100 km in real-time. The NAIRAS model addresses an important national need with broad societal, public health, and economic benefits. Commercial aircrew are classified by European and international agencies as radiation workers, yet they are the only occupational group exposed to unquantified and undocumented levels of radiation. Furthermore, the current guidelines for maximum public and prenatal exposure can be easily exceeded during a single solar storm event for commercial passengers on intercontinental or polar routes, or by frequent use of these high-latitude routes even during background conditions. The NAIRAS model will provide a new decision support system that currently does not exist, but is essential for providing the commercial aviation industry with data products that will enable airlines to achieve the right balance between minimizing flight cost while at the same time minimizing radiation risk.

NAIRAS is a physics-based model that maximizes the use of real-time input data. GCR are transported from outside the heliosphere to 1 AU using real-time measurements of ground-based neutron monitor count rates. The SEP particle spectra are determined in-situ using a combination of NOAA/GOES and NASA/ACE ion flux measurements. Both sources of cosmic rays, galactic and solar, are transported through Earth's magnetosphere using a semi-physics-based geomagnetic shielding model. The geomagnetic shielding model utilizes real-time NASA/ACE solar wind and IMF measurements. The cosmic rays are transported from the magnetosphere through the neutral atmosphere using the NASA LaRC's HZETRN deterministic transport code. The real-time, global atmospheric mass density distribution is obtained from the NOAA Global Forecasting System. Global and flight path radiation exposure visualization and decision data products have been developed, which are available at the NAIRAS website.

Future research will focus on new science questions that emerged in the the development NAIRAS prototype operational model (Mertens et al., 2010c). The science questions identified by Mertens et al. (2010c) must be addressed in order to obtain a more reliable and robust operational model of atmospheric radiation exposure. Addressing these science questions require improvements in both space weather modeling and observations.

The Automated Radiation Measurements for Aviation Safety (ARMAS) is a new initiative to address the deficiencies in observations needed to improve the reliability and robustness of operational aircraft radiation exposure assessment. The ultimate goal of the ARMAS initiative is to integrate onboard radiation instruments into a global fleet of aircraft so that the radiation measurements can be downlinked in real-time and assimilated into the NAIRAS predictions of radiation exposure. A subsidiary goal of ARMAS is to provide a testbed to evaluate the accuracy and reliability of new generations of smaller, cheaper hardware technologies in measuring the atmospheric ionizing radiation field. The ARMAS initiative enables the NAIRAS model to adopt the successful meteorological paradigm for reliable and robust weather forecast, which is physics-based models combined with real-time data assimilation of meteorological fields. The NAIRAS/ARMAS approach is a space weather version of terrestrial weather forecasts. These efforts will occupy the NAIRAS team for the next decade or more.

Other research topics unrelated to biological risk from cosmic rays will also be addressed. Effort will also be directed toward predicting the risk that cosmic rays may pose to the operation of microelectronics instrumentation onboard aircraft, and the potential influence cosmic rays may have on the chemistry and climate of planetary atmospheres.

# 6. References

AMS (2007). Integrating space weather observations & forecasts into aviation operations, *Technical report*, American Meteorlogical Society Policy Program & SolarMetrics.

Anderson J. L., Waters, M. A, Hein, M. J., Schubauer-Berigan, M. K. & Pinkerton, L. E. (2011). Assessment of occupational cosmic radiation exposure of flight attendants using questionnaire data, *Aviat Space Environ Med*, 82: 1049–54.

Aspholm, R., Lindbohm, M. L., Paakkulainen, H., Taskinen, H., Nurminen, T. & Tiitinen, A. (1999). Spontaneous abortions among finnish flight attendants, *J. Occupational & Environmental Medicine* 41(6): 486–491.

Badavi, F. F., Nealy, J. E., de Angelis, G., Wilson, J. W., Clowdsley, M. S., Luetke, N. J., Cuncinotta, F. A., Weyland, M. D. & Semones, E. J. (2005). Radiation environment and shielding model validation for cev design, *Space 2005*, number AIAA 2005-6651, Am. Inst. of Aeronaut. and Astronaut., Long Beach, California.

Badavi, F. F., Tramaglina, J. K., Nealy, J. E., & Wilson, J. W. (2007a). Low earth orbit radiation environments and and shield model validation for iss, *Space 2007*, number AIAA 2007-6046, Am. Inst. of Aeronaut. and Astronaut., Long Beach, California.

Badhwar, G. D. & O'Neill, P. M. (1991). An improved model of galactic cosmic radiation for space exploration missions, *22nd International Cosmic Ray Conference*, number OG-5.2-13, pp. 643–646.

Badhwar, G. D. & O'Neill, P. M. (1992). An improved model of galactic cosmic radiation for space exploration missions, *Nuclear Tracks Radiat. Meas.*, 20: 403–410.

Badhwar, G. D. & O'Neill, P. M. (1993). Time lag of twenty-two year solar modulation, *23nd International Cosmic Ray Conference*, Vol. 3, pp. 535–539.

Badhwar, G. D. & O'Neill, P. M. (1994). Long term modulation of galactic cosmic radiation and its model for space exploration, *Adv. Space Res.* 14: 749–757.

Badhwar, G. D. & ONeill, P. M. (1996). Galactic cosmic radiation model and its applications, *Adv. Space Res.* 17: 7–17.

Baker, D. N., Mason, G. M., Figueroa, O., Colon, G., Watzin, J. G. & Aleman, R. M. (1993). An overview of the solar, anomalous, and magnetospheric particle explorer (sampex) mission, *IEEE Trans. Geosci. Remote Sen.* 31(3): 531–541.

Band, P. R., Spinelli, J. J., Ng, V. T. Y., Moody, J. & Gallagher, R. P. (1990). Mortality and cancer incidence in a cohort of commercial airline pilots, *Aviat. Space Environ. Med.* 61: 299–302.

Barish, R. J. (1990). Health physics concerns in commercial aviation, *Health Phys.* 59: 199–204.

Barish, R. J. (2004). In-flight radiation exposure during pregnancy, *Obstet. Gynecol.* 103: 1326–1330.

Beck, P., Latocha, M., Rollet, S. & Stehno, G. (2005). Tepc reference measurements at aircraft altitudes during a solar storm, *Adv. Space Res.* 36: 1627–1633.

BEIR V (1990). *National Researach Council. Health effects of exposure to low levels of ionizing radiation*, Washington, DC, National Academy Press.

Bramlitt, E. T. (1985). Commercial aviation crewmember radiation doses, *Health Phys.* 49: 945–948.

Brandt, S. (1999). *Data Analysis, Statistical and Computational Methods for Scientists and Engineers*, Springer-Verlag, New York.

Buja, A., Lange, J. H, Perissinotto, E., Rausa, G., Grigoletto, F., Canova, C. & Mastrangelo, G. (2005) . Cancer incidence among male military and civil pilots and flight attendants: an analysis on published data, *Toxicol Ind Health* 21: 273–82.

Buja, A., Mastrangelo, G., Perissinotto, E., Grigoletto, F., Frigo, A. C., Rausa, G., Marin, V., Canova. C. & Dominici, F. (2006). Cancer incidence among female flight attendants: a meta-analysis of the published data, *J Womens Health* 15: 98–105.

Chen, J., Lewis, B. J., Bennett, L. G. I., Green, A. R. & Tracy, B. L. (2005). Estimated neutron dose to embryo and foetus during commercial flight, *Radiat. Prot. Dosim.* 114(4): 475–480.

Clem, J., Clements, D. P., Esposito, J., Evenson, P., Huber, D., LHeureux, J., Meyer, P. & Constantin, C. (1996). Solar modulation of cosmic electrons, *Astrophys. J.* 464: 507.

Clucas, S. N., Dyer, C. S. & Lei, F. (2005). The radiation in the atmosphere during major solar particle events, *Adv. Space Res.* 36: 1657–1664.

Copeland, K., Sauer, H. H., Duke, F. E. & Friedberg, W. (2008). Cosmic radiation exposure on aircraft occupants on simulated high-latitude flights during solar proton events from 1 january 1986 through 1 january 2008, *Adv. Space Res.* 42: 1008–1029.

DOC (2004). Intense space weather storms october 19 - november 07, 2003, *Service assessment*, U.S. Department of Commerce, National Oceanic and Atmospheric Administration, National Weather Service, Silver Spring, Maryland.

Dyer, C., Hands, A., Lei, F., Truscott, P., Ryden, K. A., Morris, P., Getley, I., Bennett, L., Bennett, B. & Lewis, B. (2009). Advances in measuring and modeling the atmospheric radiation environment, *IEEE Trans. Nucl. Sci.* 56(8).

Dyer, C. & Lei, F. (2001). Monte carlo calculations of the influence on aircraft radiation environments of structures and solar particle events, *IEEE Trans. Nucl. Sci.* 48(6): 1987–1995.

Dyer, C., Lei, F., Hands, A., Clucas, S. & Jones, B. (2005). Measurements of the atmospheric radiation environment from cream and comparisons with models for quiet time and solar particle events, *IEEE Trans. Nucl. Sci.* 52(6): 2326–2331.

Dyer, C., Lei, F., Hands, A. & Truscott, P. (2007). Solar particle events in the qinetiq atmospheric radiation model, *IEEE Trans. Nucl. Sci.* 54(4): 1071–1075.

Ellison, D. C. & Ramaty, R. (1985). Shock acceleration of electrons and ions in solar flares, *Astrophys. J.* 298: 400–408.

Engelmann, J. J., Ferrando, P., Soutoul, A., Goret, P., Juliusson, E., Koch-Miramond, L., Lund, N., Masse, P., Peters, B., Petrou, N., & Rasmussen, I. L. (1990). Charge composition and energy spectra of cosmic-ray nuclei for elements from be to ni. results from heao-3-c2, *Astron. Astrophys.*, 233: 96–111.

European Radiation Dosimetry Group (EURADOS) (1996). Exposure of air crew to cosmic radiation. A report of EURADOS Working Group 11, *EURADOS Report 1996.01*. In: McAulay, I. R, Bartlett, D. T, Dietze, G. & et al. editors. European Commission Report Radiation Protection 85. Luxembourg: Office for Official Publications of the European Communities.

Fanton, J. W. & Golden, J. G. (1991). Radiation-induced endometriosis in Macaca mulatta, *Radiation Res*, 126: 141–146.

Fedder, J. G. L. J. A. & Mobarry, C. M. (2004). The lyon-fedder-mobarry (lfm) global mhd magnetospheric simulation code, *J. of Atmos. and Solar-Terrestrial Phys.* 66(15-16): 1333–1350.

Ferrari, A., Pelliccioni, M. & Pillon, M. (1997a). Fluence to effective dose conversion coefficients for neutrons up to 10 tev, *Radiat. Prot. Dos.* 71(3): 165–173.

Ferrari, A., Pelliccioni, M. & Pillon, M. (1997b). Fluence to effective dose and effective dose equivalent conversion coefficients for protons from 5 mev to 10 tev, *Radiat. Prot. Dos.* 71(2): 85–91.

Foelsche, T. (1961). Radiation exposure in supersonic transports, *Technical Report TN D-1383*, NASA.

Foelsche, T. & Graul, E. H. (1962). Radiation exposure in supersonic transports, *Atompraxis* 8: 365–380.

Foelsche, T., Mendell, R. B., Wilson, J. W. & Adams, R. R. (1974). Measured and calculated neutron spectra and dose equivalent rates at high altitudes: Relevence to sst operations and space research, *Technical Report TN D-7715*, NASA.

Friedberg, W., Faulkner, D. N., Snyder, L., Jr., E. B. D. & OBrien, K. (1989). Galactic cosmic radiation exposure and associated health risk for air carrier crewmembers, *Aviat. Space Environ. Med.* 60: 1104–1108.

Gaisser, T. (1990). *Cosmic Rays and Particle Physics*, Cambridge University Press.

Getley, I. L., Duldig, M. L., Smart, D. F. & Shea, M. A. (2005a). Radiation dose along north america transcontinental flight paths during quiescent and disturbed geomagnetic conditions, *Space Weather* 3(S01004): doi:10.1029/2004SW000110.

Getley, I. L., Duldig, M. L., Smart, D. F. & Shea, M. A. (2005b). The applicability of model based aircraft radiation dose estimates, *Adv. Space Res.* 36: 1638–1644.

Gold, R. E. & et al. (1998). Electron, proton, and alpha monitor on the advanced composition and explorer satellite, *Space Sci. Rev.* 86: 541–562.

Gopalswamy, N., Yashiro, S., Liu, Y., Michalek, G., Vourlidas, A., Kaiser, M. L. & Howard, R. A. (2005). Coronal mass ejections and other extreme characteristics of the 2003 october - november solar eruptions, *J. Geophys. Res.* 110(A09S15): doi:10.1029/2004JA010958.

Grajewski, B., Waters, M. A., Yong, L. C., Tseng, C.-Y., Zivkovich, Z. & Cassinelli II, R. T. (2011). Airline pilot cosmic radiation and circadian disruption exposure assessment from logbooks and company records, *Ann. Occup. Hyg.* 55(5): 465–475.

Haggerty, D. K., E. C. R. G. C. H. & Gold, R. E. (2006). Quantitative comparison of ace/epam data from different detector heads: Implications for noaa rtse users, *Adv. Space Res.* 38: 995–1000.

Hammer, G. P, Blettner, M & Zeeb, H. (2009). Epidemiological studies of cancer in aircrew, *Radiat Prot Dosimetry* 136: 232–9.

Heinrich, W., Roesler, S. & Schraube, H. (1999). Physics of cosmic radiation fields, *Radiat. Prot. Dosim.* 86: 253–258.

ICRP (2008). *ICRP Publication 103: 2007 Recommendations of the International Commission on Radiological Protection*, ISBN 0-7020-3048-1, Elsevier.

ICRP (1991). *ICRP Publication 60: 1990 Recommendations of the International Commission on Radiological Protection*, Vol. 21(1-3), Pergamon Press.

ICRU (1986). *ICRU Report 40: The quality factor in radiation protection*, International Commission on Radiation Units and Measurements.

Jiang, T. N., Lord, B. I. & Hendry, J. H. (1994). Alpha particles are extremely damaging to developing hemopoiesis compared to gamma radiation, *Radiat. Res.* 137: 380–384.

Kahler, S. W. (2001). Origin and properties of solar energetic particles in space, *in* P. Song, H. J. Singer & G. L. Siscoe (eds), *Space Weather*, American Geophysical Union, Washington, DC.

Kalnay, E. & et al. (1996). The ncar/ncep 40-year reanalysis project, *Bull. Amer. Meteor. Soc.* 77: 437–470.

Kress, B. T., Hudson, M. K., Perry, K. L. & Slocum, P. L. (2004). Dynamic modeling of geomagnetic cutoff for the 23-24 november 2001 solar energetic particle event, *Geophys. Res. Lett.* 31(L04808): doi:10.1029/2003GL018599.

Kress, B. T., Mertens, C. J. & Wiltberger, M. (2010). Solar energetic particle cutoff variations during the 28-31 october 2003 geomagnetic storm, *Space Weather* 8(S05001): doi:10.1029/2009SW000488.

Lambiotte, J. J., Wilson, J. W. & Filipas, T. A. (1971). Proper-3c: A nucleon-pion transport code, *Technical Report TM X-2158*, NASA.

Langlais, B. & Mandea, M. (2000). An igrf candidate geomagnetic field model for epoch 2000 and a secular variation model for 2000-2005, *Earth Planets Space* 52: 1137–1148.

Lauria, L., Ballard, T. J., Caldora, M., Mazzanti, C. & Verdecchia, A. (2006). Reproductive disorders and pregnancy outcomes among female flight attedants, *Aviation, Space, and Envirnomental Medicine* 77(7): 533–559.

Lewis, B. J., Bennett, G. I., Green, A. R., McCall, M. J., Ellaschuk, B., Butler, A. & Pierre, M. (2002). Galactic and solar radiation exposure to aircrew during a solar cycle, *Radiat. Prot. Dosim.* 102(3): 207–227.

Lindborg, L., Bartlett, D. T., Beck, P., McAulay, I. R., Schnuer, K., Schraube, H. & (Eds.), F. S. (2004). Cosmic radiation exposure of aircraft crew: compilation of measured and calculated data. a report of eurados working group 5, European Radiation

Dosimetry Group, Lexembourg: Office for the Official Publications of the European Communities, European Communities.

Lopate, C. (2004). Private Communication.

Meier, M., Hubiak, M., Matthiä, D., Wirtz, M. & Reitz, G. (2009). Dosimetry at aviation altitudes, *Radiat. Prot. Dos.* 136(4): 251–255.

Menn, W. & et al. (2000). The absolute flux of protons and helium at the top of the atmosphere using imax, *Astrophys. J.* 533: 281–297.

Mertens, C. J., Kress, B. T., Wiltberger, M., Blattnig, S. R., Slaba, T. S., Solomon, S. C. & Engel, M. (2010a). Geomagnetic influence on aircraft radiation exposure during a solar energetic particle event in october 2003, *Space Weather* 8(S03006): doi:10.1029/2009SW000487.

Mertens, C. J., Moyers, M. F., Walker, S. A. & Tweed, J. (2010b). Proton lateral broadening distribution comparisons between grntrn, mcnpx, and laboratory beam measurements, *Adv. Space Res.* 45: 884–891.

Mertens, C. J., Tobiska, W. K., Bouwer, D., Kress, B. T., Solomon, S. C., Kunches, J., Grajewski, B., Gersey, B. & Atwell, W. (2010c). Nowcast of atmospheric ionizing radiation for aviation safety. Submitted as a white paper to The National Academies Decadal Strategy for Solar and Space Physics (Heliophysics) RFI.

Mertens, C. J., Tobiska, W. K., Bouwer, D., Kress, B. T., Wiltberger, M. J., Solomon, S. C. & Murray, J. J. (2009). Development of nowcast of atmospheric ionizing radiation for aviation safety (nairas) model, *1st AIAA Atmosheric and Space Environments Conference*, number AIAA 2009-3633, Am. Inst. of Aeronaut. and Astronaut., San Antonio, Texas.

Mertens, C. J., Wilson, J. W., Blattnig, S. R., Kress, B. T., Norbury, J. W., Wiltberger, M. J., Solomon, S. C., Tobiska, W. K. & Murray, J. J. (2008). Influence of space weather on aircraft ionizing radiation exposure, *16th Aerospace Sciences Meeting and Exhibit*, number AIAA 2008-0463, Am. Inst. of Aeronaut. and Astronaut., Reno, Nevada.

Mertens, C. J., Wilson, J. W., Blattnig, S. R., Solomon, S. C., Wiltberger, M. J., Kunches, J., Kress, B. T. & Murray, J. J. (2007a). Space weather nowcasting of atmospheric ionizing radiation for aviation safety, *45th Aerospace Sciences Meeting and Exhibit*, number AIAA 2007-1104, Am. Inst. of Aeronaut. and Astronaut., Reno, Nevada.

Mertens, C. J., Wilson, J. W., Walker, S. A. & Tweed, J. (2007b). Coupling of multiple coulomb scattering with energy loss and straggling in hzetrn, *Adv. Space Res.* 40: 1357–1367.

Mewaldt, R. A., Cohen, C. M. S., Labrador, A. W., Leske, R. A., Mason, G. M., Desai, M. I., Looper, M. D., Mazur, J. E., Selesnick, R. S., & Haggerty, D. K. (2005). Proton, helium, and electron spectra during the large solar particle events of october-november 2003, *J. Geophys. Res.* 110(A09S10): doi:10.1029/2005JA011038.

NAS/NRC (1980). *National Academy of Sciences/National Research Council: Health effects of exposures to low levels of ionizing radiation*, National Academy Press, Washington, DC. Committee on the Biological Effects of Ionizing Radiation, BEIR V.

NCRP (1993). *National Council on Radiation Protection and Measurements: Limitations of exposure to ionizing radiation*, Vol. 116, National Council on Radiation Protection and Measurements.

NCRP (2009). *National Council on Radiation Protection and Measurements: Ionizing Radiation Exposure of the Population of the United States*, NCRP Report No. 160, National Council on Radiation Protection and Measurements.

Nealy, J. E., Cucinotta, F. A., Wilson, J. W., Badavi, F. F., Dachev, T. P., Tomov, B. T., Walker, S. A., Angelis, G. D., Blattnig, S. R. & Atwell, W. (2007). Pre-engineering spaceflight validation of environmental models and the 2005 hzetrn simulations code, *Adv. Space Res.* 4: 1593–1610.

Neher, H. V. (1961). Cosmic-ray knee in 1958, *J. Geophys. Res.* 66: 4007–4012.

Neher, H. V. (1967). Cosmic-ray particles that changed from 1954 to 1958 to 1965, *J. Geophys. Res.* 72: 1527–1539.

Neher, H. V. (1971). Cosmic rays at high latitudes and altitudes covering four solar maxima, *J. Geophys. Res.* 76(7): 1637–1651.

Neher, H. V. & Anderson, H. R. (1962). Cosmic rays at balloon altitudes and the solar cycle, *J. Geophys. Res.* 67: 1309–1315.

NOAA (2009). National oceanic and atmospheric administration, space weather prediction center. Available from http://www.swpc.noaa.gov/ftpdir/indices/SPE.txt.

O'Brien, K., Friedberg, W., Smart, D. F. & Sauer, H. H. (1998). The atmospheric cosmic- and solar energetic particle radiation environment at aircraft altitudes, *Adv. Space Res.* 21: 1739–1748.

O'Brien, K., Smart, D. F., Shea, M. A., Felsberger, E., Schrewe, U., Friedberg, W. & Copeland, K. (2003). World-wide radiation doseage calculations for air crew memebers, *Adv. Space Res.* 31(4): 835–840.

Ogilvy-Stuart, A. L. & Shalet, S. M. (1991). Effect of radiation on the human reproductive system, *Environ Health Perspect Suppl*, 101(Suppl 2): 109–116.

O'Neill, P. M. (2006). Badhwar-oneill galactic cosmic ray model update based on advanced composition explorer (ace) energy spectra from 1997 to present, *Adv. Space Res.* 37: 1727–1733.

Onsager, T. G. & et al. (1996). Operational uses of the goes energetic particle detectors, in goes-8 and beyond, *in* E. R. Washwell (ed.), *SPIE Int. Soc. Opt. Eng.*, Vol. 2812, pp. 281–290.

Parker, E. N. (1965). The passage of energetic charged particles through interplanetary space, *Planet. Space Sci.* 13: 9–49.

Picone, J. M., Hedin, A. E., Drob, D. P. & Aikin, A. C. (2002). Nrlmsis-00 empirical model of the atmosphere: Statistical comparisons and scientific issues, *J. Geophys. Res.* 107(A12): 1468. doi:10/1029/2002JA009430.

Reitz, G., Schnuer, K. & Shaw, K. (1993). Editorial - workshop on radiation exposure of civil aircrew, *Radiat. Prot. Dosim.* 48: 3.

Schraube, H., Mares, V., Roesler, S. & Heinrich, W. (1999). Experimental verification and calculation of aviation route doses, *Radiat. Prot. Dosim.* 86(4): 309–315.

Slaba, T. C., Blattnig, S. R. & Aghara, S. K. (2010a). Coupled neutron transport for hzetrn, *Radiat. Meas.* 45: 173–182.

Slaba, T. C., Blattnig, S. R. & Badavi, F. F. (2010c). Faster and more accurate transport procedures for hzetrn, *J. Comput. Phys.* 229: 9397–9417.

Slaba, T. C., Blattnig, S. R. & Badavi, F. F. (2010d). Faster and more accurate transport procedures for hzetrn, *Technical Report TP-2010-216213*, NASA.

Slaba, T. C., Blattnig, S. R., Clowdsley, M. S., Walker, S. A. & Badavi, F. F. (2010b). An improved neutron transport algorithm for hzetrn, *Adv. Space Res.* 46: 800–810.

Slaba, T. C., Qualls, G. D., Clowdsley, M. S., Blattnig, S. R., Simonsen, L. C., Walker, S. W., & Singleterry, R. C. (2009). Analysis of mass averaged tissue doses in max, fax, and cam, and caf, *Technical Report TP-2009-215562*, NASA.

Smart, D. F. & Shea, M. A. (1994). Geomagnetic cutoffs: A review for space dosimetry calculations, *Adv. Space Res.* 14(10): 10,787–10,796.

Smart, D. F. & Shea, M. A. (2005). A review of geomagnetic cutoff rigidities for earth-orbiting spacecraft, *Adv. Space Res.* 36: 2012–2020.

Störmer, C. (1965). *The Polar Aurora*, Oxford at the Clarendon Press.

Tai, H., Bichsel, H., Wilson, J. W., Shinn, J. L., Cucinotta, F. A. & Badavi, F. F. (1997). Comparison of stopping power and range databases for radiation transport study, *Technical Report TP-3644*, NASA.

Toffoletto, F. R., Sazykin, S., Spiro, R. W., Wolf, R. A. & Lyon, J. G. (2004). Rcm meets lfm: initial results of one-way coupling, *J. of Atmos. and Solar-Terrestrial Phys.* 66(15-16): 1361–1370.

Townsend, L. W., Stephens, D. L., Hoff, J. L., Zapp, E. N., Moussa, H. M., Miller, T. M., Campbell, C. E. & Nichols, T. F. (2006). The carrington event: Possible doses to crews in space from a comparable event, *Adv. Space Res.* 38: 226–231.

Townsend, L. W., Zapp, E. N., Jr., D. L. S. & Hoff, J. L. (2003). Carrington flare of 1859 as a prototypical worst-case solar energetic particle event, *IEEE Trans. Nucl. Sci.* 50(6): 2307–2309.

Tsyganenko, N. A. (1989). Determination of magnetic current system parameters and development of experimental geomagneitc field models based on data from imp and heos satellite, *Planet Space Sci.* 37: 5–20.

Tsyganenko, N. A. (2002). A model of the near magnetosphere with dawn-dusk asymmetry: 1. mathematical structure, *J. Geophys. Res.* 107(A8): 1179. doi:10.1029/2001JA000219.

Tsyganenko, N. A. & Sitnov, N. I. (2005). Modeling the dynamics of the inner magnetosphere during strong geomagnetic storms, *J. Geophys. Res.* 110: A03208. doi:10.1029/2004JA010798.

Tylka, A. J., Cohen, C. M. S., Dietrich, W. F., Lee, M. A., Maclennan, C. G., Mewaldt, R. A., Ng, C. K. & Reames, D. V. (2005). Shock geometry, seed populations, and the origin of variable elemental composition at high energies in large gradual solar particle events, *ApJ.* 625: 474–495.

Tylka, A. J. & Lee, M. A. (2006). Spectral and compositional characteristics of gradual and impuslive solar energetic particle events, in solar eruptions and energetic particles, *in* N. Gopalswamy, R. Mewaldt & J. Torsi (eds), *Solar Eruptions and Energetic Particles*, Vol. Geophysical Monograph 165, American Geophysical Union, Washington, DC.

UNSCEAR (1988). *UNSCEAR 1988 Report to the General Assembly: Sources, effects, and risks of ionizing radiation*, number E.88.IX.7, United Nations Scientific Committee on the Effects of Atomic Radiation, United Nations, New York.

Upton, A. C., Chase, H. B., Hekhuis, G. L., Mole, R. H., Newcombe, H. B., Robertson, J. S., Schaefer, H. J., Synder, W. S., Sondhaus, C. & Wallace, R. (1966). Radiobiological aspects of the supersonic transport, *Health Phys.* 12: 209–226.

VanAllen, J. A. (1968). *Physics of the Magnetosphere*, Vol. 10, Springer-Verlag, New York, New York, chapter Particle Description of the Magnetosphere.

Wallance, R. G. & Sondhaus, C. A. (1978). Cosmic ray exposure in subsonic air transport, *Aviation Space, and Environ. Med.* 74: 6494–6496.

Wang, W., Wiltberger, M., Burns, A. G., Solomon, S. C., Killeen, T. L., Maruyama, N. & Lyon, J. G. (2004). Initial results from the coupled magnetosphere-ionosphere-thermosphere model: thermosphere-ionosphere responses, *J. of Atmos. and Solar-Terrestrial Phys.* 66: 1425–1441.

Waters, M., Grajewski, B., Pinkerton, L. E., Hein, M. J. & Zivkovich, Z. (2009). Development of historical exposure estimates of cosmic radiation and circadian rhythm disruption for cohort studies of Pan- Am flight attendants, *Am J Ind Med* 52: 751–61.

Waters, M., Bloom, T. F. & Grajewski, B. (2000). The NIOSH/FAA working womens health study: Evaluation of the cosmic-radiation exposures of flight attendants, *Health Phys.* 79(5): 553–559.

Wilson, J. W. (1977). Analysis of the theory of high-energy ion transport, *Technical Report TN D-8381*, NASA.

Wilson, J. W. (2000). Overview of radiation environments and human exposures, *Heath Phys.* 79(5): 470–494.

Wilson, J. W., Badavi, F. F., Cucinotta, F. A., Shinn, J. L., Badhwar, G. D., Silberberg, R., Tsao, C. H., Townsend, L. W. & Tripathi, R. K. (1995a). Hzetrn: Description of a free-space ion and nucleon transport and shielding computer program, *Technical Report TP 3495*, NASA.

Wilson, J. W., Joes, I. W., Maiden, D. L. & Goldhagan, P. (eds) (2003). *Analysis, results, and lessons learned from the June 1997 ER-2 campaign*, NASA CP-2003-212155, NASA Langley Research Center.

Wilson, J. W., Lambiotte, J. J., Foelsche, T. & Filippas, T. A. (1970). Dose response functions in the atmosphere due to incident high-energy protons with applications to solar proton events, *Technical Report TN D-6010*, NASA.

Wilson, J. W., Mertens, C. J., Goldhagan, P., Friedberg, W., Angelis, G. D., Clem, J. M., Copeland, K. & Bidasaria, H. B. (2005b). Atmospheric ionizing radiation and human exposure, *Technical Report TP-2005-213935*, NASA.

Wilson, J. W., Miller, J. & Cucinotta, F. A. (eds) (1997). *Shielding strategies for human space exploration*, NASA Conference Publication 3360, NASA Johson Space Center.

Wilson, J. W. & Townsend, L. W. (1988). Radiation safety in commercial air traffic: A need for further study, *Health Phys.* 55: 1001–1003.

Wilson, J. W., Townsend, L. W., Nealy, J. E., Chung, S. Y., Hong, B. S., Buck, W. W., Lamkin, S. L., Ganapol, B. D., Khan, F. & Cucinotta, F. A. (1989). Bryntrn: A baryon ttansport model, *Technical report*, NASA.

Wilson, J. W., Townsend, L. W., Schimmerling, W., Khandelwal, G. S., Khan, F., Nealy, J. E., Cucinotta, F. A., Simonsen, L. C., Shinn, J. L., & Norbury, J. W. (1991). Transport methods and interactions for space radiation, *Technical Report RP-1257*, NASA.

Wilson, J. W., Tripathi, R. K., Cucinotta, F. A., Shinn, J. L., Badavi, F. F., Chun, S. Y., Norbury, J. W., Zeitlin, C. J., Heilbronn, L. & Miller, J. (1995b). Nucfrg2: An evalution of the semiempirical nuclear fragmentation database, *Technical Report TP-3533*, NASA.

Wilson, J. W., Tripathi, R. K., Mertens, C. J., Blattnig, S. R., Clowdsley, M. S., Cucinotta, F. A., Tweed, J., Heinbockel, J. H., Walker, S. A. & Nealy, J. E. (2005c). Verification and validation of high charge and energy (hze) transport codes and future development, *Technical Report TP-2005-213784*, NASA.

Wilson, J. W., Tweed, J., Walker, S. A., Cucinotta, F. A., Tripathi, R. K., Blattnig, S. & Mertens, C. J. (2005a). A benchmark for laboratory exposures with 1 a gev iron ions, *Adv. Space Res.* 35: 185–193.

Wilson, J. W. & et al. (2006). International space station: A testbed for experimental and computational dosimetry, *Adv. Space Res.* 37: 1656–1663.

Xapsos, M. A., Barth, J. L., Stassionopoulos, E. G., Messenger, S. R., Walters, R. J., Summers, G. P. & Burke, E. A. (2000). Chracterizing solar proton energy spectra for radiation effects applications, *IEEE Trans. Nucl. Sci.* 47(6): 2218–2223.

# Part 2

# Effects on Materials

# 7

# Influence of Ionizing Radiation and Hot Carrier Injection on Metal-Oxide-Semiconductor Transistors

Momčilo Pejović[1], Predrag Osmokrović[2],
Milica Pejović[1] and Koviljka Stanković[2]
*[1]University of Niš, Faculty of Electronic Engineering, Niš,*
*[2]University of Belgrade, Faculty of Electrical Engineering, Belgrade,*
*Serbia*

## 1. Introduction

### General characteristics of MOS components

Favourable characteristics of silicon dioxide (further referred to as oxide or $SiO_2$) as an almost irreplaceable dielectric in MOS (Metal-Oxide-Semiconductor) components have contributed to a considerable extent to the great success of the technology of manufacturing integrated circuits during the last decades. However, it has been proved that instabilities of electric charge in the gate oxide and at $Si - SiO_2$ interface, which inevitably occur due to the influence of ionizing radiation ($\gamma$ and X-radiation, electrons, ions) (IR) and hot carrier injection (which includes Fowler-Nordheim high electric field Stress, avalanche hole injection, avalanche electron injection, and so on (HCI) during the operation of MOS transistors, lead to instabilities of their electric parameters and characteristics and present a serious problem to the reliability of MOS integrated circuits [1-4]. Studying these instabilities, particularly causes and mechanisms responsible for their occurrence, is very significant as it may help the manufacturer define appropriate technological parameters with the aim of increasing their reliability. Considering the rate of development and requirement that the components should be highly reliable, the investigation of their instabilities in normal working conditions is practically impossible. In order to determine, as quickly as possible, mechanisms that cause instabilities, accelerated reliability tests are used and they comprise of the application of strong electric fields (so-called electric stress) in static and impulse modes at room temperature and/or at elevated temperatures. In this context, reliability comprises probability that the component may fulfil its target function during a certain period and under certain conditions, i.e. that it may preserve the electric characteristics and the value of electric parameters under certain exploitation conditions during a certain period. Reliability is defined as a probabilistic category as no failure of the component can be predicted with certainty. Namely, if a group of components manufactured in almost the same conditions is subjected to stress, a smaller number of them will fail earlier (early failure), while a larger number of them will fail later. In order to remove from such a group of components those with the possibility of early failure, a selection procedure is applied to all the components. This procedure consists of a series of tests the most significant of which is the burn-in test. This test is carried out in two steps: short-term burning (10 – 30 hours) in normal and somewhat intense working conditions, in order to remove components liable to failure due to their manufacturing defects, and sufficiently long burning in tough working conditions in order to remove components with manifested functional failures. A significant fact that should be emphasized in relation to MOS components is that, although the high quality of components guarantees a certain level of their reliability, many defects may remain hidden and manifest during the application thus leading to the degradation of the components and their failure.

In recent times, the development of MOS integrated circuits shows a tendency of increasing the complexity and constantly decreasing dimensions, which comprises a need for increasingly thin gate oxides. In addition, discrete power MOS components have been also developed. At the beginning of their development, it was deemed that they would very soon replace bipolar transistors in power electronics. However, technological processes used in the manufacture of these transistors were more complex in relation to bipolar transistors so that a real progress in power electronics was made when VDMOS (Vertical-Double-Diffused MOS) transistors were developed because the manufacture of them was much simpler. Due to their favourable electric characteristics (great operative speed, low switch-on resistance, high switching voltage, etc.), the VDMOS transistors have found their application in the fields such as automotive industry [5]. Switching characteristics of VDMOS transistors at high frequencies (higher than 100 kHz [6]) have also proved very good owing to which these transistors are applied as switching components for high-frequency sources of supply used in medical electronics, engine control systems, switching sources of supply in telecommunication researches, etc.

In addition to its favourable properties, $SiO_2$ used in the manufacture of MOS components and integrated circuits has some shortcomings as well. Namely, instabilities of electric charge in the gate oxide and in interface traps, which are inevitable due to the presence of the IR or HCI processes, during the operation of MOS components and MOS integrated circuits, also lead to instabilities of their electric parameters and characteristics. Due to this, instabilities represent one of the most serious problems in relation to a reliable operation of MOS integrated circuits [2, 3] particularly when these components are used in radiation surroundings.

Researches have shown [3] that MOS components are very sensitive to different types of ionizing radiation and that their sensitivity significantly depends on the manufacturing technology. Additionally, some technological processes such as lithography with the use of a ray of electrons, X-rays, and processes in plasma, etc. included in the technological sequence for the production of integrated circuits with the aim of decreasing the dimensions of their components, may be significant sources of defects in the oxide [3].

The attention of today's researches on the impact of the IR and HCI processes on MOS components is directed in two ways. One way comprises the production of components with the highest possible resistance to these types of processes, while the other way leads towards the production of MOS transistors sensitive to IR in order to produce sensors and dosimeters of ionizing radiation [3]. It should be emphasized that the first operating principles of PMOS sensors and dosimeters of ionizing radiation were published as early as 1974 [7]. Unlike the influence of IR and HCI on the reliability of commercial MOS transistors which is the subject of many researches worldwide, PMOS transistors as IR dosimeters are investigated to a much lesser extent.

## 2. Origin and characteristics of defects and electrical charges in oxide and at Si-SiO₂ interface

Electrical charges in the gate oxide and at the Si-SiO₂ interface may be divided into four groups as follows [2, 8, 9]: *mobile ions, charges at the traps, fixed centres charges and charges at interface traps*. However, if a division of charges based on their influence on the current-voltage characteristics of MOS transistors is made, only charges in the gate oxide (including mobile ions, charges at the trap centres, and fixed charges) and charges at interface traps may be distinguished. The main classification of charges in the gate oxide and at interface traps of MOS transistors is shown schematically in Figure 1.

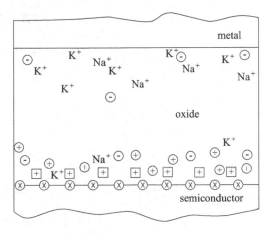

Na⁺, K⁺- mobile ions
$\oplus$ , $\ominus$ - charge at trap centers
$\boxed{+}$     - fixed charge
$\otimes$     - charge at interface traps

Fig. 1. Classification of charge at $Si-SiO_2$ interface and in the gate oxide of MOS transistor.

*Mobile ions* in the gate oxide are mainly positive ions of alkali metals such as Na⁺ and K⁺, as well as hydrogen ions. They may be incorporated into the oxide during the mere processing of the wafer, or they may reach the oxide, coming from the surrounding surfaces, during the processes of passivation and packing [9, 10]. The existence of these ions in the gate oxide is associated with the fact that during high temperature processes, they easily diffuse into the oxide, from contaminated surfaces. Mobile ions are not distributed uniformly in the oxide, but most of them are located in the semiconductor or metal whereby many more ions are located in the metal as the attractive force towards the metal is much higher than the attractive force towards the semiconductor.

The most important instability mechanism related to electric charges in the gate oxide is the migration of positive ions to the oxide-metal interface and oxide-semiconductor interface under the influence of the positive polarization of the gate and elevated temperature. A series of treatments are taken with the MOS technology in order to minimize these instabilities. Besides the compulsory annealing at high temperature and high level of cleanliness of the technological space and materials used, some modifications in the oxide manufacturing technology have been made in order to minimize instability effects.

Furthermore, a generally accepted procedure for minimizing mobile ions is the introduction of chlorine into the oxide, which is achieved by introducing a certain quantity of HCl or $Cl_2$ into the oxidation atmosphere or by passing oxygen through trichloroethylene.

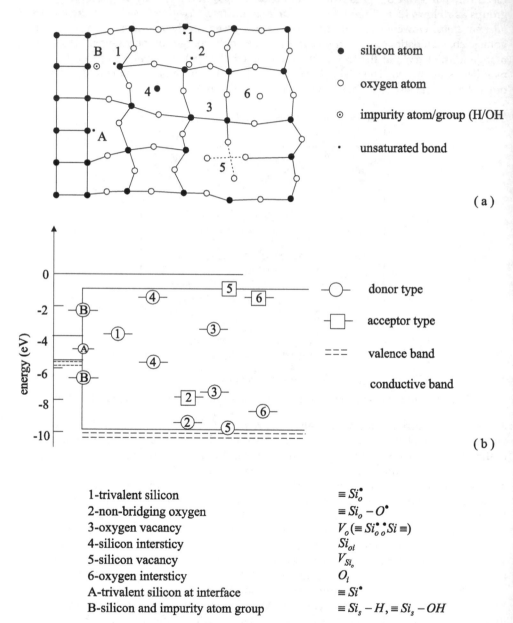

1-trivalent silicon $\qquad\qquad\qquad \equiv Si_o^\bullet$
2-non-bridging oxygen $\qquad\qquad \equiv Si_o - O^\bullet$
3-oxygen vacancy $\qquad\qquad\quad V_o(\equiv Si_o^{\bullet\bullet} \,_o^\bullet Si \equiv)$
4-silicon intersticy $\qquad\qquad\quad Si_{oi}$
5-silicon vacancy $\qquad\qquad\quad\; V_{Si_o}$
6-oxygen intersticy $\qquad\qquad\quad O_i$
A-trivalent silicon at interface $\qquad \equiv Si^\bullet$
B-silicon and impurity atom group $\quad \equiv Si_s - H, \equiv Si_s - OH$

Fig. 2. Most common defects in oxide and at $Si - SiO_2$ interface: (a) two dimensional model of chemical bonds; (b) energy band model.

*Charges at the trap centres* are defects and admixed atoms in the oxide which have permissible energy levels in the forbidden zone of the oxide. They are mainly located near the $Si - SiO_2$ interface, while their number inside the oxide is negligible. Figure 2 shows the model of chemical bonding and diagram of energy zones that illustrate the most significant defects in the oxide and at the $Si - SiO_2$ interface [4, 11]. Although this is not adequately shown in Figure 2a, it should be taken into account that the $Si - SiO_2$ interface does not spread through only one atom layer, but rather through several of them. As Figure 2b shows, due to a great difference in the widths of the forbidden zones of silicon (1.12 eV) and oxide (9 eV), discontinuity of energy zones occurs at the interface, and charge carriers encounter a high energy barrier (around 3.2 eV for electrons and 4.7 eV for holes) which in normal conditions prevents any movement of electrons and holes from the semiconductor into the oxide.

The origin of the traps varies. They may be formed both on $\equiv$Si-O- bonds and on defects in the oxide (oxygen vacancies, $\equiv$Si-H and $\equiv$Si-OH bonds, other admixed atoms, etc.). This means that there are different energy levels of the traps depending on the type and quality of the oxide. The traps may be neutral or charged either positively or negatively. During exploitation in certain conditions, they may become electrically charged by capturing electrons or holes, and neutralized by releasing the same, a consequence of which is the instability of the density of effective electric charge captured on them. Therefore, the defects are mainly donor-type defects meaning that they may be positively charged or neutral, so that total electric charge in the oxide is positive. It is important to note that holes may be captured more easily than electrons. Thus the ratio between the number of captured holes and the total number of holes injected into the oxide is approximately one, while, on the average, out of $10^5$ injected electrons only less than one is captured. Nevertheless, instabilities of charges at the traps caused by the electron capture may be equal to instabilities caused by the capture of holes or even more pronounced. The reason for this lies in the fact that electrons may be more easily injected into the oxide, which increases the possibility of their capture.

The most significant mechanisms that lead to the occurrence of free carriers in the oxide may be classified into three groups: a) tunnelling of electrons and holes from the semiconductor or metal into the oxide, 2) injection of hot electrons and holes from the semiconductor into the oxide, and 3) generation of electron-hole pairs in the mere gate oxide which may be a consequence of various factors such as ionizing radiation effects, photo-generation or application of a strong electric field. To the traps in the oxide or to the valence or conductive zone of the oxide, holes from the valence zone of the semiconductor or electrons from the conductive zone of the semiconductor or metal may be tunnelled. The processes of tunnelling electrons or holes from the traps into the conductive or valence zone of the oxide, or directly into the semiconductor are possible, which by analogy leads to instability of the charge density at the traps.

There are four mechanisms related to the injection of hot carriers from the semiconductor into the gate oxide. These are: 1) substrate hot electron injection, 2) channel hot electron injection, 3) drain avalanche hot carrier injection (whereby the injection of holes is preferred when the normal component of the field is directed towards the oxide, while the injection of electrons is preferred when the field is directed towards the semiconductor), and 4) hot electron injection caused by secondary ionization by holes in the substrate. It should be noted that such changes in the density of charges at the traps are not the only form of instability that occur during the exploitation of MOS transistors. Namely, new traps may be formed which are neutral at the beginning, but which may become charged by capturing electrons and holes.

*Fixed charge* is located near the $Si-SiO_2$ interface at distances smaller than 8 nm and its density is constant regardless of changes in the threshold voltage of MOS components. This charge is always positive as it mainly consists of unsaturated bonds of silicon near the $Si - SiO_2$ interface. Fixed charge depends on the orientation of crystals and it is considerably higher with crystal orientation (111) than with crystal orientation (100). The quantity of fixed charge depends on the Technological conditions of oxidation (atmosphere and temperature). The oxide burn-in in nitrogen or argon at temperatures higher than 600°C decreases the density of fixed charge. This is one of the reasons for which the oxide burn-in has been introduced as a standard procedure in the manufacture of CMOS integrated circuits. It should be emphasized that instability of fixed charges is rarely manifested separately from changes in other charges in the oxide. Instabilities of this type are commonly followed by simultaneous changes in charges at interface traps and oxide traps. This is why in literature fixed charge is mentioned in a much broader sense, although this term also refers to charge in the traps particularly considering the instability of charge in them. Since the origin of fixed charge and traps is frequently the same, it may be assumed that they differ in instability mechanisms. Namely, instability of charge in the traps is a consequence of capturing and releasing free electrons and holes in the oxide, while the instability of fixed charge reflects in its increase as a result of appropriate chemical reactions. One of such reactions is the dissociation of the $Si - H$ bond near the $Si - SiO_2$ interface. Under the influence of modest negative polarization of the gate and increased temperature, the $Si - H$ group reacts with the $Si - O$ group, whereby a interface trap is formed on the Si atom from the semiconductor, while released hydrogen and oxygen form an OH group which diffuses towards the interface. After the transfer of one electron into the semiconductor, the $Si$ atom in the oxide remains positively charged and this represents the fixed charge. However, it is often impossible to determine a strict boundary between fixed charge and charge at the traps.

*Charges at interface traps* represent a significant group of electric charges in the oxide. Interface traps are located at the $Si - SiO_2$ interface the energy levels of which are in the forbidden zone of the semiconductor. These traps develop as a consequence of the semiconductor structure breaking because surface atoms remain with one unsaturated bond. This means that on clean areas of the semiconductor the number of interface traps is approximately equal to the number of surface atoms. However, thermal oxidation reduces the number of interface traps by several orders of magnitude as a great number of silicon atom bonds become saturated by bonding to oxygen atoms. From the aspect of the origin of interface traps, there is a complete analogy between them and fixed charge and thus with the trap centres. The difference is that fixed charge and the trap centres are in the oxide, while interface traps are at the $Si - SiO_2$ interface and they are easily accessible to free carriers from the semiconductor. The exchange of charge between interface traps and the semiconductor is relatively fast, so that the density of charge in interface traps at a given moment depends on the distribution of the interface traps energy levels and the position of the Fermi level in the semiconductor forbidden zone. Although the process of thermal oxidation considerably decreases the density of interface traps, an additional decrease is needed for a proper operation of MOS components. Today, a procedure commonly used is the oxide burn-in after metallization performed in nitrogen or in a mixture of nitrogen and hydrogen. In this manner, hydrogen is directly introduced in the oxide, or it is formed after the reaction of water and aluminium. This procedure decreases both the density of interface traps and density of fixed charge and therefore it is applied in almost all standard technologies. Another favourable circumstance is that oxides that increase in the presence of hydrogen chloride also contain, in addition to chlorine that neutralizes mobile ions, hydrogen that decreases the density of interface traps.

The mechanisms of instability of charges in interface traps are a consequence of the formation of new interface traps. Almost all of the described mechanisms may be monitored through the increase in the density of interface traps. The mechanism of fixed charge formation has already been described, whereby interface traps are formed simultaneously. The ion drift towards the Si – SiO$_2$ interface may lead to an increase in the density of interface traps. It has been proved that the processes of electric charge injection in the oxide most easily form new interface traps that may be achieved by the IR and HCI processes. It is important to point out that owing to the passage of electrons through the Si – SiO$_2$ interface new interface traps may be formed whereby electrons will not be captured by the traps, which means that the density of charge in them will not change.

As it has already been mentioned, interface traps exchange charge with the semiconductor very rapidly, while this is not the case with the trap centres. On the other hand, a considerable number of defects in the oxide are located in the close vicinity of the interface and they may, like interface traps and regardless of the fact that their microstructure is the same as that of defects deeper in the oxide, exchange charge with silicon. However, unlike real interface traps, these defects are separated from charges in silicon by an energy barrier so that the exchange of charge between them and silicon is carried out through carrier tunnelling whereby the duration of such a process depends to a great extent on the interface distance and the barrier height, and it may range from 1 μs to several years. The long period over which these defects exchange charge with silicon indicates that there is no strict boundary between the effects of charge in the oxide and the effects of interface traps on the electric parameters of MOS transistors. This is why in literature these defects are called boundary traps [12-15]. They may act both as the traps in the oxide and as interface traps. Such nature of the boundary traps may lead to problems in the interpretation of electric measuring results and to difficulties in separating the effects of interface traps and electric charge in the oxide since, depending on measuring conditions (values of the polarization voltage, signal frequency), the boundary traps may behave as interface traps or/and traps in the oxide. By combining appropriate measuring techniques and analysis it is possible to separate the effects of electric charge in the boundary traps [16, 17], which is particularly significant for the characterization of MOS components with a thin gate oxide, where all defects in the oxide are located near the interface and where these defects with their density values are comparable to or even exceed interface traps [14]. In case of thick oxides, as with power MOS transistors, the standard division into the traps in the oxide and interface traps is generally acceptable. In this case, the participation of boundary centres is that at the electric characterization of components will be registered as interface traps or the traps in the oxide, to a great extent depends on the measuring speed and the frequency of the signal that is used during the measurement.

## 3. Classification of the traps according to their effects on electric characteristics

According to the above, it may be concluded that all charges may be divided into two groups: charges in the oxide and charges at interface traps. Changes of charge densities in the oxide $\Delta N_{ot}$ and interface traps $\Delta N_{it}$ are directly responsible for changes of sub-threshold characteristics of MOS transistors during the IR and HCI processes. The participation of a certain type of defects in the values of $\Delta N_{ot}$ and $\Delta N_{it}$ primarily depends on their electric effects on the charge carriers in the channel of MOS transistors, as well as on their location in space. If the defects may capture carriers from the channel, due to which the sub-threshold characteristic slope decreases, it is deemed that they behave as interface traps and that value $\Delta N_{it}$ depends on them. If the defects act on carriers in the channel by attracting or rejecting them with Coulomb forces (depending on the mark of their electric charge), sub-threshold characteristics will be changed due to the change of $\Delta N_{ot}$. It should be emphasized that the influence of defects developed in the IR and HCI processes on $\Delta N_{ot}$ and $\Delta N_{it}$ is determined primarily by their electric effects and after by their location. This indicates that some defects located at the $Si - SiO_2$ interface may behave as defects in the oxide (influence $\Delta N_{ot}$), while some defects located in the oxide may behave as interface traps (influence $\Delta N_{it}$) [3, 4, 18]. Although the influence of the location of defects on sub-threshold characteristics is smaller than their electric impact, it should not be neglected. For example, during recording sub-threshold characteristics (which is a fast process), only part of these defects may capture carriers from the channel.

The most common techniques for determining densities of the traps in the oxide and at the $Si - SiO_2$ interface are subthreshold-midgap (SMG) [19] and charge pumping (CP) techniques [20, 21] (a detailed description of these techniques is given in Section 6). Based on the position of the traps detected by these techniques, a new classification of the centres became necessary as with the use of the SMG technique both $\Delta N_{ot}$ (SMG) and $\Delta N_{it}$ (SMG) may be determined, while with the use of the CP technique only $\Delta N_{it}$ (CP) may be determined. The use of these two techniques is suitable as there is a great difference in effective frequencies (from several Hz for the SMG technique to 1 MHz for the CP technique). As a slower technique, SMG is used for determining two types of centres: fixed traps (FT) with a density of $\Delta N_{ft} \equiv \Delta N_{ot}$ (SMG) and switching traps (ST) with a density of $\Delta N_{st} \equiv \Delta N_{it}$ (SMG) [22, 23].

Fixed traps (FT) are traps in the oxide that do not exchange charges with Si substrate during recording sub-threshold characteristics by the SMG technique. They are commonly located deeper in the oxide, although they may also be near or at the $Si - SiO_2$ interface. A significant characteristic of FT is that they may be permanently annealed [22, 23].

Switching traps (ST) may be divided into faster switching traps (FST) and slower switching traps (SST). ST may exchange charges with Si substrate during the measurement. The speed of ST depends on their distance from the $Si - SiO_2$ interface [24]. This indicates that all defects developed in the IR or HCI processes, which are located near this interface and are capable of exchanging charges with the substrate, become part of ST. Furthermore, some ST may be located deeper in the oxide so that there is not sufficient time for them to exchange, during the the measurement, electric charges with carriers in the channel, meaning that $\Delta N_{ft}$ may contain a certain number of ST. FST which are located at the mere interface and which represent true interface traps (with density $\Delta N_{it}$, i.e. meaning that $\Delta N_{it}$(CP) $\equiv \Delta N_{fst}$). SST (with density $\Delta N_{sst}$) located in the oxide, near the $Si - SiO_2$ interface are frequently called slow states (SS) [21], anomalous positive charge (APC) [25], switching oxide traps (SOT) [26] or border traps (BT) [24].

Finally, it should be pointed out that during the measurement by the SMG technique, ST exchange charges during the measurement meaning that in case of an n-channel MOS transistor (NMOS) these centres capture charges from the channel but releases them when the measurement is finished. It has been shown that ST may be temporarily annealed (compensation process) or permanently annealed (neutralization process), however not during the measurement but by exposing components to elevated temperatures.

## 3.1 Defects in SiO$_2$ and Si – SiO$_2$ interface caused by IR and HCI processes
### 3.1.1 Formation of electron-hole pairs in SiO$_2$

IR and HCI processes lead to the formation of similar defects in SiO$_2$ since the electrons play a crucial role in the energy transfer to the oxide in both cases, and the difference is in the electron energy. Namely, the secondary electrons released in the oxide by the gamma photons (IR), and the hot electrons (HCI), and, eventually, the secondary electrons released by the hot electrons play a main role in defect creations. However, it has been shown [22, 27-31] that a difference in electron energy has a significant influence on the created defect types. For instance, the highly energetic secondary electrons (the IR case) produce significantly less negatively charged FT than the low-energy hot electrons (the HCI case).

In the case of IR, the gamma photons interact with the electrons in the SiO$_2$ molecules mainly via Compton effect, releasing secondary electrons and holes, i.e. gamma photons break Si$_0$ – O and Si$_0$ – Si$_0$ covalent bonds in the oxide [3] (the index $_0$ is used to denote the oxide). The released electrons (so called "secondary electrons") which are highly energetic, may be recombined by holes at the place of production, or may escape recombination. The secondary electrons that escape recombination with holes travel some distance until they leave the oxide, losing their kinetic energy through the collisions with other secondary electrons or, what is more probable, with the bonded electrons in Si$_0$ – O and Si$_0$ – Si$_0$ covalent bonds in the oxide, releasing more secondary electrons (the latter bond represents an oxygen vacancy).

Each secondary electron, before it has left the oxide or been recombined by the hole, can break a lot of covalent bonds in the oxide producing a lot of new secondary highly energetic electrons, since its energy is usually much higher than an impact ionizing process energy (energy of 18 eV is necessary for the creation of one electron-hole pair [3], i.e. for electron ionization).

In the case of IR processes, gamma radiation breaks both covalent bonds in SiO$_2$ between oxygen atoms and weak $\equiv$ Si$_0$ – H and $\equiv$ Si$_0$ – OH bonds in the oxide. These processes develop in the following reactions [32, 33]

$$O_3 \equiv Si_0 - O - Si_0 \equiv \xrightarrow{h\nu} \equiv Si_0 - O^\bullet + e^- + h^+, \tag{1}$$

$$\equiv Si_0 - H \equiv \xrightarrow{h\nu} Si_0^\bullet + H^\bullet + e^- + h^+, \tag{2}$$

$$\equiv Si_0 - OH \equiv \xrightarrow{h\nu} Si_0^\bullet + OH^\bullet + e^- + h^+. \tag{3}$$

As it can be noted, these reactions lead to the formation of electron-hole pairs (e$^-$ - h$^+$). A part of the electrons formed by reactions (1)-(3) break covalent bonds in the oxide (reaction (1)). It is obvious that the secondary electrons play a more important role in bond breaking than highly energetic photons, as a consequence of the difference in their effective masses, i.e. in their effective cross section. The electrons leaving the place of production escape the oxide rapidly (for several picoseconds), but the holes remain in the oxide.

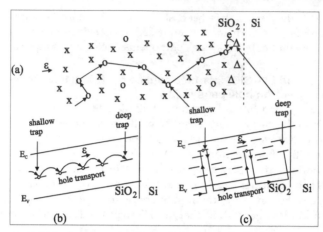

Fig. 3. Space diagram: (a) hole transport through the oxide bulk in the case of positive gate bias. "x" represents unbroken bonds and "o" represents broken bonds (trapped holes at shallow traps), respectively, and "Δ" represents the hole trap precursors near the interface (precursor of a deep trap). Energetic diagram: the hole transport (b) by tunneling between two localized traps and (c) by the oxide valence band.

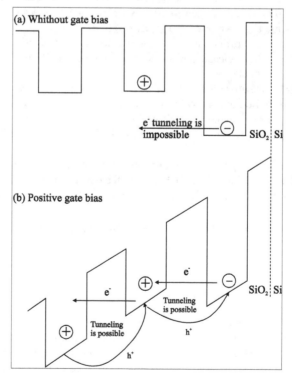

Fig. 4. The electron tunneling between adjacent centers: (a) shallow centre and (b) deep centre.

The holes released in the oxide bulk are usually only temporary, but not permanently captured at the place of production, since there are no energetically deeper centres in the oxide bulk. The holes move towards one of the interfaces ($SiO_2$ – Si or gate-$SiO_2$), depending on the oxide electric field direction, where they are trapped at energetically deeper trap hole centres (see Figures 3 and 4). Moreover, even in the zero gate voltage case, the electrical potential due to a work function difference between the gate and substrate is high enough for partial or complete holes moving towards the interface. In addition, some electrons could be trapped at the electron trapping centres of electron capture, but this is not very probable in the IR case [30].

In the case of HCI (e.g. Fowler-Nordheim tunnelling into the gate oxide), the hot electrons, having significantly lower energies than the secondary electrons created by IR, are injected either from the substrate or from the gate. Two cases should be considered: (1) thin oxide ($d_{ox} \leq 10$ nm) and (2) thick oxide ($d_{ox} > 10$ nm). In the former case, the hot electrons, tunnelling into the oxide conduction band, pass the oxide without collisions, since their travel distance is short and, consequently, the probability of the collision with the electrons in the covalent oxide bonds is low [34, 35]. The hot electrons reach the gate, if they are injected from the substrate, or reach the substrate, if they are injected from the gate, where they generate electron-hole pairs. The released (secondary) electrons are attracted by the gate or the substrate electrode, depending on the location of their release, while the generated holes are injected into the oxide.

In the case of thick oxide ($d_{ox} > 10$ nm), a hot electron passing through the oxide has to collide with the electrons bonded in the oxide, giving a certain amount of energy to them and performing the impact ionization [34, 36, 37]. The hot electrons, whose energies are usually higher than 9 eV representing the $SiO_2$ band gap, can break the $Si_0$ – O and $Si_0$ – $Si_0$ covalent bonds in the oxide by impact ionization and create electron-hole pairs.

The hot and secondary electrons after certain collisions with each other, and what are more probable, with bonded electrons, escape the oxide rapidly, but the holes transport towards one of the interfaces. The high electric field forces the holes to leave their production places, since there is no significant concentration of energetically deep trapping hole created in the oxide bulk, but these centres exist near the interfaces (see Figure 3). Besides, the hot and secondary electrons can produce more electron-hole pairs before they leave the oxide. In addition, some hot/secondary electrons could be trapped at electron centres creating negative charge, which is a very probable mechanism [31].

### 3.1.2 The defects created by electrons in the impact ionization process

The hot (HCI) and secondary electrons (IR and HCI) passing through the oxide bulk collide with bonded electrons in the most numerous bond: the non-strained silicon-oxygen bond, $\equiv Si_0 - O - Si_0 \equiv$ (it could be also represented as $\equiv Si_0 \cdot \cdot O \cdot \cdot Si_0 \equiv$, where $\cdot \cdot$ represents two electrons making a covalent bond), by reaction [23]

$$\equiv Si_0 - O - Si_0 \equiv \xrightarrow{e^-} \equiv Si_0 - O^\bullet + {}^+Si_0 \equiv +2e^-, \tag{4}$$

$$Si_0 - O - Si_0 \equiv \xrightarrow{e^-} \equiv Si_0 - O^+ + {}^\bullet Si_0 \equiv +2e^-, \tag{5}$$

where $\equiv$, the indices $_0$ and $\bullet$ denote three $Si_0$ – O bonds ($O_3 \equiv Si_0$ – O), silicon atom in the oxide and an electron, respectively.

Reaction (4) represents the most probable reaction in the oxide bulk, since these are the most numerous centres. The formed $\equiv Si_0 - O^{+\bullet}Si_0 \equiv$ complex is energetically very shallow, representing the temporary hole centre (the trapped holes can easily leave it [11]). The creation of $\equiv Si_0 - O^{+\bullet}Si_0 \equiv$ complex, given in reaction (5) is less probable (the $\bullet$ denotes the unpaired electron, i.e. "uncoupled spin").

Nevertheless, the strained silicon-oxygen bond ($\equiv Si_0 - O - Si_0 \equiv$), mainly distributed near the interfaces, can also be easily broken by the passing hot/secondary electrons, usually creating the amphoteric non-bridging-oxygen (NBO) centre $\equiv Si_0 - O^{\bullet}$, and positively charged $E'$ centre, $\equiv Si_0^+$ [38] known as an $E'_s$ centre [39] (the same process as in reaction (4)). A NBO centre is an amphoteric defect that could be more easily negatively charged than positively by trapping an electron:

$$\equiv Si_0 - O^{\bullet} + e^- \rightarrow \equiv Si_0 - O^-. \tag{6}$$

Obviously, the NBO centre, as an energetically deeper centre, is the main precursor of the negatively charged traps (defect) in the oxide bulk and interface regions.

A hot/secondary electron passing through the oxide can also collide with an electron in the strained oxygen vacancy bond $\equiv Si_0 - Si_0 \equiv$, i.e. $\equiv Si_0^{\bullet\bullet}Si_0 \equiv$, which is a precursor of an $E'_\gamma$ centre ($\equiv Si_0^{\bullet}$) [40], breaking this bond and knocking out an electron:

$$\equiv Si_0^{\bullet\bullet}Si_0 \equiv \rightarrow \equiv Si_0^+ + {}^{\bullet}Si_0 \equiv + 2e^- \tag{7}$$

Moreover, this is a conventional structural model of the $E'_\gamma$ centre: a hole trapped at an oxygen monovacancy ($\equiv Si_0^{+\bullet}Si_0 \equiv$). The oxygen vacancy bonds are mainly distributed in the vicinity of interfaces.

The mentioned reactions could occur anywhere in the oxide: near the interfaces and in the oxide bulk. The trapped charge can be positive (oxide trapped holes) and negative (oxide trapped electrons) and the former is more important, since the holes trapping centres are numerous, including three types ($E'_s$, $E'_\gamma$ and NOB centres), compared with one electron capture (NBO). The holes and electrons captured near the Si – SiO$_2$ interface have the greatest effect on MOSFET characteristics, since they have the strongest influence on the channel carriers.

### 3.1.3 Hole transport

The holes trapped at $\equiv Si_0^+$ centres form oxygen vacancies, and strained silicon-oxygen bonds are energetically deep and steady, at which the holes can remain for a longer time period, i.e. their filling with electrons is more difficult in relation to some shallowly captured holes. These centres exist near both interfaces, especially near the Si – SiO$_2$ interface. The holes created and trapped at the bulk defects (reaction (4)), representing energetically shallow centres, are forced to move towards one of the interface under the electric field, where they are trapped at deeper traps, since there are a lot of oxygen vacancies, as well as a lot of strained silicon-oxygen bonds near the interfaces, grouping all trapped positive charge there. The holes leave the energetically shallow centres in the oxide spontaneously and they are transported to the interface by a hopping process using either shallow centres in the oxide (Figure 3(b); the holes "hop" from one to another centre) or centres in the oxide valence band (Figure 3(c)) [41]. Figure 3 displays the holes transport in the space for the positive gate bias (a) and the energetic diagram for the possible mechanisms of this space process ((b) and (c)).

Figure 4 shows a possible hole (electron) tunnelling mechanism between adjacent centres: a shallow centre and deep centre. Figure 4(a) shows the case without gate bias, for the Si – SiO$_2$ interface. In this case, the holes, i.e. electrons tunnelling between these centres, are not possible. The holes are factious species representing the electron unoccupied places, and the hole movement is in fact the movement of bonded electrons, however the electron movement is the movement of free electrons.

When the MOS transistor is positively biased (Figure 4(b)), the bonded electron can tunnel from the deep centre to the shallow centre. It represents the hole tunnelling from shallow to deep centres, being permanently trapped at a deep centre. The electron, which is now in the shallow centre, can easily tunnel from this shallow centre to the next adjacent shallow centre, enabling the holes transport towards the interface. The pictures are similar for the negative gate bias [31].

### 3.1.4 The defects created by the holes

The holes can be created either in the oxide bulk (thick oxide) or in the gate/substrate and injected into the oxide bulk (thin oxide). Moving throughout the oxide, the holes can react with the hydrogen defect precursors ($\equiv Si_0 - H$ and $\equiv Si_0 - OH$) and create the following defects [42].

$$\equiv Si_0 - H + h^+ \rightarrow \equiv Si_0^{\bullet} + H^+, \tag{8}$$

$$\equiv Si_0 - H + h^+ \rightarrow \equiv Si_0^+ + H^\circ, \tag{9}$$

$$\equiv Si_0 - OH + h^+ \rightarrow \equiv Si_0 - O^{\bullet} + H^+, \tag{10}$$

$$\equiv Si_0 - OH + h^+ \rightarrow \equiv Si_0 - O^+ + H^\circ, \tag{11}$$

The hot and secondary electrons may also interact, in a similar way, with these defect precursors. However, these precursors are not so important for the hot and secondary electrons, since their concentration is significantly lower than the concentration of the precursors representing the non-strained silicon-oxygen bonds. Moreover, it could not be expected that the holes break bonds in the non-strained silicon-oxygen bonds. Because of that, the precursors described in reactions (8-11) are only important for the holes that are transported through the oxide since the hydrogen is relatively weakly bounded and the holes can easily break these bonds. Reactions (8-11) are particularly important for the creation of the interface trap since they produce the hydrogen ion H$^+$ and hydrogen atom H$^\circ$, which take part in defect creation at the Si – SiO$_2$ interface. The existence of these precursors in the oxide and, particularly, at the Si – SO$_2$ interface is very reasonable, since annealing in the hydrogen ambient is a standard step during the manufacture of numerous MOS devices.

When the holes reach the interface, they can break both the strained oxygen vacancy bonds $\equiv Si_0 - Si_0 \equiv$, forming $E'_\gamma$ centres [38]

$$\equiv Si_0 - Si_0 \equiv + h^+ \rightarrow \equiv Si_0^{\bullet} + \equiv Si_0^+ \tag{12}$$

and the strained silicon-oxygen bonds $\equiv Si_0 - O - Si_0 \equiv$, creating the amphoteric NBO centres $\equiv Si_0 - O^{\bullet}$ and $E_s^{'}$ ($\equiv Si_0^+$) centres:

$$\equiv Si_0 - O - Si_0 \equiv + h^+ \rightarrow Si_0 - O^{\bullet} \equiv Si_0^+ \tag{13}$$

It could be assumed that the strained Si - O bonds, $\equiv Si_0 - O - Si_0 \equiv$ and oxygen vacancy, $\equiv Si_0 - Si_0 \equiv$, represent the main defect precursors in the oxide bulk and at the interfaces. However, these precursors exist mainly near the interface. It should be noted that $E_{\gamma}^{'}$, $E_s^{'}$ (reactions (12) and (13)) and NBO centre (reaction (13)) represent energetically deeper hole and electron trapping centres, respectively. The energetic levels of the defects created after the holes at $E_{\gamma}^{'}$ and $E_s^{'}$ centres and electrons at NBO centre, respectively, have been trapped, can be various. In the chemical sense same defects show different behaviours depending on the whole bond structure: the angles and distances between the surrounding atoms. It could be assumed that these defects represent fixed traps (FT) and slow switching traps (SST) [29, 30]. They can be positively and negatively charged, as well as neutral.

In the literature, there are many controversies about the oxide defects types induced by the IR and HCI processes and there are different names for the same or similar defects (see [23] for more details).

### 3.1.5 Gate oxide/substrate (SiO$_2$ – Si) interface

The defects at the Si — SiO$_2$ interface, known as the fast switching traps (FST), or true interface traps, represent an amphoteric defect $Si_3 \equiv Si_s^{\bullet}$ a silicon atom $\equiv Si_s^{\bullet}$ at the Si — SiO$_2$ interface back bonded to three silicon atoms from the substrate $\equiv Si_3$ and the FST are usually denoted as $\equiv Si_s^{\bullet}$ or $Si^{\bullet}$. They can be directly created by incident photons/hot electrons tunnelling from the substrate or the gate [43, 44], however this amount can be neglected. The direct creation of FST is only emphasized in the case of hot electrons for the positive gate bias applied to thin oxides ($d_{ox} < 10$ nm), where the electrons, tunnelling in the conduction band of oxide, pass the oxide without any collisions (the probability for collision process is small), accelerating themselves and reaching the interface with energy enough for an interface defect creation. Besides direct creation, the FST are mainly created by trapped holes ($h^+$ model) [45-48] and by hydrogen released in the oxide (hydrogen-released species model - $H$-model) [49-51].

The $h^+$ model proposes that a hole trapped near the interface create FST, suggesting that an electron-hole recombination mechanism is responsible [46]. Namely, when holes are trapped near the interface and electrons are subsequently injected from the substrate, recombination occurs. The interface trap may be formed from the energy released by this electron-hole recombination. It was supposed [23] that FST can be created by the reaction:

$$\equiv Si_0^+ + e^+ + \equiv Si_s - Si_s \rightarrow \equiv Si_0^{\bullet} + \equiv Si_0^+ + \equiv Si_0^{\bullet} . \tag{14}$$

The main shortcoming of the h$^+$ model is its impossibility in explaining the delayed creation of FST [51]. However, this fact cannot completely disqualify this model, and we can suppose that part of interface traps are created by it.

The H model proposes that $H^+$ ions, released in the oxide by trapped holes (reactions (8) and (10)), drift towards the $Si - SiO_2$ interface under the positive electric field. When an $H^+$ ion arrives at the interface, it picks up an electron from the substrate, becoming a highly reactive hydrogen atom $H°$ [52]:

$$H^+ + e^- \rightarrow H°. \tag{15}$$

Also, according to the $H$ model, the hydrogen atoms $H°$ released in reactions (9) and (11) diffuse towards the $Si - SiO_2$ interface under the existing concentration gradient.

Highly reactive $H°$ atoms react without an energy barrier at the interface producing FST throughout the following reactions [53-55]:

(i) The creation of interface trap $Si_s^{\bullet}$, when $H°$ reacts with an interface trap precursor $Si_s - H$ [52]

$$Si_s - H + H° \rightarrow Si_0^{\bullet} + H_2. \tag{16}$$

or an interface trap precursor $Si_s - OH$:

$$Si_s - OH + H° \rightarrow Si_0^{\bullet} + H_2O. \tag{17}$$

Many investigations have shown that the $P_{b_0}$ and $P_{b_1}$ centres which exist at the (100) interface represent interface traps [56-58]. The $P_{b_0}$ defect has the following structure: $Si_s \equiv Si_{s_0}^{\bullet}$, but the structure of $P_{b_1}$ defect, here denoted as $\equiv Si_{s_1}^{\bullet}$, is not known (the remaining three bonds of the $\equiv Si_{s_1}^{\bullet}$ defect can be bonded to various species [59]).

(ii) The passivation of the interface traps, when $H°$ reacts with previously formed (fabrication process or IR and HCI processes induced) interface trap [53, 60]

$$Si_s^{\bullet} + H° \rightarrow Si_s - H. \tag{18}$$

(iii) Dimerization of hydrogen, when $H°$ reacts with another $H°$ also existing near the interface [61, 62]

$$H° + H° \rightarrow H_2 \tag{19}$$

Reactions (16) and (17) are most probable at the start of the interface trap creation, since both have a large number of $Si_s - H(OH)$ precursors, and the effective cross sections for reactions (16) and (17) are higher than for reactions (18) and (19) [52].

## 4. Annealing of MOS components after IR and HCI processes

For the electric characteristics of MOS components, changes of charge in the oxide and changes in interface traps are significant both during the IR and HCI processes and after the end of these processes. Some of the mechanisms occurring after the end of the IR and HCI processes remain active, and new mechanisms also develop and last for a longer period of time. The generation of charge in the oxide takes place as long as an IR or HCI process goes on. When these processes end, the density of charge in the oxide starts to decrease, while the generation of interface traps continues and this process may occur for a longer period of time [61, 63-73]. A process during which positive charge density in the oxide decreases with a simultaneous increase in interface traps density after an IR or HCI process, in which a large quantity of positive charge was incorporated in the oxide, is called annealing. This term is not fully adequate although it is frequently used, as only charge in the oxide is commonly "annealed" while the density of interface traps may increase. MOS components annealing may be performed spontaneously (in air atmosphere, at room temperature, without polarization), or at lowered or elevated temperatures, or in the presence of various intensity and direction electric field.

A number of experimental results [24, 74-80] have shown that the density of interface traps during annealing mainly does not change, while the density of trapped charge in the oxide decreases. Several models are proposed for the trapped charge annealing [75, 77]. It has been observed that during annealing part of trapped charge is permanently burnt-in (neutralized), while the other part may only be compensated, which leads to the effect of "inverse burn-in" (apparent increase in the density of trapped charge during annealing at gate negative polarization), which is attributed, by a group of authors, to the existence of the so-called Switching Oxide Trap (SOT) centres [77, 78]. Another group of authors attribute it to the existence of the so-called Anomalous Positive Charge (APC) centres [24, 79, 80]. Both groups explain inverse burn-in by the exchange of charge with substrate whereby such exchange, in case of SOT, is performed by electron tunnelling as the appropriate levels of these centres lay at the height of the substrate conductive zone [78], while APC are, similarly to interface traps, capable of performing direct exchange with the substrate as their energy levels are at the level of the forbidden zone [24, 79, 80]. Regardless of their different approaches, both groups have practically contributed to establishing the fact that SOT and APC are most probably the same centre, i.e. $E'$ centre that, depending on the conditions of the experiment and measurement may exhibit various properties (burn-in, inverse burn-in or exchange of charges) and may be registered as a interface trap or the so-called boundary centre of capture.

A characteristic of these models of charge burn-in is that they are related to captured charge in the oxide and that they mainly do not exchange charge at interface traps, i.e. that they consider changes in the density of captured charge during annealing separately and independently of changes in the density of interface traps. This may be deemed the main disadvantage of this group of models, particularly because there are numerous experimental results that may be very accurately described by the above-described $h^+$ and H models. It is important to point out that according to $h^+$ and H models, the mechanisms of the formation of defects in the oxide and at the $Si - SiO_2$ interface take place not only during the IR or HCI processes, but after their cessation as well. This practically means that these two models are more comprehensive and more acceptable for explaining the annealing process than the charge burn-in model. It is important to emphasize here that most experimental results indicate the dominant role of hydrogen and that the $H$ model in many cases gives a real picture of the processes in the oxide and at the $Si - SiO_2$ interface. Therefore, the processes of the captured charge density decrease and simultaneous increase in the density of interface traps, which take place within a shorter or longer time after the cessation of the IR or HCI processes, may be best described by the H model. However, this model may not describe the annealing processes in case in which the generation of interface traps begins and ends during mere tension so that their density do not increase during the annealing. In addition, none of the models mentioned so far may explain the phenomenon of a decrease in the density of interface traps during the annealing for a long period [81-83] and a latent increase in the density of interface traps [28,84]. Furthermore, none of the mentioned models may explain the phenomenon of an increase and then decrease in the density of interface traps during annealing for a longer period of time [85-88]. On the other hand, the latent generation of interface traps that occurs during annealing is manifested through rapid increase in their density after reaching apparent saturation, and simultaneous rapid decrease in the density of trapped charge, which is characteristic of MOS transistors the oxide of which contains a high concentration of oxygen and bounded hydrogen [87]. It is deemed that it is not the hydrogen originating from the mere oxide that is responsible for the latent increase in the density of interface traps, but rather the hydrogen which is, during the IR or HCI processes, released in the form of molecules in the adjusted layers (polysilicon gate, protective oxide). The hydrogen molecule diffuses towards the gate oxide and through it further towards the oxide-semiconductor interface in the vicinity of which it reacts with a positively charged centre of capture, according to the following reaction:

$$\equiv Si_0^+ + H_2 \rightarrow \equiv Si_0 - H + H^+. \tag{20}$$

As it can be seen from the last reaction, the captured charge is neutralized and hydrogen ions are released which then diffuse towards the Si $-$ SiO$_2$ interface and participate in the formation of a interface trap (reactions (16) and (17) [89]). The late increase in the density of interface traps is attributed to the long-lasting diffusion of hydrogen molecules which is particularly difficult through the polysilicon gate, as well as to a possibility that oxygen vacancies slow down the movement of hydrogen in the oxide [67, 89].

This explains the latent increase in the density of interface traps but cannot explain their decrease that occurs after the latent increase. Owing to this, based on experimental results of the annealing of irradiated power VDMOS transistors, the so-called Hydrogen-Water (H $-$ W) model has been proposed, which can explain the behaviour of the densities of interface traps and captured charge during annealing including both the latent increase and the decrease in the density of interface traps [28, 85, 86]. According to this model, all free hydrogen (present in the oxide before the IR or HCI processes or released during these processes) is utilized for the so-called conventional generation of interface traps that ends with the cessation of the IR or HCI processes and/or in the initial stage of annealing immediately after the cessation of these processes. In the further stage of annealing, for initiating the latent generation of interface traps it is necessary and sufficient that a certain quantity (even a small one) of hydrogen ions appears at the interface. The origin of these ions will be discussed later. When these late ions reach the Si $-$ SiO$_2$ interface under the influence of the electric field, they takeover electrons from the substrate (reaction 15) and form neutral atoms of hydrogen that may participate in one of the following processes:

(i) formation (depassivation) of interface traps (reaction (16)); (ii) pasivation of interface traps (reaction (18)); and (iii) dimerization (reaction (19)). Reaction (18) is more probable than the other two reactions because the concentration of $\equiv Si_s - H$ precursor is higher than the concentration of interface traps ($\equiv Si_s^{\bullet}$), as well as because the cross section of reactants is larger in this reaction in comparison to the other two, so that the density of interface traps begins to increase. In reactions (16) and (19), hydrogen molecules are released and therefore their concentration at the interface increases (concentration gradient occurs) due to which these molecules diffuse towards the inside of the oxide whereby on their way many of them react on with positively charged traps. The following reaction commonly occurs:

$$\equiv Si_0{}^+ + H_2 \rightarrow \equiv Si_0 - H + H^+. \tag{21}$$

Hydrogen ions formed in this reaction drift towards the interface, takeover electrons from the substrate and the processes (i) - (iii) are repeated. In this manner, for further formation of interface traps an additional source of $H^+$ ions is provided, and at the same time, burn-in of captured charge is performed, in compliance with results according to which the latent increase in the density of interface traps corresponds to the decrease in the density of captured charge.

Water molecules are deemed responsible for the decrease in the density of interface traps that occurs after the latent increase. Water molecules are mainly bounded (physically or chemically) in the thermal and/or protective CVD oxide. During annealing, they may be released (more expressed at higher temperatures) and after that they slowly diffuse towards the Si – SiO$_2$ interface where they finally react with interface traps according to the following reaction:

$$\equiv Si_s^{\bullet} + H_2O \rightarrow \equiv Si_s - OH + H^o. \tag{22}$$

As it can be seen from the last reaction, the final contribution of water molecules after long annealing is the passivation of interface traps. Namely, in reaction (22) one interface trap is passivated and a neutral hydrogen atom formed in the same reaction may react with (i) another neutral hydrogen atom (reaction (19)) which will not change the result of reaction (22), meaning that one interface trap remains passivated, (ii) with the $\equiv Si_s^{\bullet}$ defect (reaction (18)) with which one more interface trap is passivated, or (iii) with the $\equiv Si_s - H$ precursor (reaction (16), the only one with which the effect of reaction (22) is annulled). In this manner, the H – W model covers the latent increase and, afterwards, the decrease in the density of interface traps. Based on this model, it may be concluded that a small quantity of any type of hydrogen particles (ions, neutral atoms or molecules) may trigger the latent generation of interface traps. Although the authors of this model [28] mention a possibility that this generation is triggered by hydrogen ions captured in the traps in the oxide, or by hydrogen molecules formed in reaction (22) from the water molecule after its slow diffusion through the oxide, they still believe that for triggering the latent increase in the density of interface traps, hydrogen molecules originating from the protective CVD oxide and/or polysilicon gate are most probably responsible. The cause of their late arrival to the oxide-semiconductor interface may be the reduced speed of diffusion through polysilicon, even through the gate oxide (as a consequence of a decrease in the diffusion constant in the oxide owing to the IR processes), as well as a possibility that hydrogen molecules may have reduced speed by being captured at oxygen vacancies from which they are subsequently released.

## 5. Influence of charges in the oxide and interface traps on MOS transistors parameters

Charges in the oxide, as well as interface traps have a significant influence on the characteristics of MOS transistors. The transfer characteristic of MOS transistors in the saturation region may be described by the following expression [2, 4]:

$$I_D = \beta \left( V_G - V_T \right)^2 / 2, \tag{23}$$

where $\beta$ is an amplifying factor and may be expressed as

$$\beta = \frac{\mu WC_{ox}}{L} .$$ 
(24)

In expressions (23) and (24), $I_D$ is drain current, $V_G$ is voltage at the gate, $V_T$ is threshold voltage, $\mu$ is electric mobility in the channel, $C_{ox} = \varepsilon_{ox}/d_{ox}$ is the capacitance of the gate oxide per area unit ($\varepsilon_{ox}$ and $d_{ox}$ represent dielectric constant and oxide thickness, respectively), while $W$ and $L$ are the width and effective length of the channel, respectively. Based on these expressions it may be noticed that charges in the oxide and in interface traps may influence the electric characteristics of MOS transistors only through the influence on threshold voltage and mobility of carriers in the channel. Positive charge in the oxide influences carriers in silicon by its electric field, attracting electrons towards the interface and rejecting holes from it. This leads to a change in threshold voltage which is decreased with NMOS, and increased (by absolute value) with PMOS transistors. At the same time, electrons approaching the interface with NMOS transistors should increase, while holes moving away with PMOS transistors should decrease the dispersal of carriers on the uneven areas of the interface, by which the mobility of carriers would be decreased with NMOS, and increased with PMOS transistors. However, it is deemed that charges in the oxide have a low influence on the carrier mobility and that their impact is not electrostatic but it is rather a consequence of the carrier capturing process [4, 90]. Namely, carrier capture in interface traps will not change the total amount of charge in the channel region, which means that no electric effect is produced. The influence of interface traps may be explained in the following way: interface traps capture a certain number of electrons or holes induced by a change in the surface potential caused by the gate voltage, due to which the formation of a channel requires higher gate voltage comprising an increase in the threshold voltage both with NMOS and PMOS transistors. At the same time, capturing carriers at the interface traps with both types of MOS transistors leads to a decrease in the number of conductive carriers in the channel (to a decrease in current) which is manifested in a decrease in the slope of transfer characteristics, a decrease in the amplifying factor and thus a decrease in the mobility of carriers $\mu$. Namely, the greater the density of interface traps, the greater the number of captured carriers in relation to a total number of induced carriers, which is manifested in an apparently greater mobility degradation. Through such a conclusion, the real situation (decreased number of mobile carriers in the channel) is replaced with an apparent one (unchanged number of mobile carriers with reduced mobility), but the effect on the conductivity in the channel (i.e. on the transistor current) remains absolutely the same. Due to this, the use of a simpler model given in expressions (23) and (24), whereby the so-called effective (or apparent) mobility is used instead of the real one, is much more purposeful than complicated consideration of the real situation which would require the knowledge about the accurate function of the interface trap energy levels distribution in the forbidden zone of silicon. This claim is supported by the definition of the mobility of carriers according to which this mobility is not a physical value but only the coefficient of proportionality in the expression for the dependence of drift rate on electric field. Therefore, the value of mobility is not constant but it is rather necessary to adjust it in order to maintain the applicability of Ohm's law to the transport of current in the semiconductor [4]. Owing to this, the concept of effective mobility is common in modelling the effects of various factors on transport processes in the semiconductor, such as doping effects, effects of electric field,

high levels of injection, surface dispersal, crystal orientation, as well as effects of charge in
the oxide and effects of interface traps [4, 91-93]. Further in this text, the term mobility will
be used, but it will comprise the effective value.

The considerations given so far suggest that with NMOS transistors a decrease in mobility
should be expected with an increase in the density of either interface traps or charge in the
oxide. Considering the dominant influence of interface traps, a decrease in mobility in
general should be also expected for PMOS transistors, even in the case in which the density
of charge in the oxide increases somewhat more rapidly than the density of interface traps.
In literature, however, there are examples of an increase in mobility, one of which was
recorded in a study of PMOS transistors with a polysilicon gate [94]. In the same
experiment, PMOS transistors with an aluminium gate were irradiated and an expected
decrease in mobility was recorded, while an expected increase in threshold voltage was
almost the same with both types of transistors. By additional analysis, a great increase in the
density of charge in the oxide was determined, approximately equal with both types of
samples while at the same time an increase in the density of interface traps was much
greater with the samples with aluminium gates in relation to the samples with polysilicon
gates. An increase in mobility with PMOS transistors with polysilicon gates was explained
by their high resistance to the formation of interface traps, i.e. by a disproportionately great
increase in the density of charge in the oxide in relation to the increase in the density of
interface traps during the IR processes. This result is not in accordance with the conclusion
that charge in the oxide has low influence on mobility. It may be deemed even controversial
as similar results with PMOS transistors with polysilicon gates [95, 96] showed the opposite
result, i.e. reduced mobility. However, regardless of the contradiction of these results, it may
be concluded that charge in the oxide has significant influence on the mobility of carriers in
the channel, although such influence is lower than the influence of interface traps. This is
indicated by the results of researches with PMOS dosimetric transistors which show that a
considerable increase in the density of charge in the oxide leads to an increase in mobility
[97].

Many experimental results have shown that the influence of charge in the oxide and the
influence of interface traps on the threshold voltage of MOS transistors may be modelled by
the following expression [98]:

$$V_T = V_{T0} \mp \frac{qN_{ot}}{C_{ox}} + \frac{qN_{it}}{C_{ox}} \tag{25}$$

where $V_{T0}$ is the threshold voltage of an ideal MOS transistor without any charge in the
oxide and without interface traps, while the second and third terms include their influence,
whereby the sigh "-" in front of the second term refers to an NMOS transistor, while the sign
"+" refers to a PMOS transistor. Therefore, the influence of charge in the oxide ($qN_{ot}$) is
included as voltage drop that creates an electric field on the oxide, i.e. $qN_{ot}/C_{ox}$ . On the
other hand, it is claimed that the influence of interface traps on carriers in the channel
reflects in their capture, so a that part of carriers induced in the channel by the gate voltage
remain immobile. If, for given gate voltage, the number of captured carriers from the
channel per area unit is $qN_{it}$ , then part of the gate voltage amounting to $qN_{it}/C_{ox}$ is spent
not on the formation of mobile carriers but rather on the formation of captured ones. By this
amount the gate voltage required for the formation of a channel is increased (according to

absolute value), which means that the influence of interface traps on threshold voltage is $qN_{it}/C_{ox}$ .

In order to give consideration to the influence of charge in the oxide and of interface traps on mobility, several models may be used [99-101] which are based on the empirical expression for the dependence of effective mobility on charge in the oxide and the concentration of admixtures in the channel of NMOS transistors given in the paper [102]

$$\mu = \frac{\mu_0}{1 + \alpha N_{ot}}, \qquad (26)$$

where $\mu_0$ = 3440 – 164 1og $N_A$ [cm²/(Vs)] is mobility without charge in the oxide ($N_A$ is the concentration of acceptor impurities expressed in cm³) and $\alpha_{ot}$ = - 1.04 · 10⁻¹² + 1.93 log $N_A$ [cm²]. Based on later researches [98] it has been shown that the above expression should be modified taking into account the contribution of interface traps as well. Therefore, mobility may be expressed in the following way:

$$\mu = \frac{\mu_0}{1 + \alpha_{it} N_{it} \pm \alpha_{ot} N_{ot}}, \qquad (27)$$

where coefficients $\alpha_{it}$ and $\alpha_{ot}$ depend on technology, and due to greater influence of charge on interface traps $\alpha_{it} > \alpha_{ot}$. The sign "+" in front of the third term in the denominator refers to NMOS transistors, while the sign "-" refers to PMOS transistors. This is in accordance with the observation above that charge in the oxide, being almost always positive, may lead to an increase in mobility with PMOS transistors (although this happens rarely owing to the dominant influence of charge on interface traps).

Due to the increase in interface traps, the leakage current of the inversely polarized connection increases as well. Leakage current consists of free electrons and holes generated in the depletion region of the inversely polarized junction, which are transferred through the field into the n-type or p-type semiconductor. The generation of free holes in the presence of interface traps takes place through the transfer of electrons from the valence band to the energy level of interface traps near the Fermi level, while the generation of free electrons takes place through the transfer of electrons from the energy level of interface traps near the Fermi level to the conductive band of the semiconductor. Therefore, electrons pass from the valence band into the conductive band through interface traps. This process leads to the occurrence of leakage current of the inversely polarized junction. The leakage current directly depends on the density of interface traps.

The low-frequency noise of MOS transistors also increases with the increase in the density of interface traps. This noise appears because interface traps capture carriers that move through the channel retaining them for a certain period of time and then emitting them back into the channel. This is a process of accidental nature and it modulates the channel current which may be particularly observed at low frequencies and high densities of interface traps.

The avalanche breakdown of the inversely polarized drain-substrate junction is a consequence of the presence of charge in the gate oxide originating from mobile ions, fixed charge, and charge in the traps. The sum of all these charges gives effective charge per area unit which is almost always positive. This charge transforms the drain-substrate p⁺-n junction in PMOS transistors into the p⁺-n⁺ junction meaning that breakdown voltage is reduced. In NMOS transistors, the substrate is of p-type, and charge in the oxide depletes it leading to a reduction of the field at the drain-substrate n⁺-p junction and thus to an increase in avalanche breakdown voltage.

It is important to emphasize that it is impossible to give an appropriate typical quantitative picture of the described instabilities. There are differences not only in the values of changes in corresponding parameters, but also in the occurring instability mechanisms, and as it has been already mentioned these differences may lead to changes in an observed parameter even in opposite directions. The next important difference is in time dependencies of respective instabilities meaning that even if the same initial instability is manifested with this group of MOS transistors, the instabilities may significantly differ after some time. In addition, time dependences of respective parameters instabilities may have very complex forms as they are frequently defined by simultaneous activities of several mechanisms. Such differences in the behaviour of MOS transistor parameters are conditioned by differences in the manner of the gate oxide formation and by conditions of the component exploitation, and they may lead to qualitatively different forms of instabilities of the MOS transistor parameters.

## 6. Techniques of determining electric charge density in oxide and interface traps

Change of electric charge density in gate oxide and interface traps, which occur during IR and HCI processes, are studied through the application of certain techniques. An important role in that procedure is played by changes in electric parameters, primarily threshold voltage and amplification factor (considering that the influence of electric charge in oxide and interface traps on electric characteristics are transferred through them), which are determined on the basis of recorded I-V characteristics.

It is already known that a change in threshold voltage of MOS transistors due to IR and HCI processes can be expressed in the following way:

$$\Delta V_T = \Delta V_{ot} + \Delta V_{it} , \tag{28}$$

where $\Delta V_{ot}$ and $\Delta V_{it}$ are contributions to the change in threshold voltage due to the electric charge in oxide and in interface traps, respectively. By using the expression (25), the change in surface electric charge density in oxide $\Delta N_{ot}$ and the change in interface traps density $\Delta N_{it}$ can be determined as

$$\Delta N_{ot} = \pm \frac{C_{ox}}{q} \Delta V_{ot} , \; \Delta N_{it} = \frac{C_{ox}}{q} \Delta V_{it} , \tag{29}$$

where the signs "+" and "-" refer to p-channel and n-channel MOS transistors, respectively. $\Delta V_T$ (expression (28)) represents an experimentally determined value of threshold voltage $\Delta V_T = V_T - V_T(0)$ and the corresponding change in the electric charge densities in oxide $\Delta N_{ot} = N_{ot} - N_{ot}(0)$ and the interface traps $\Delta N_{it} = N_{it} - N_{it}(0)$ (expression (29)), where $V_T(0)$, $N_{ot}(0)$ and $N_{it}(0)$ are values of corresponding values prior to, and $V_T$, $N_{ot}$ and $N_{it}$ after IR or HCI processes.

There are several techniques used to determine electric charge density, and/or to separate the influence of electric charge density in gate oxide and interface traps, and each of these techniques has its own advantages and deficiencies [3,103]. Here we will describe the two techniques most commonly used today. These are subthreshold midgap (SMG) technique which is used to determine electric charge density in gate oxide and interface traps, and charge pumping (CP) technique which is used only to determine change in interface traps density.

## 6.1 Subthreshold midgap technique

The subthreshold midgap (SMG) technique is based on the analysis of change in the subthreshold transfer characteristics of MOS transistors [19]. Theoretically, the subthreshold characteristic (dependence of drain current on the gate voltage for the given voltage on drain $V_{DS}$) can be described by the expression [2]:

$$I_D = \mu \frac{W}{L} \cdot \frac{N_A L_D kT}{\sqrt{2}} \left(\frac{n_i}{N_A}\right)^2 \exp\left(\frac{q\varphi_s}{kT}\right) \cdot \left[1 - \exp\left(-\frac{qV_{DS}}{kT}\right)\right] \cdot \sqrt{\frac{kT}{q\varphi_s}}, \tag{30}$$

where $N_A$ is the concentration of impurities in the p-channel area, $L_D$ is Debye length, $\varphi_s$ is surface potential which represents the voltage function at the gate [2]. The last expression clearly expresses exponential dependence of drain current on the surface potential, and thereby on the gate voltage as well. The subthreshold characteristics of n-channel MOS transistor prior to and after being subjected to the IR process, as well as the necessary elements to explain the technique are given on the Figure 5 [19].

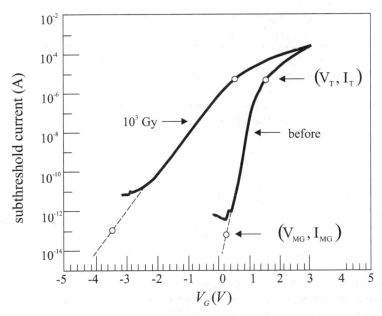

Fig. 5. Subthreshold current as a function of gate voltage, before irradiation and after gamma radiation dose of $10^3$ Gy.

From the viewpoint of subthreshold characteristic, an increase in interface traps density, as already mentioned before, under the influence of IR or HCI processes, is manifested through the change in its slope. Namely, as an increase in gate voltage is followed by capture of carriers in interface traps, the "response" of surface potential will be decreased, and along with it, the drain current as well, to the changes in gate voltage, and therefore the voltage of subthreshold characteristic will be milder.

In case when Fermi level of semiconductors on the Si – SiO$_2$ interface is located in the middle of the forbidden zone, the total electric charge on interface traps, regardless of the disposition within the forbidden zone of the substrate, is equal to zero. This means that the shift on the voltage axis between two subthreshold characteristics in that case is solely the consequence of change in electric charge density in oxide. The gate voltage in which Fermi level on the Si – SiO$_2$ interface is located in the middle of the forbidden zone is marked with $V_{MG}$ (midgap voltage), and is calculated as abscissa of the point ($V_{MG}$, $I_{MG}$) on the subthreshold characteristics. The above-mentioned point is determined on the basis of its ordinate, i.e. $I_{MG}$ current (midgap current) which is calculated on the basis of the expression (30) by replacing appropriate values of surface potential $\varphi_s = \varphi_F$ and experimentally determined mobility $\mu$. As the mobility of carriers depends on several parameters such as temperature, admixture concentrations, degree of component degradation (and thereby also the electric charge density in the oxide and interface traps), it must be determined experimentally for each component and for each radiation dose or duration of HCI process. It is determined from the slope of the curve $\sqrt{I_D} = f(V_G)$ [19], as the drain current of MOS transistors in saturation is given through the following expression:

$$I_D = \mu \frac{W}{2L} C_{ox} (V_{GS} - V_T)^2 . \qquad (31)$$

The $I_{MG}$ current in $V_{MG}$ voltage is very small (order of magnitude of picoampere or smaller, depending on the size of transistor), i.e. it is often significantly smaller than the lowest measured level of subthreshold current, determined by the leakage current. For this reason, the point ($V_{MG}$, $I_{MG}$) is located on the extension of the most linear part of subthreshold characteristics marked by the dashed line on Figure 5. Therefore, change in electric charge density in oxide, formed under the influence of IR or HCI process is generated as [19]

$$\Delta N_{ot} = \frac{C_{ox}}{q} (V_{MG}(0) - V_{MG}) \qquad (32)$$

where $V_{MG}(0)$ is the value of $V_{MG}$ voltage prior to IR or HCI processes.

Other two important points on the Figure 5 are the points ($V_T$, $I_T$), which are located on subthreshold characteristics prior to and after the radiation. These points could be determined in a similar way as the points ($V_{MG}$, $I_{MG}$), and/or on the basis of the expression (30) and the definition of threshold voltage, and/or on the basis of the expression (30) and the definition of surface potential for the case of strong inversion. However, these points can be determined with highest reliability on the basis of knowledge about their abscissas, i.e. used generated values of threshold voltage, which avoid any ambiguity in the definition of threshold and/or start of strong inversion. The threshold voltage is most frequently determined as section of extrapolated dependence of the square root of drain current in MOS transistor in saturation from the threshold voltage at the gate (expression (31)), with $V_G$ axis.

Presence of interface traps changes the slope of subthreshold characteristics, i.e. the difference between the voltage $V_T$ and $V_{MG}$ which represents the measure of "stretch-out" and is marked as $V_{S0}$ (stretch-out voltage)

$$V_{S0} = V_T - V_{MG} , \qquad (33)$$

Accordingly, an increase in electric charge density on interface traps which causes change in subthreshold characteristics inclination, i.e. $V_{S0}$ voltage is now more simply determined on the basis of the expression [19]

$$\Delta N_{it} = \frac{C_{ox}}{q}\left(V_{S0} - V_{S0}(0)\right) , \qquad (34)$$

where $V_{S0}(0)$ is the value of stretchout voltage prior to, and $V_{S0}$ after IR or HCI process.

### 6.1.1 Charge-pumping technique

As opposed to the SMG technique, the charge-pumping (CP) technique does not give changes in charge densities in oxide and in interface traps, but is used solely to determine interface traps density while charge density in oxide can be subsequently determined on the basis of the expression (29) under the condition that the change in threshold voltage is known [20, 21, 104].

The charge-pumping effect can be explained on the basis of the diagram shown in Figure 6 [104]. The source and the drain of the transistor are short-circuited, and the p-n junction of source and drain with the substrate are inversely polarized with $V_R$ voltage. In the absence of signal at the gate, under the influence of inverted polarization at the junction source-substrate and drain-substrate, the inverted saturation current of these connections will flow. When a train of rectangular pulses of sufficiently high amplitude is brought to gate (with pulse generator), a change of current direction in the substrate occurs. The intesity of that current is proportional with the pulse frequency, and "pumping" of the same amount of electric charge towards the substrate. As current cannot flow through oxide, the electric charge in the substrate come through p-n junctions of source and drain. In this way, in the case of n-channel MOS transistors, a channel is formed under the gate in positive pulse half-period, whereby electrons are captured on interface traps. During the negative half-period, when the channel area turns into the state of accumulation, mobile electrons from the channel are returned to the source and drain, and the captured electrons are recombined

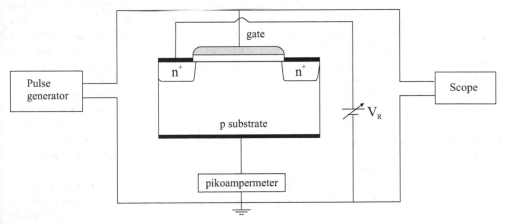

Fig. 6. Schematic diagram of charge pumping measurement.

with holes from the accumulated layer, thereby generating CP current which is proportional with the number of recombination centers, i.e. interface traps:

$$I_{CP} = q \cdot A_G \cdot f \cdot N_{it} \tag{35}$$

where $A_G$ is gate surface, and $f$ is pulse frequency. In order to avoid recombination with the channel electrons, it is necessary to ensure their return to the source and drain before overflow of cavities from the substrate occurs, which is accomplished by using reverse polarization of p-n junction or using a train of trapezoid pulses or triangular pulses with sufficient time for rise time $t_r$ and fall time $t_f$ pulse. However, part of the electrons whose capture is shallowest, are in the meantime thermally emitted into conductive band of the substrate, reducing the width of interface traps energy range measured by the CP technique, so that CP current is generated by interface traps within the range [21]

$$\Delta E = -2kT \ln\left( v_{th} n_i \sqrt{\sigma_n \sigma_p} \frac{|V_T - V_{FB}|}{|\Delta V_G|} \sqrt{t_r t_f} \right) \tag{36}$$

which is 0.5 eV from the middle of the forbidden band. In the expression (36) $v_{th}$ is thermal velocity, $\sigma_n$ and $\sigma_p$ are cross-section surfaces of carrier captures, $n_i$ is self-concentration of carriers in the semiconductor, and $\Delta V_G$ is pulse height

Using of expression (35), interface traps density can be calculated on the basis of the measured maximum value of CP current. Maximum CP current is directly proportional to the pulse frequency, and a small-size transistor with usual state density needs a frequency of at least several kHz to enable the charge-pumping current level reach the order of magnitude of picoampere. Due to this, CP measuring is most often conducted with frequencies in the range between 100 kHz and around 1 MHz, whereby only real (FST) interface traps are registered (in some frequencies, CP is also contributed by part of SST which also captures electrons from the channel [105]).

As the CP technique requires a separate outlet for the substrate, it could be concluded that it is not applicable for power VDMOS transistors, in which the p-bulk is technologically connected to the source. However, thanks to the very structure of these transistors [106], the CP technique is applicable in a somewhat altered form; as shown in Figure 7 [25, 107]. It should be pointed out that VDMOS power transistor represents a parallel connection of a large number of cells (elementary transistor structures) with a large surface, which is especially suitable for the CP technique (higher level of current which is easier to measure).

The role of source and drain, as minority carrier in power VDMOS transistors (Figure 7) are taken over by p⁻ body areas of two adjacent cells, while the role of substrate (source of majority carriers) is played by n⁻ epitaxial layer between the cells which is directly connected to the n⁺ area of the drain. In this case, the CP current is generated through recombination through interface traps which are located on the interface oxide – n⁻ substrate, and VDMOS structure acts as a PMOS transistor. In this way, the CP technique practically perform characterization of interface above the epitaxial layer, and not above the channel area, which must be taken into consideration when calculating interface traps density with use of the expression (35), whereby an adequate value of the active surface above the epitaxial layer should be taken as $A_G$. The oxide above the channel and the oxide above the epitaxial layer have the same characteristics and thickness, as they are made simultaneously in the production process, which also refers to the polysilicon gate. On the other hand, it is known that the surface interface traps do not depend on the type of semiconductor on which the oxide is formed. This means that, within the qualitative analysis, it can be considered that the conclusions attained on the basis of the CP measurements also refer to the interface above the channel surface.

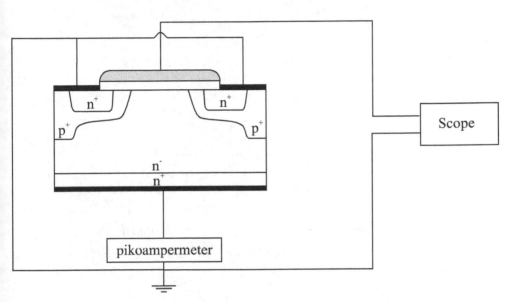

Fig. 7. Schematic diagram of charge pumping measurement for power VDMOS transistor.

## 7. Some results of IR and HCI processes and later annealing of MOS transistors

### 7.1 Behavior of CMOS transistors with Al gate from integrated circuits of the CD4007UB type in the course of IR process and subsequent annealing

The results of radiation with gamma rays, X-rays and electrons, as well as of subsequent annealing of MOS transistors with Al-gate from the integrated circuits CD4007UB are displayed in several papers [108-115], in which CMOS (NMOS and PMOS) transistors were used. This made it possible to directly evaluate the instability of electric parameters and characteristics of these integrated circuits on the basis of the data on instability of electric parameters and characteristics of these transistors.

Figures 8 and 9 display changes in the threshold voltage of CMS transistors during the radiation with gamma rays and later annealing on the temperature of 115°C for 0 V gate polarizations [111, 112]. As it can be seen, the threshold voltage in NMOS transistor (Figures 8) decreases down to the value of the absorbed dose of gamma radiation of 200 Gy, and then rises again, while in PMOS transistors (Figure 9], it continually rises. Change in threshold voltage also takes place during the annealing at elevated temperature, but in such a way as to approximate their values to the ones present prior to the radiation. It has been shown [112] that the annealing process can be accelerated by positive polarization on the gate.

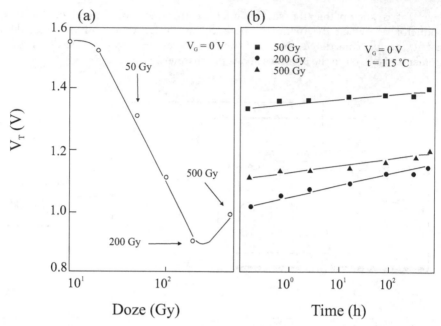

Fig. 8. Threshold voltage shift ($\Delta V_T$) for NMOS transistor during gamma irradiation, $V_G = 0$ V (a) and thermal annealing, $V_G = 0$ V (b).

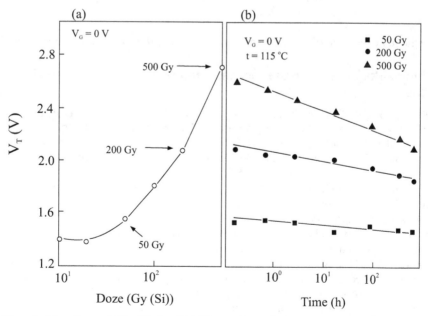

Fig. 9. Threshold voltage shift $\Delta V_T$ for PMOS transistor during gamma irradiation, $V_G = 0$ V (a) and thermal annealing, $V_G = 0$ V (b).

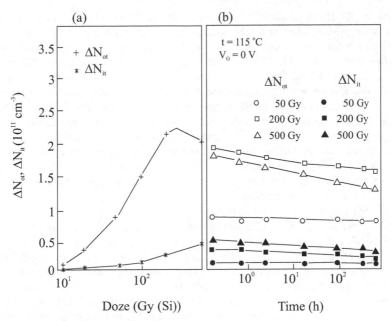

Fig. 10. The oxide trap ($\Delta N_{ot}$) and interface traps ($\Delta N_{it}$) density of NMOS transistor during gamma irradiation, $V_G$ = 0 V (a) and thermal annealing, $V_G$ = 0 V (b).

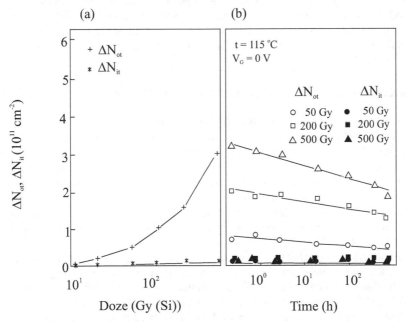

Fig. 11. The oxide trap ($\Delta N_{ot}$) and interface traps ($\Delta N_{it}$) density of PMOS transistor during gamma irradiation, $V_G$ = 0 V (a) and thermal annealing, $V_G$ = 0 V (b).

Figures 10 and 11 show changes in charge density in gate oxide and interface traps during radiation with gamma rays and later annealing in the temperature of 115°C for 0 V gate polarizations for the same transistors for which the change in threshold voltage (Figures 8 and 9) was monitored [111, 112]. The method described in the paper [90] was used to determine these densities. From the behavior of the curves in Figures 10 and 11, it was concluded that the annealing of radiated transistor is carried out through annealing of radiation defects described in detail in chapter 3. Namely, both the density of charge in the gate oxide and the density of interface traps show the tendency of decreasing during the burn in. It has been shown that the annealing of defects caused by gamma radiation depends not only on the conditions of transistor annealing, but on the conditions of their radiation as well, which can also be detected in Figures 10 and 11. Namely, the level of radiation exposure dictates the degree of annealing and the speed of the process of burn in the defects, so that the annealing is smaller in higher degrees of radiation exposure, considering the higher density of defects, but their annealing is more intense. This fact is supported by radiating a CMOS transistor with electrons and X-radiation whose energies amount to 10 MeV and later annealing with low-energy UV radiation [109].

In order to be able to consider the temperature annealing or annealing or annealing by low-energy UV radiation of gamma radiation degraded of electric characteristics of CMOS transistors as successful, it is necessary to ensure conditions in which the degraded electric parameters will become stable after annealing. It has been shown [110] that these conditions can be fulfilled by annealing on the temperature of 115° C and positive polarization at the 10 and 15 V gate. Check of characteristics stability was performed by continuing the annealing in an increased temperature and without polarization of the gate, and it was observed that the threshold voltage remains stable. Besides being definitive, the annealing process is accelerated by polarization at the 10 and 15 V gate. With lower voltage values in the continued temperature treatment of radiation-exposed transistors, the threshold voltage starts decreasing, which can most probably be attributed to the effect of the so-called "inverted" annealing.

## 7.2 Behavior of power VDMOS transistor during the IR process and the subsequent annealing

Electric parameters behavior of MOS transistors with polysilicon gate during gamma radiation was studied for many years [4]. For the commercial components can be said that they are mostly known. Figure 12 shows typical changes in threshold voltage during radiation of n-channel power VDMOS power transistors of the type EF1N10 at room temperature, and with gate voltages $V_G = 0$ and 10 V [116]. It can be seen that the threshold voltage decreases along with the increase in radiation dose and that the changes are more pronounced with higher values of gate voltage. It was also shown that the mobility is decreased during the radiation, and the changes are greater when the gate voltage is 10 V than when it is not applied.

Change in the oxide traps charge density density $\Delta N_{ot}$ and interface traps $\Delta N_{it}$ during the IR process for the same transistors in which the change in threshold voltage was monitored (Figure 12), determined by use of SMG technique, are displayed in Figures 13 and 14, respectively [116]. An increase in $\Delta N_{ot}$ and $\Delta N_{it}$ values during gamma radiation can be observed, and these changes are greater in the case $V_G = 10$ V. It has also been observed that the increase in the oxide trap density is substantially greater than the increase in interface traps density.

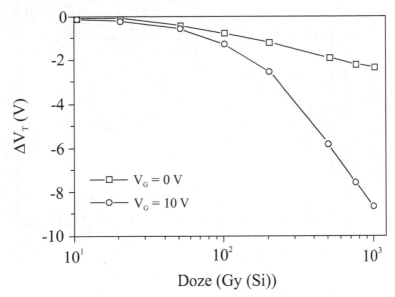

Fig. 12. Threshold voltage shift ( $\Delta V_T$ ) of n-channel power VDMOS transistor during gamma irradiation for $V_G$ = 0 V and $V_G$ = 10 V.

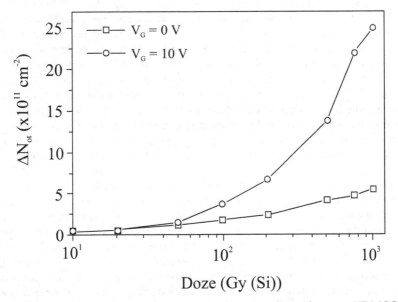

Fig. 13. The oxide traps charge density density ( $\Delta N_{ot}$ ) of n-channel power VDMOS transistor during gamma irradiation for $V_G$ = 0 V and $V_G$ = 10 V.

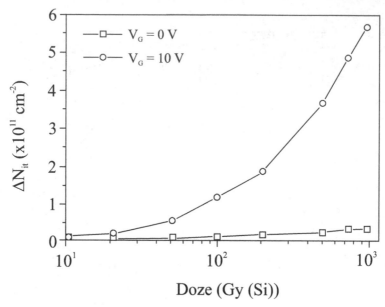

Fig. 14. The interface trap density ( $\Delta N_{it}$ ) of n-channel power VDMOS transistor during gamma irradiation for $V_G$ = 0 V and $V_G$ = 10 V.

Several papers [22, 27, 28, 85, 86, 88, 117-124] displayed the results of annealing the radiation-exposed power VDMOS transistors in room temperature and elevated temperatures. Figure 15 displays changes in threshold voltage of VDMOS power transistor of the type EFL1N10 when, during the gamma radiation, there was no voltage at the gate, and during the annealing it amounted to $V_G$ = 10 V, while the annealing was conducted at room temperature, 55° C and 140° C. Figure 16 refers to the same annealing conditions, except that the voltage at the gate during the gamma radiation amounted to $V_G$ = 10 V [27]. As it can be seen from these figures, a greater and faster change in threshold voltage $\Delta V_T$ occurs in case of higher temperature, and occurrence of super-recovery was detected only in transistors for which radiation was conducted when there was no polarization at the gate, and when their annealing was conducted on the temperature of 140° C.

The influence of temperature and voltage at the gate on the change in density of oxide traps charge $\Delta N_{ot}$ and interface traps $\Delta N_{it}$ for the same transistors for which a change in threshold voltage was monitored (Figures 15 and 16) are displayed in Figures 17 and 18 [27], and these changes are determined by using SMG technique. It can be seen that the density of interface traps after some annealing time starts increasing, while the time period prior to the start of increase can be very long. As mentioned before, this phenomenon of increase of $\Delta N_{it}$ value after some saturation is known in literature under the title latent increase (latent generation) of interface traps. The time interval prior to the occurrence of latent increase in interface traps density depends on the annealing temperature, it being shorter if the annealing temperature is higher. Also, a direct link between the latent increase in interface trap density and the latent decrease of density of the oxide trapped charge have been detected, as well as the fact that, after the latent increase in interface traps density, comes their decrease (passivization) during exposure to elevated temperature.

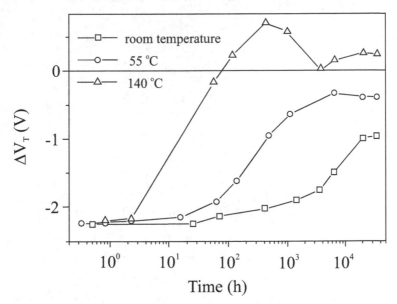

Fig. 15. Threshold voltage shift ($\Delta V_T$) of n-channel power VDMOS transistor during annealing; $V_G = 0$ V during irradiation and $V_G = 10$ V during annealing.

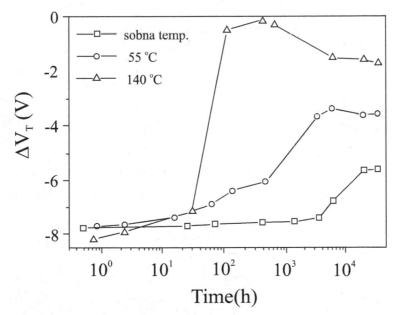

Fig. 16. Threshold voltage shift ($\Delta V_T$) of n-channel power VDMOS transistor during annealing; $V_G = 10$ V during irradiation and annealing.

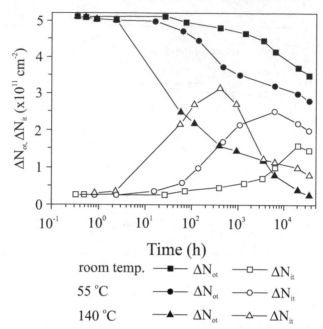

Fig. 17. The oxide traps charge density ($\Delta N_{ot}$) and interface traps ($\Delta N_{it}$) density during annealing for n-channel power VDMOS transistor irradiated for $V_G = 0$ V.

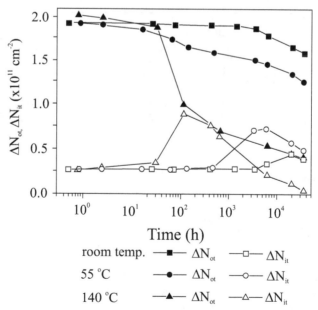

Fig. 18. The oxide traps charge density ($\Delta N_{ot}$) and interface trap ($\Delta N_{it}$) density during annealing for n-channel power VDMOS transistors irradiated at $V_G = 10$ V.

One of the important parameters which influence changes in density of oxide trapped charge $\Delta N_{ot}$ and interface traps $\Delta N_{it}$ during the annealing of power VDMOS transistors which have previously been exposed to gamma radiation is the value of voltage at the gate during the annealing. Experimental studies [119] have shown that the recovery of radiation-exposed transistors on the temperature of 140° C for the polarizations at the gate amounting to $V_G = 0, 5$ and 10 V, leads to latent increase in interface traps density and latent decrease in the density of oxide trapped charge. In the case $V_G = 0$, when only an electric field is present in the oxide due to the difference of work function of poly-Si gate and substrate, these changes are significantly smaller, when compared to the changes caused by the gate voltage of 5 and 10 V. The values of maximum densities of interface traps formed during the annealing, for $V_G = 5$ and 10 V have very few mutual differences. Also, there are certain differences in the time interval prior to the start of latent increase in $\Delta N_{it}$.

Changes in densities of captured electric charge and interface traps shown in Figures 13 through 18 are very well described by H-W model [27, 28], whose more detailed description is given in chapter 4.

### 7.2.1 Isochronal annealing of power VDMOS transistors after the IR process
The isochronal annealing implies to the annealing of MOS transistors after IR or HCI processes with variable temperature. Figure 19 shows behavior of density of the oxide trapped charge and interface traps during an isochronal annealing of radiation-exposed power VDMOS transistors of the type EFP8N15 in the temperature range between 50° C and 290° C [121]. These densities were followed by using SMG technique. Duration of annealing at every temperature amounted to 5 min, while the temperature change amounted to 10° C, and the gate voltage during the annealing amounted to $V_G = 10$ V. As it can be seen, the

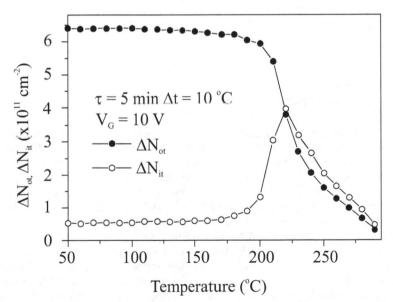

Fig. 19. The oxide traps charge density ($\Delta N_{ot}$) and interface trap ($\Delta N_{it}$) density for n-channel power VDMOS transistors during isochronal annealing determined using SMG technique.

values of $\Delta N_{ot}$ and $\Delta N_{it}$ are insignificantly changed up to the temperature of approximately 175° C, while at higher temperatures there occurs a rapid increase in interface traps densities, followed by decrease in the density of oxide trapped charge. The interface traps density starts decreasing from the temperature t=225° C. Figures 20 shows behavior of interface traps density with the temperature of isochronal annealing, generated by SMG and CP techniques. Similarities in behavior of these curves can be seen, but the values of $\Delta N_{it}$ generated by SMG technique are significantly higher, in accordance with the discussion given in chapter 6.

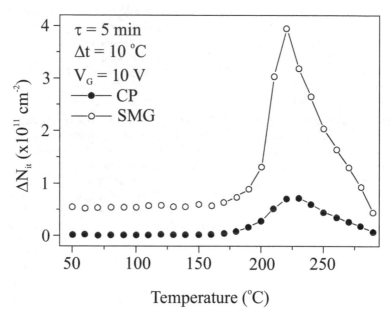

Fig. 20. The interface trap ($\Delta N_{it}$) density for n-channel power VDMOS transistor during isochronal annealing determined using SMG and CP techniques.

### 7.2.2 Influence of voltage polarization at the gate during the annealing of power VDMOS transistors which had previously been subjected to gamma radiation

Figure 21 shows changes in the density of oxide trapped charge and interface traps during the annealing of power VDMOS transistor of the type EFP8N15 which had previously been subjected to the IR process. Annealing was conducted under alternating change in the voltage polarization at the gate and temperatures of 200° C and 250° C, while each phase, for one direction of electric field lasted for 1 h [116]. The densities of the oxide trapped charge and interface traps were determined by SMG technique. It can be seen from the figure that during the first phase for $V_G$ = 10 V, there occurred a latent increase in interface traps density and latent decrease in the density of oxide trapped charge, for both temperatures. The first phase is similar to the results shown in Figures 17 and 18. During the second phase ($V_G$=-10 V), at the annealing temperature of 200° C, the value of $\Delta N_{it}$ continues to decrease, which could be expected on the basis of the H-W model, as the diffusion of water molecules is not dependent on the direction of electric field in oxide, while at the temperature of 250° C there is no change in this density, as almost all interface traps have already been passivated during the first phase. However, in the third phase ($V_G$=+10 V), the behavior of interface traps density is different from the behavior in the first phase. Namely, there is no latent increase in interface traps density, although the value of $\Delta N_{it}$ at first insignificantly raise at the very start of the annealing at the temperature of 250° C, and then continued to decrease, while at the temperature of 200° C, a mild increase was detected, and then slow decrease of this density (these changes are significantly smaller than during the first phase). On the basis of such behavior of interface traps and the application of H-W model, it was concluded that there were no hydrogen particles during the third phase (and especially no $H_2$ molecules) which could cause latent increase of $\Delta N_{it}$.

The paper [116] also shows the results of $\Delta N_{it}$ generated through the application of CP technique for power VDMOS transistors of the type EFP8N15 annealing under the same conditions as shown in Figure 21. The behavior of this density is very similar to the behavior of the density generated through the SMG technique, but the values of $\Delta N_{it}$ generated with CP technique are significantly lower, which is in accordance with the sensitivity of the method discussed in chapter 6.

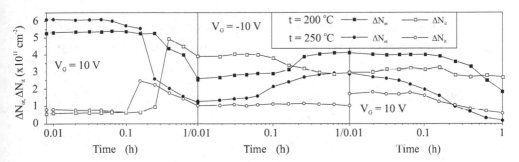

Fig. 21. The oxide traps charge density ($\Delta N_{ot}$) and interface trap ($\Delta N_{it}$) density during annealing with positive and negative polarization on the gate ($V_G = \pm 10$ V) for n-channel VDMOS transistors.

### 7.3 Behavior of power VDMOS transistor during the HCI process and subsequent annealing

Figure 22 shows the changes in threshold voltage of power VDMOS transistor EFL1N10 during the HCI process, with gate voltages of +80 V and -80 V [126]. It can be seen that the HCI process leads to significant changes in the threshold voltage $\Delta V_T$. An initial decrease in the value of $\Delta V_T$ (during the first 40 min under positive, i.e. 20 min under negative gate voltage) and subsequent increase until the end of tension implies the occurrence of the "turn-around" effect. Similar behavior was also detected in power VDMOS transistor of the type IRF510 [29].

Figure 22 also shows the change in threshold voltage during the second HCI process. Namely, after the first HCI process, transistors were annealing at the temperature of 150° C in the duration of 3000 h, and then subjected to the HCI process again. It can be seen that the behavior is similar, while the changes in $\Delta V_T$ are more pronounced in the first HCI process, for both polarization signs at the gate.

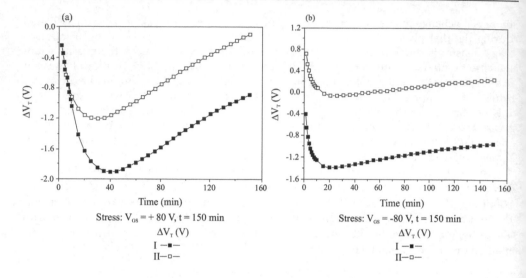

Fig. 22. Threshold voltage shift ( $\Delta V_T$ ) of n-channel power VDMOS transistors for positive ((a) $V_{GS}$ = 80 V ) and negative ((b) $V_{GS}$ = −80 V ) HCI during the first and second stress cycle.

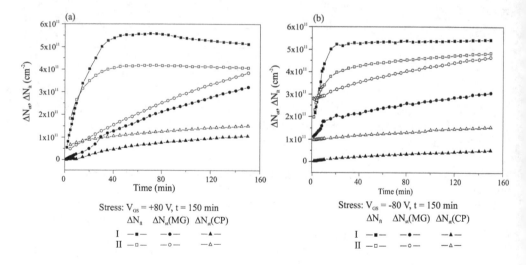

Fig. 23. The oxide traps charge density ( $\Delta N_{ot}$ ) and interface trap ( $\Delta N_{it}$ ) density of n-channel power VDMOS transistors for positive ((a) $V_{GS}$ = 80 V ) and negative ((b) $V_{GS}$ = −80 V ) HCI during the first and second stress cycle.

Figure 23 shows the changes in oxide trapped charge density in the gate oxide $\Delta N_{ot}$ determined by SMG technique and the changes in interface traps density $\Delta N_{it}$ (SMG) determined by SMG technique and $\Delta N_{it}$ (CP) determined by CP technique for the same transistors that were subjected to the HCI procedures during the monitoring of change in threshold voltage (Figure 22) [126]. From the figure it can be seen that the behaviors of $\Delta N_{it}$ (SMG) and $\Delta N_{it}$ (CP) are qualitatively the same, while there are significant quantitative differences between them. On the basis of the discussion given in section 3, the quantitative concurrences should not be expected, as the share of ST in the densities $\Delta N_{it}$ (SMG) and $\Delta N_{it}$ (CP) is different. Besides that, SMG and CP techniques in power VDMOS transistors register interface traps in two different areas of gate surface. SMG technique registers current in the p⁻ -area of the channel, while CP technique registers interface traps in the n⁻ -epi area of power VDMOS transistors.

Park et al [127] studied this problem in detail in a large number of similar samples and concluded that the initial values of threshold voltage in these two interfaces are significantly different. Another cause of quantitative difference in the values of $\Delta N_{it}$ (SMG) and $\Delta N_{it}$ (CP) is the fact that these two techniques record different areas of the silicon forbidden band. Namely, SMG technique records the defects in the area located at about 0.45 eV of the upper part of the silicon forbidden band, while CP technique records the defects at the same distance, but in the lower part of this area [127].

The initial increase $\Delta N_{ot}$ shown in Figure 23 is a consequence of the formation of positively charged FT at the early stage of HCI process [126]. $\Delta N_{it}$ achieves saturation at the later stage of this process, as the number of electrons captured by NBO centers increases. The increase in $\Delta N_{it}$ value for the entire duration of the HCI process is necessitated by the reaction of H⁺ ions which are released in the oxide (H model) and the holes drifting towards the Si – SiO₂ interface and forming interface traps ($h^+$ model). Greater changes of $\Delta N_{it}$ (SMG) occur because this technique comprises all ST (ST=SST+FST), while $\Delta N_{it}$ (CP) comprises only FST.

Figures 24 shows the changes in threshold voltage of $\Delta V_T$, while Figures 25 shows the changes in the density of oxide trapped charge $\Delta N_{ot}$ and the change in interface traps

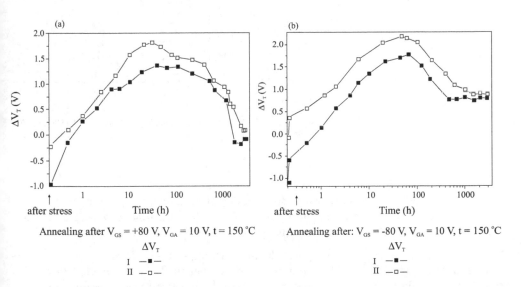

Fig. 24. Threshold voltage shift ( $\Delta V_T$ ) of n-channel power VDMOS transistor during the first and second thermal post HCI annealing cycle following (a) positive and (b) negative HCI.

densities determined by SMG technique ($\Delta N_{it}$ (SMG)) and CP technique ($\Delta N_{it}$ (CP)) during the annealing of power VDMOS transistor of the type EFL1N10 which had previously been subjected to the HCI process under gate voltages of +80 V and -80 V, annealing at the temperature of 150° C, and then subjected to HCI process again [126]. It can be seen that $\Delta V_T$ rises during the first 10 hours of recovery, then it decreases and achieves saturation in case of duration of 1.000 h of annealing, and that the behavior of $\Delta V_T$ is very similar in both polarization cases. The only difference is that $\Delta V_T$ achieves saturation sooner in the annealing preceded by HCI process with a negative gate polarization.

From Figure 25 it can be seen that there are no differences caused by the opposite sign of a gate polarization during the HCI process. Nor do the trends of the monitored changes differ during the first and the second recovery. As expected, the density of $\Delta N_{ot}$ decreases during annealing. $\Delta N_{it}$ (SMG) initially rises, while its decrease would occur during a longer annealing period, until the saturation is achieved. Very similar behavior also applies to the values of $\Delta N_{it}$ (CP), but this change is significantly smaller than the change $\Delta N_{it}$ (SMG) for the reasons discussed previously. The behavior of $\Delta N_{ot}$, $\Delta N_{it}$ (SMG) and $\Delta N_{it}$ (CP) during annealing is very well described by the H-W model which details are given in chapter 5.

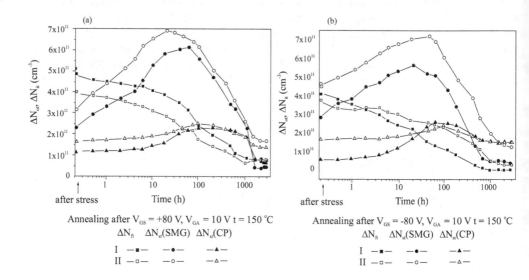

Fig. 25. The oxide traps charge density ( $\Delta N_{ot}$ ) and interface trap ( $\Delta N_{it}$ ) density of n-channel power VDMOS transistors during the first and second thermal post HCI annealing cycle following (a) positive and (b) negative HCI.

In the paper [22] shows comparative results of annealing of the n-channel power VDMOS transistor of the type EFL1N10, where a certain number of them is subjected to gamma radiation, and others to the HCI (Fowler-Nordheim high electric field stress) process. The changes in $\Delta N_{ot}$ , $\Delta N_{it}$ (SMG) and $\Delta N_{it}$ (CP) were monitored. It was noticed that, during annealing of transistors that had previously been subjected to gamma radiation there is a latent increase in interface trap density, both with SMG and CP technique. For transistors that had previously been subjected to the HCI process, a latent increase in interface trap density during the annealing occurs when only CP technique is used to determine their density. This shows that there is a difference in the nature of capture centers emerged during the IR and HCI processes. Finally, it should be pointed out that the behavior of $\Delta N_{it}$ (SMG) and $\Delta N_{it}$ (CP) after exposure to IR process given in the paper [127, 128], is identical to the behavior of $\Delta N_{it}$ (SMG) given in the paper [126] in case of the HCI process. The experiment was conducted with the samples that are structurally the same, but not identically processed as the samples in the work [126]. This comparison only illustrates the complicated nature of defect behavior occurring during the gamma radiation and the HCI process.

### 7.4 Application of PMOS transistor with Al-gate as a sensor and dosimeter of ionizing radiation

As early as 1974, there emerged an idea on the possibility of application of MOS transistors for the detection of the absorbed dose of ionizing radiation [7]. It was followed by the design and production of radiation-sensitive PMOS transistor with Al gate which is known under the title RADFET (radiation sensitive field effect transistors) which could be used as both sensor and dosimeter of ionizing radiation (gamma and X-radiation) [7]. These dosimeter transistors have so far found partial application in modern aircrafts [129, 130] (in which higher values of absorbed doses are measured) in medicine (in radiology, where they have the role of radiation sensors) [131, 132], in nuclear industry [132] and military [133, 134]. However, no PMOS dosimeter for small doses of radiation measuring, which could be used as a personal dosimeter has yet been realized.

The basic parameter for PMOS dosimetric transistor is the threshold voltage $V_T$, on basis of which the absorbed radiation dose is determined. The change in threshold voltage, as already said above, is a consequence of oxide trapped charge density $\Delta N_{ot}$ and interface traps $\Delta N_{it}$ caused by the IR process. In the case of PMOS transistors, the increase in these densities leads to the increase in $V_T$, as opposed to NMOS transistors, in which an increase of $\Delta N_{ot}$ reduces $V_T$, and increase of $\Delta N_{it}$ increases its value.

In general, the change in threshold voltage $\Delta V_T$ with an absorbed dose of radiation $D$ is given by the following expression:

$$\Delta V_T = A \cdot D^n ,\tag{37}$$

where $A$ is the constant, and n is the degree of linearity. For $n = 1$, the constant $A$ represents the sensitivity $S$ of dosimetric transistor

$$S = \frac{\Delta V_T}{D} ,\tag{38}$$

Figure 26 represents the single-point method, which is primarily used to determine the threshold voltage of PMOS dosimetric transistors [135]. It consists of the establishing constant current through the channel $I_D$ (the value of 10 μA is usually taken) and measuring of $V_0$ voltage which corresponds with this current. It is also considered that the change in this voltage ($\Delta V_0$) corresponds with the change in threshold voltage ($\Delta V_T$).

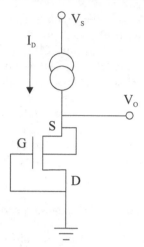

Fig. 26. The threshold voltage determination using single point method.

The second important parameter of PMOS dosimetric transistors is the recovery of threshold voltage after the radiation, which is known under the term fading f(t). It can be expressed in the following way:

$$f(t) = \frac{V_T(0) - V_T(t)}{V_T(0) - V_{T0}} = \frac{V_T(0) - V_T(t)}{\Delta V_T(0)} ,$$ (39)

where $V_{T0}$ is the threshold voltage prior to radiation, $V_T(0)$ immediately after irradiation, $V_T(t)$ after annealing time $t$ and $\Delta V_T(0)$ is the threshold voltage shift immediately after irradiation of PMOS dosimetric transistor.

Fading can also be determined on the basis of the values $\Delta N_{ot}$ and $\Delta N_{it}$ [136, 137]

$$f(t) = \frac{\Delta N'_{ot}(t) + \Delta N'_{it}(t)}{\Delta N_{ot}(0) + \Delta N_{it}(0)} ,$$ (40)

in which $\Delta N'_{ot}(t) = \Delta N_{ot}(0) - \Delta N_{ot}(t)$ ("the annealing part" of the oxide trapped charge) and $\Delta N'_{it}(t) = \Delta N_{it}(0) - \Delta N_{it}(t)$ ("the annealing part" of the interface traps), $\Delta N_{ot}(0)$ and $\Delta N_{it}(t)$ are the densities of oxide trapped charge and interface traps after the annealing for the time $t$, respectively, while $\Delta N_{it}(0)$ and $\Delta N_{it}(t)$ are adequate densities after the radiation, i.e. at the start of annealing.

The papers [136-143] show the results of exposure to gamma radiation and subsequent annealing of PMOS dosimetric transistors manufactured in the company Ei-Microelectronics, Nis, Serbia, while the papers [135, 143-145] show the results of radiation and annealing of PMOS dosimetric transistors manufactured in Tyndall National Institute, Cork, Ireland. Figure 27 shows change in threshold voltage with an increase in absorbed dose of gamma radiation for the transistors with oxide thickness of 1.23 μm [136]. It can be seen, that this dependence can be displayed in a coordinate system with log-log scale, with a straight line for all applied voltages at the gate. Also, changes in threshold voltage are greater in the case of positive voltages at the gate, although sensitivity grows with the increase of the absolute value $V_G$ in both polarization modes (the smallest value applies to the zero polarization). This means that the increase of $\Delta V_T$ takes place regardless of the direction of the electric field in oxide, but the size of these changes is dependent on it.

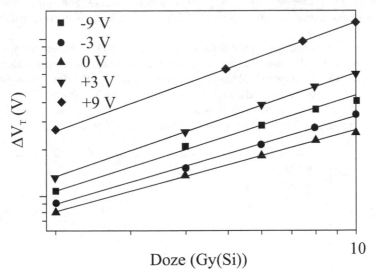

Fig. 27. Threshold voltage shift ($\Delta V_T$) during irradiation of pMOS transistor, with 1.23 μm thick oxide.

Figure 28 shows change in threshold voltage as a function of absorbed dose of gamma radiation for different oxide thicknesses, for the gate voltage of $V_G = 3$ V [139]. It can be seen

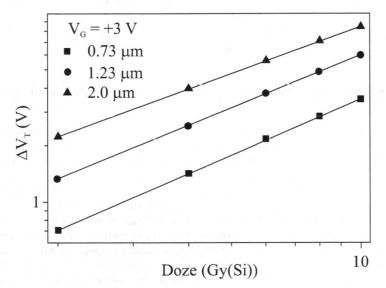

Fig. 28. Threshold voltage shift ($\Delta V_T$) during irradiation of pMOS transistors with different oxide thickness and gate polarization of +3 V.

that the sensitivity grows along with the thickness of gate oxide, and as the tested PMOS dosimetric transistors have thermal oxide with the same thickness, these results show a significant role of CVD oxide in the change of threshold voltage. The gate voltage for these transistors was the same, i.e. the electric field in the thickest oxide was smallest, which affected their sensitivity.

The paper [135] shows changes in the threshold voltage of PMOS dosimetric transistors for the oxide thicknesses of 100 mm for the values of absorbed gamma radiation doses between 50 and 300 Gy, while the paper [143] stated this dependence for the dose range between 100 Gy and 500 Gy for gate voltages during radiation $V_G=0$ and 5 V (Figure 29). It was found that there is a linear dependence between the threshold voltage change and the absorbed radiation dose when the dose range amounted between 10 Gy and 500 Gy (Figure 29).

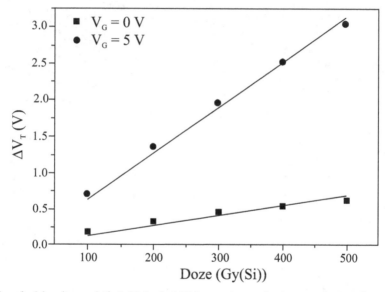

Fig. 29. Threshold voltage shift ($\Delta V_T$) of pMOS transistors during irradiation for $V_G$ = 0 V and $V_G$ = 5 V.

Change of fading, as the other important parameter for PMOS dosimetric transistors, during annealing at room temperature is shown in figures 30 and 31 [116] for the transistors with oxide thickness of 1.23 and 2.0 μm. The radiation and recovery were conducted with the same voltage values at the gate. It can be seen that the transistors having oxide thickness of 2.0 μm show a more pronounced negative fading. On the basis of fading behavior of transistors with the same oxide thickness, but with different gate polarization, it is obvious that no concrete conclusion which would apply to all radiation-exposed transistors can be given. Lack of knowledge about the explicit form of dependence of fading behavior since the recovery time is not a deficiency of these PMOS transistors (which cannot be said for the sensitivity), as the practical application requires solely that it be smaller than some previously set value (usually, fading should be no less than ± 10% after three months of annealing at room temperature). On the basis of these results, it can be seen that for the oxide thickness of 2.0 μm, when the sensitivity is at its peak, the fading has a small value, and therefore in the case $V_G=9$ V, the sensitivity is $S = 2$ V/Gy for the absorbed dose $D = 10$ Gy(Si) and fading 5.9 % for annealing time of 3500 h

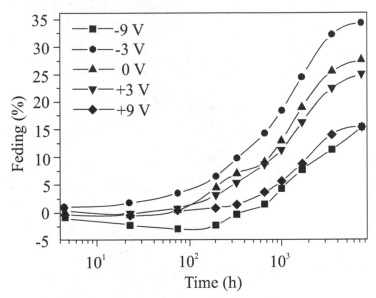

Fig. 30. Fading of irradiated pMOS transistors, with gate oxide thickness of 1.23 μm, during room temperature/bias annealing.

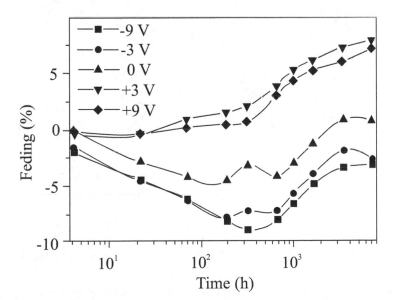

Fig. 31. Fading of irradiated pMOS transistors, with gate oxide thickness of 2.0 μm, during room temperature/bias annealing.

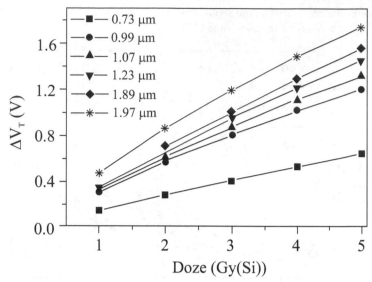

Fig. 32. Threshold voltage shifts ( $\Delta V_T$ ) during irradiation at dose rate $1.2 \cdot 10^{-3}$ Gy/s for different gate oxide thickness of pMOS transistors.

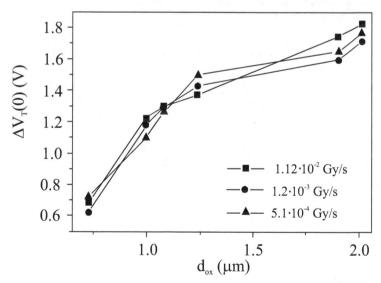

Fig. 33. Threshold voltage shifts ( $\Delta V_T$ ) for different pMOS transistors irradiated up to 5 Gy(Si).

Figure 32 shows changes in the threshold voltage from a dose of gamma radiation of PMOS dosimetric transistors with different gate oxide thicknesses (the thickness of thermal oxide was 0.3 μm, and the thickness of CVD oxide were different) during radiation with the dose of 1.2 ·10⁻³ Gy [138]. It can be seen that the sensitivity grows along with the gate oxide thickness. This is clearly displayed in Figure 33 which shows changes in threshold voltage immediately after radiation, i.e. prior to annealing at room temperature for three dose speeds. On the basis of these figures, it can be concluded that the main influence on sensitivity is exerted by the total oxide thickness, and then the thickness of CVD oxide. Namely, the transistors with oxide thickness of 1.97 μm are more sensitive that a transistor with oxide thickness of 1.89 μm, although they have smaller thickness of CVD oxide (the same applies to the oxide thicknesses of 1.07 μm and 0.99 μm). A cause for such behavior can be the thickness and the type of thermal oxide.

As opposed to the results shown in Figures 32, in which the sensitivity of the tested PMOS transistors is given in the range between 1 and 5 Gy, Figures 34 shows their sensitivity in the dose range between 0.003 and 1 Gy for the case when there was no gate polarization during radiation [116].

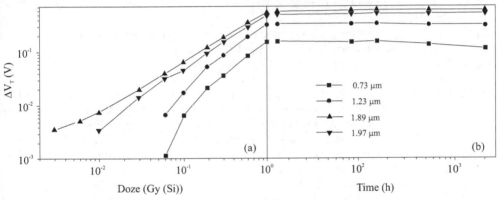

Fig. 34. Threshold voltage shifts ( $\Delta V_T$ ) of pMOS transistor during irradiation with dose rate of $7.51 \cdot 10^{-5}$ Gy/s (a) and annealing at room temperature (b), for different values of gate oxide thickness.

As it can be seen from the figure 34, the greatest sensitivity is displayed by the transistors with oxide thickness of 1.89 $\mu m$, which can register radiation doses in the order of magnitude cGy. As this transistor has the greatest thickness of CVD oxide, it can be assumed that this oxide would have a significant role in the area of small doses. The changes in threshold voltage during annealing of up to 2000 hours are insignificant for any oxide thickness (Figure 34b). Results with transistors of this oxide thickness show that their sensitivity increases with the increase of gate voltage during the gamma radiation, especially when the doses are higher than 0,01 Gy.

## 8. References

[1] J.R. Davis, Instabilities in MOS Devices, London: Gordon an Breach Science Publishers, 1981.
[2] S.M. Sze, Physics of Semiconductors Devices, New York: Wiley and Sons, 1981.

[3] T.P. Ma and P.V. Dressendorfer, Ionizin Radiation Effects in MOS Devices and Circuits, New York: Willey and Sons, 1989.

[4] S. Dimitrijev, Understanding Semiconductor Devices, New York: Oxford University Press, 2000.

[5] S. Gamerith and M. Polzl, Negative bias temperature stress on low voltage p-channel DMOS transistors and the role of nitrogen, Microelectron. Reliab., Vol. 42, pp. 1439-1443, 2002.

[6] B.J. Baliga, Switching lots of watts at high speed, IEEE Spectrum, Vol. 18, pp. 42-53, 1981.

[7] A.G. Holmes-Siedle, The space charge dosimeter- general principle of new method of radiation dosimetry, Nucl. Instrum. and Methods., Vol. NI-121, pp. 169-179, 1974.

[8] B.E. Deal, Standardized terminology for oxide charges associated with thermally oxidized silicon, IEEE Trans. Electron. Dev., Vol. ED-27, pp. 606-608, 1980.

[9] J.R. Davis, Instabilities in MOS Devices, London: Gordon and Breach Science Publishers, 1981.

[10] R.J. Kriegler, Ion instabilities in MOS structures, Proc. Int. Reliab. Phys. Symp., pp. 250-258, 1974.

[11] C.T. Sah, Origin of interface states and oxide charges generated by ionizing radiation, IEEE Trans. Nucl. Sci., Vol. NS-23, pp. 1563-1567, 1976.

[12] D.M. Fleetwood, Border traps in MOS devices, IEEE Trans. Nucl. Sci., Vol. NS-39, pp. 269-271, 1992.

[13] D.M. Fleetwood, P.S. Winokur, R.A. Reber Jr., T.L. Meisenheimer, J.R. Schwank, M.R. Shaneyfelt, L.C. Riewe, Effects of oxide traps, interface traps and border traps on metal-oxide-semiconductor devices, J. Appl. Phys., Vol. 73, pp. 5058-5074, 1993.

[14] D.M. Fleetwood, M.R. Shaneyfelt, W.L. Warren, J.R. Schwank, T.L. Meisenheimer, P.S. Winikur, Border traps: Issues for MOS radiation response and long-term reliability, Microelectron. Reliab., Vol. 35, pp. 403-408, 1995.

[15] D.M. Fleetwood and N.S. Saks, Oxide, interface and border traps in thermal, $N_2O$ and $N_2O$ -nitrogen oxides, J. Appl. Phys., Vol. 79, pp. 1583-1594, 1996.

[16] D.M. Fleetwood, P.S. Winokur, L.C. Riewe, R.A.. Reber Jr., Bulk oxide traps and border traps in metal-oxide-semiconducter capacitors, J. Appl. Phys., Vol. 84, pp. 6141-6148, 1996.

[17] D.M. Fleetwood, Revised model of thermally stimulated current in MOS capacitors, IEEE Trans. Nucl. Sci., Vol. NS-44, pp. 1826-1833, 1997.

[18] C.T. Wang, Hot carrier design considerations for MOS devices and circuits, New York: Van Nostrand Reinhold, 1992.

[19] J.P. McWhorter and P.S. Winokur, Simple technique for separating the effects of interface traps and trapped-oxide charge in metal-oxide semiconductor transistor, Appl. Phys. Lett., Vol. 48, pp. 133-135, 1986.

[20] M.A.B. Elliot, The use charge pumping currents to measure interface trap densities in MOS transistors, Solid-State Electron., Vol. 19, pp. 241-247, 1976.

[21] G. Groeseneken, H.E. Maes, N. Beltran and R.F. De Keersmaecker, A Reliable approach to charge-pumping measurements in MOS Transistors, IEEE Trans., Electron Dev., Vol. ED-31, pp. 42-53, 1984.

[22] G.S. Ristić, M.M. Pejović, A.B. Jakšić, Comparision between postirradiation annealing and post high electric field stress annealing od of n-channel power VDMOSFETs, Appl. Surface Science, Vol. 220, pp. 181-185, 2003.

[23] G.S. Ristić, Influence of ionizing radiation and hot carrier injection on metal oxide-semiconducter transistors, J. Phys. D: Appl. Phys., Vol. 41, 02300 (19 pp), 2008.

[24] L.P. Trombetta, F.J. Feigh, R.J. Zeto, Positive charge generation in metal-oxide-semiconductor capacitors, J. Appl. Phys., Vol. 69, pp. 2512-2521, 1991.

[25] O. Habaš, Z. Prijić, D. Pantić, N. Stojadinović, Charge-pumping characterization of $SiO_2/Si$ interface virgin and irradiated power VDMOSFETs, IEEE Trans. Electr. Dev., Vol. ED-43, pp. 2197-2208, 1996.

[26] T. Uchino, M. Takashaki and T. Yoko, Model of oxygen-deficiency-related defects in $SiO_2$ glass, Phys. Rev. B, Vol. 62, pp. 2983-2986, 2000.

[27] M. Pejović and G. Ristić, Creation and passivation of interface traps in irradiated MOS transistors during annealing at different temperatures, Solid State Electr. Vol. 41, pp. 715-720, 1997.

[28] M. Pejović, G. Ristić and A. Jakšić, Formation and passivation of interface traps in irradiated n-channel power VDMOSFETs during thermal annealing, Appl. Surf. Sci., Vol. 108, pp. 141-148, 1997.

[29] G.S. Ristić, M.M. Pejović and A.B. Jakšić, Fowler-Nordheim high electric field stress of power VDMOSFETs, Solid-State Electron., Vol. 49, pp. 1140-1152, 2005.

[30] G.S. Ristić, M.M. Pejović and A.B. Jakšić, Defect behaviour in n-channel power VDMOSFETs during HEFS and thermal post-HEFS annealing, Appl. Surf. Sci., Vol. 252, pp. 3023-3032, 2006.

[31] G.S. Ristić, M.M. Pejović and A.B. Jakšić, Physico-chemical processes in metal-oxide-semiconductors with thick gate oxide during high electric field stress, J. Non. Cryst. Solids, Vol. 353, pp. 170-179, 2007.

[32] D.L. Griscom, Diffusion of radiolytic molecular hydrogen as a mechanism for the post-irradiation buildup of interface states in $SiO_2$-Si structures, J. Appl. Phys., Vol. 58, pp. 2524-2533, 1985.

[33] A.G. Revesz, Defect structure and irradiation behaviour of non-crystaline $SiO_2$, IEEE Trans. Nucl. Sci., Vol. NS-18, pp. 113-116, 1971.

[34] D.J. DiMaria, E. Cartier, D.A. Buchanan, Anode hole inection and trapping in silicon dioxide, J. Appl. Phys., Vol. 80, pp. 304-317, 1996.

[35] T.P. Chem, S.Li, S. Fung, C.D. Beling, K.F. Lo, Post-stress interface trap generation induced by oxide stress with FN injection, IEEE Trans. Electron Devices, Vol. ED-45, pp. 1972-1977, 1998.

[36] D.J. DiMaria, E. Cartier and D. Arnold, Impact ionization, trap creation, degradation and breakdown in silicon dioxide films on silicon, J. Appl. Phys., Vol. 73, pp. 3367-3384, 1993.

[37] D.J. DiMaria, Defect production, degradation and breakdown of silicon dioxide film, Solid-State Electron., Vol. 41, pp. 957-965, 1997.

[38] D.L. Griscom, Optical properties and structure of defects in silica glass, J. Ceram. Soc. Japan, Vol. 99, pp. 923-941, 1991.

[39] R. Helms and E.H. Poindexter, The silicon-silicon-dioxide system: its microstructure and imperfections, Rep. Progr. Phys., Vol. 57, pp. 791-852, 1994.

[40] R.A. Weeks, Paramagnetic resonance of lattice defects in irradiated quartz, J. Appl. Phys., Vol. 27, pp. 1376-1381, 1956.

[41] H.E. Boesch Jr., F.B. Mclean, J.M. McGarrity, G.A. Ausman Jr., Hole transport and charge relaxation in irradiated $SiO_2$ MOS capacitors, IEEE Trans. Nucl. Sci., NS-22, pp. 2163-2167, 1975.

[42] J.F. Conley, P.M. Lenahan, Electron spin resonance analysis of EP center interactions with $H_2$ : Evidence for a localized EP center structure, IEEE Trans. Nucl. Sci., Vol. 42, pp. 1740-1743, 1995.

[43] D.A. Buchanan, A.D. Marwick, D.J. DiMaria, L. Dori, Hot-electron-induced hydrogen redistribution and defect generation in metal-oxide-semiconductor capacitors, J. Appl. Phys., Vol. 76, pp. 3595-3608, 1994.

[44] D.J. DiMaria, D.A. Buchanan, J.H Stathis, R.E. Stahlbush, Interface states induced by the presence of trapped holes near the silicon-silicondioxide interface, J. Appl. Phys., Vol. 7, pp. 2032-2040, 1995.

[45] S.K. Lai, Two carrier nature of interface-state generation in hole trapping and radiation damage, Appl. Phys. Lett., Vol. 39, pp. 58-60, 1981.

[46] S.K. Lai, Interface trap generation in silicon dioxide when electrons are captured by trapped hols, J. Appl. Phys., Vol. 54, pp. 2540-2546, 1983.

[47] S.T. Chang, J.K. Wu, S.A. Lyon, Amphoteric defects at $Si-SiO_2$ interface, Appl. Phys. Lett., Vol. 48, pp. 662-664, 1986.

[48] S.J. Wang, J.M. Sung, S.A. Lyon, Relationship between hole trapping and interface state generation in metal-oxide-silicon structure, Appl. Phys. Lett., Vol. 52, pp. 1431-1433, 1988.

[49] F.B. Mclean, A framework for understanding radiation-induced interface states in $SiO_2$ MOS structures, IEEE Trans. Nuclear Sci., Vol. NS-27, pp. 1651-1657, 1989.

[50] N.S. Saks, C.M. Dozier, D.B. Brown, Time dependence of interface trap formation in MOSFETs following pulsed irradiation, IEEE Trans. Nucl. Sci., Vol. NS-35, pp. 1168-1177, 1988.

[51] N.S. Saks, D.B. Brown, Interface trap formation via the two-stage $H^+$ process, IEEE Trans. Nucl. Sci., Vol. NS-36, pp. 1848-1857, 1989.

[52] D.L. Griscom, D.B. Brown, N.S. Saks, Nature of radiation-induced point defects in amorphous $SiO_2$ and their role in $SiO_2$-on-Si structures, The Physics and Chemistry of $SiO_2$ and the $Si-SiO_2$ interface, ed C.R. Helms and B.E. Deal (New York: Plenum) 1988.

[53] K.L. Brower , S.M. Myers, Chemical kinetics of hydrogen and (111) $Si-SiO_2$ interface defect, Appl. Phys. Lett., Vol. 57, pp. 162-164, 1990.

[54] J.H. Statis, E. Cartier, Atomic hydrogen reactions with $P_b$ centers at the (100) $Si-SiO_2$ interface, Phys. Rev., Lett., Vol. 72, pp. 2745-2748, 1994.

[55] E.H. Poindexter, Chemical reactions of hydrogenous species in the $Si-SiO_2$ system, J. Non Cryst. Solids, Vol. 187, pp. 257-263, 1995.

[56] D. Wuillaume, D. Goguenheinm, J. B, Borgoin, Nature of defect generation by electric field stress at $Si-SiO_2$ interface, Appl. Phys. Lett., Vol. 58, pp. 490-492, 1991.

[57] E.H. Poindexter, P.J. Caplan, B.E. Deal, R.R. Razouk, Interface states and electron spin resonance centers in thermally oxidized (111) and (100) silicon wafers, J. Appl. Phys., Vol. 52, pp. 879-883, 1981.

[58] E.H. Poindexter, G.J. Gerardi, M.E. Ruecked, P.J. Caplan, N.M. Johnson, D.K. Biegelsen, Electronic traps and $P_b$ centres at $SiO_2/Si$ interface: Bend-gap energy distribution, J. Appl. Phys., Vol. 56, pp. 2844-2853, 1984.

[59] A. Stesmans, V.V. Afanasiev, Electron spin resonance features of interface defects in thermal (100) $SiO_2/Si$, J. Appl. Phys., Vol. 83, pp. 2449-2457, 1998.

[60] E. Cartie, J.H. Stathis, D.A. Buchanan, Passivation and depassivation of silicon dangling bonds at the $SiO_2/Si$ inteface by atomic hydrogen, Appl. Phys. Lett., Vol. 63, pp. 1510-1512, 1993.

[61] D.L. Griscom, Diffusion of radiolytic molecular hydrogen as a mechanism for the post-irradiation buildup of interface states in $SiO_2$ -on-Si structures, J. Appl. Phys., Vol. 58, pp. 2524-2533, 1985.

[62] M.L. Reed, J.D. Plummer, $SiO_2/Si$ interface trap production by low-temperature thermal processing, Appl. Phys. Lett., Vol. 51, pp. 514-516, 1987.

[63] N.S. Saks, S.M. Dozier, D.B. Brown, Time dependence of interface traps formation in MOSFETs following pulsed irradiation, IEEE Trans. Nucl. Sci., Vol. NS-35, pp. 1168-1177, 1988.

[64] D.B. Brown, N.S. Saks, Time dependence of radiation-induced interface trap formation in MOS devices as a function of oxide thickness and applied field, J. Appl. Phys., Vol. 70, pp. 3734-3747, 1991.

[65] R.E. Stahlbush, B.J. Mrstik, R.K. Lawrence, Post-irradiation behaviour of the interface state density and trapped positive sharge, IEEE Trans. Nucl. Sci., Vol. NS-37, pp. 1641-1649, 1990.

[66] B.J. Mrstik, R.W Rendul, Si-$SiO_2$ interface state generation during X-ray irradiation exposure to hydrogen ambient, IEEE Trans. Nucl. Sci., Vol. NS-38, pp. 1101-1110, 1991.

[67] R.E. Stalbush. A.H. Edwards, D.L. Griscom, B.J. Mrstik, Post-irradiation cracking of $H_2$ and formation of interface states in irradiated metal-oxide-semiconductor field-effect trnsistros, J. Appl. Phys., Vol. 73, pp. 658-667, 1993.

[68] N.S. Saks, R.B. Klein, R.E. Stahlbush, B.J. Mrstik, R.W. Rendell, Effects of post-stress hydrogen annealing of MOS oxides after Co-60 irradiation or Fowler-Nordheim injection, IEEE Trans. Nucl. Sci., Vol. NS-40, pp. 1341-1349, 1993.

[69] I.S. Al-Kofahi, J.F. Zhang, G. Groeseneken, Continuing degradation of the Si-$SiO_2$ interface after hot hole stress, J. Appl. Phys., Vol. 81, pp. 2686-2692, 1997.

[70] J.F. Zhang, I.S. Al-Kofahi, G. Groeseneken, Behaviour of hot stress Si-$SiO_2$ interface at elevated temperature, J. Appl. Phys., Vol. 83, pp. 843-850, 1998.

[71] G. Van den Bosch, G. Groeseneken, H.E. Maes, R.B. Klein, N.S. Saks, Oxide and interface degradation resulting from substrate hot-hole injection in metal-oxide-semiconductor field-effect transistors at 295 and 77 K, J. Appl. Phys., Vol. 75, pp. 2073-2080, 1994.

[72] C.Z. Zhao, J.F. Zhang, G. Groesenecen, R. Degraeve, J.N. Ellis, C.D. Beech, Interface state generation after hole injection, J. Appl. Phys., Vol. 90, pp. 328-336, 2001.

[73] J.F. Zhang, C.Z. Zhao, G. Groeseneken, R. Degraeve, Analysis of the kinetics for interface state generation following hole injection, J. Appl. Phys., Vol. 93, pp. 6107-6116, 2003.

[74] J.R. Schwank, P.S. Winikur, P.J. McWhorter, F.W. Sexton, P.V . Dressendorfer, D.C. Turpin, Physical mechanisms contributing to devices rebound, IEEE Trans. Nucl. Sci., Vol. NS-31, pp. 1434-1438, 1984.

[75] T.R. Oldham, A.J. Lelis, F.B. McLean, Spatial dependence of trapped holes determined from tunneling analysis and measured annealing, IEEE Trans. Nucl. Sci., Vol. NS-33, pp. 1203-1210, 1986.

[76] P.J. McWhorter, S.L. Miller, W.W. Miller, Modeling the annealing of radiation-induced trapped hols in a varying thermal environmental, IEEE Trans. Nucl. Sci., Vol. NS-37, pp. 1682-1689, 1990.

[77] A.L. Lelis, H.E. Boesch Jr., T.R. Oldham, F.B. Melean, Reversibility of trapped hole annealing, IEEE Trans. Nucl. Sci., Vol. NS-35, pp. 1186-1191, 1988.

[78] A.L. Lelis, T.R. Oldham, H.E. Boesch Jr., F.B. McLean, The nature of the trapped hole annealing process, IEEE Trans. Nucl. Sci., Vol. NS-36, pp. 1808-1815, 1989.

[79] R.K. Freitag, D. B. Brown, C.M. Dozier, Exsperimental evidence of two species of radiation induced trapped positive charge, IEEE Trans. Nucl. Sci., Vol. ED-40, pp. 1316-1322, 1993.

[80] R.K. Freitag, D. B. Brown, C.M. Dozier, Evidence for two types of radiation-induced trapped positive sharge, IEEE Trans. Nucl. Sci., Vol. NS-41, pp. 1828-1834, 1994.

[81] G. Singh, K.F. Galloway, T.J. Russel, Temperature-induced rebound in power MOSFETs, IEEE Trans. Nucl. Sci., Vol. NS-34, pp. 1366-1369, 1987.

[82] D.M. Fleetwood, Long-term annealing study of midgap interface-trap charge neutrality, Appl. Phys. Lett., Vol. 60, pp. 2883-2885, 1992.

[83] D.M. Fleetwood, M.R. Shaneyfelt, L.C. Riewe, P.S. Winokur, R.A. Reber, Jr., The role of border traps in MOS high-temperature postirradiation annealing response, IEEE Trans. Nucl. Sci., Vol. 40, pp. 1323-1334, 1993.

[84] J.R. Schwank, D.M. Fleetwood, M.R. Shaneyfelt, P.S. Winokur, C.L. Axness, L.C. Riewe, Latent interface-trap buildup and its implications for hardness assurance, IEEE Trans. Nucl. Sci., Vol. NS-39, pp. 1953-1963, 1992.

[85] G.S. Ristić, M.M. Pejović, A.B. Jakšić, Modeling of kinetics of creation and passivation of interface trap in metal-oxide-semicondictor transistors during postirradiation annealing, J. Appl. Phys., Vol. 83, pp. 2994-3000, 1998.

[86] G.S. Ristić, M.M. Pejović and A.B. Jakšić, Analysis of postirradiation annealing of n-channel power vertical-double-diffused metal-oxide semiconductor transistors, J. Appl. Phys., Vol. 87, pp. 3468-3477, 2000.

[87] D.M. Fleetwood, W.L. Waren, J.R. Schwank, P.S. Winokur, M.R. Shaneyfelt, L.C. Reewe, Effects of interface traps and border traps on MOS postirradiation annealing response, IEEE Trans. Nucl. Sci., Vol. NS-42, pp. 1698-1707, 1995.

[88] M. Pejović, A. Jakšić and G. Ristić, The behaviour of radiation-induced gate-oxide defects in MOSFETs during annealing at 140 ºC, J. Non. Cryst. Solids, Vol. 240, pp. 182-192, 1998.

[89] J. Shwank, D.M. Fleetwood, M.R. Shaneyfelt, P.S. Winokur, Latent thermally activated interface-trap generation in MOS devices, IEEE Electr. Devices Lett., Vol. ED-13, pp. 203-205, 1992.

[90] S. Dimitrijev, N. Stojadinivić, Analysis of CMOS transistors instabilities, Solid-State Electronics, Vol. 30, pp. 991-1003, 1989.

[91] B. J. Baliga, Modern Power Devices, New York: Johan Wiley and Sons, 1987.

[92] Y. Tsividis, Operation and modeling of the MOS transistors, New York: McGraw-Hill, 1987.

[93] N. Arora, MOSFET models for VLSI circuit simulation, Wien-New York: Springer Verlag, 1993.

[94] S. Golubović, S. Dimitrijev, D. Župac, M. Pejović, N. Stojadinović, Gamma radiation effects in CMOS transistors, 17th European Solid State Device Research Conf., ESSDERC 87, Bologna, Italy, pp. 725-728, 1987.

[95] D Župac, K.F. Galloway, R.D. Schrimpf, P. Augier, Effects of radiation-induced oxide-trapped charge on inversion layer hole mobility at 300 and 77 K, Appl. Phys. Lett., Vol. 60, pp. 3156-3158, 1992.

[96] D. Župac, K.F. Galloway, R.D. Schrimpf, P. Augier, Radiation induced mobility degradation in p-channel double-diffused metal-oxide-semiconducter power transistors at 300 and 77 K, J. Appl. Phys., Vol. 73, pp. 2910-2915, 1993.

[97] N. Stojadinović, M. Pejović, S. Golubović, G. Ristić, V. Davidović, S. Dimitrijev, Effect of radiation-induced oxide-trapped sharge on mobility in p-channel MOSFETs, Electron. Lett., Vol. 31, pp. 497-498, 1995.

[98] N. Stojadinović, S. Golubović, S. Djorić, S. Dimitrijev, Modeling radiation-induced mobility degradation in MOSFETs, Phys. Stat. Sol. (a), Vol. 169, pp. 63-66, 1998.

[99] K.F. Galloway, M. Gaitan, T.J. Russel, A simple model for separating interface and oxide charge effects on MOS device characteristics, IEEE Trans. Nucl. Sci., Vol. NS-31, pp. 1497-1501, 1984.

[100] F.C. Hsu and S. Tam, Relationship between MOSFET degradation and hot-electron-induced interface-state generation, IEEE Electr. Dev. Lett., Vol. EDL-5, pp. 50-52, 1984.

[101] A. Bellaaouar, G. Sarrabayrouse, P. Rossel, Influence of ionizing irradiation on the channel mobility of MOS trnsistors, Proc, IEE, Vol. 132, pp. 184-186, 1985.

[102] S.C. Sun, J.D. Plumer, Electron mobility in inversion and accumulation layers on thermally oxidized silicon surface, IEEE Trans. Electr. Dev., Vol. ED-27, pp. 1497-1507, 1980.

[103] D.K. Schroder, Semiconductor Material and Device Characterization, New-York, Wiley, 1990.

[104] J.S. Brugler, P.G.A. Jespers, Charge pumping in MOS devices, IEEE Trans. Electr. Dev., Vol. ED-16, pp. 297-302, 1969.

[105] R.E. Paulsen, R.R. Siergiej, M.L. French, M.H. White, Observation of near-interface oxide traps with the charge pumping technique, IEEE Electron Dev. Lett., Vol. 13, pp. 627-629, 1992.

[106] A. Jakšić, M. Pejović, G. Ristić, S. Raković, Latent interface-trap generation in commercial power VDMOSFETs, IEEE Trans. Nucl. Sci., Vol. NS-45, pp. 1365-1371, 1998.

[107] S.C. Witezak, K.F. Galloway, R.D. Schrimpf, J.R. Brews, G. Prevost, The determination of Si-SiO$_2$ interface trap density in irradiated four-terminal VDMOSFETs using charge pumping, IEEE Trans. Nucl. Sci., Vol. NS-43, pp. 2558-2564, 1996.

[108] S. Dimitrijev, S. Golubović, D: Župac, M. Pejović, N. Stojadinović, Analisis of gamma-radiation induced instability mechanisms in CMOS transistors, Solid-State Electr., Vol. 32, pp. 349-353, 1989.

[109] S. Golubović, M. Pejović, S. Dimitrijev, N. Stojadinović, UV- radiation annealing of the electron and X-irradiation demaged CMOS transistors, Phys. Stat. Sol. (a), Vol. 129, pp. 569-575, 1992.

[110] M. Pejović, G. Ristić, S. G. Golubović, A comparision between thermal annealing and UV-radiation annealing of γ-irradiation NMOS trnsistors, Phys. Stat. Sol. (a), Vol. 140, pp. K53-K57, 1993.

[111] M. Pejović, S. Golubović, G. Ristić, M. Odalović, Annealing of gamma-irradiated Al-gate NMOS transistors, Solid-State Electron., Vol. 37, pp. 215-216, 1994.

[112] M. Pejović, S. Golubović, G. Ristić, M. Odalović, Temperature and gate-bias effect on gamma-irradiated Al-gate metal-oxide-semiconductor transistors, Jpn. J. Appl. Phys., Vol. 33, pp. 986-990, 1994.

[113] S. Golubović, G. Ristić, M. Pejović, S. Dimitrijev, The role of interface traps in rebound mechanisms, Phys. Stat. Sol. (a), Vol. 143, pp. 333-339, 1994.

[114] M. Pejović, S. Golubović, G. Ristić, Temperature-induced rebound in Al-gate NMOS transistors, IEE Proc.-Circuits Devices Syst., Vol. 142, pp. 413-416, 1995.

[115] M. Pejović, A. Jaksić, G. Ristić, B. Baljošević, Processes in n-channel MOSFETs during postirradiation thermal annealing, Radiat. Phys. Chem., Vol. 49, pp. 521-525, 1997.

[116] G.S. Ristić, Radiation and postirradiation effects in power VDMOS transistors and PMOS dosimetric transistors, PhD, University of Niš, Faculty of Electronic Engineering, 1998.

[117] A. Jakšić, G. Ristić, M. Pejović, Analysis of the processes in power MOSFETs during γ-ray irradiation and subsquent thermal annealing, Phys. Stat. Sol. (a), Vol. 155, pp. 371-379, 1996.

[118] A. Jakšić, M. Pejović, G. Ristić and S. Raković, Latent interface-trap generation in commercial power VDMOSFETs, Abstract of 4th European Conf. RADECS 97, pp. A5-A6, 1997.

[119] G.S. Ristić, M.M. Pejović, A.B. Jakšić, Numerical simulation of creation-passivation kinetics of interface traps in irradiated n-channel power VDMOSFETs during thermal annealing with various gate bias, Microelectronics Engin., Vol. 40, pp. 51-60, 1998.

[120] G.S. Ristić, A.B. Jakšić, M.M. Pejović, Latent interface-trap builup in power VDMOSFETs: new experimental evidence and numerical simulation, European Conf. of Radiation and its Effects on Components and Systems, RADECS 99, Abbaye de Fontevraud, France, pp. H14-H17, 1999.

[121] A.B. Jakšić, M.M. Pejović, G.S. Ristić, Isothermal and isochronal annealing experiments on irradiated commercial power VDMOSFETs, IEEE Trans. Nucl. Sci., Vol. NS-47, pp. 659-666, 2000.

[122] A.B. Jakšić, G.S. Ristić, M.M. Pejović, New experimental evidence of latent interface-trap buildup in power VDMOSFETs, IEEE Trans. Nuclear Sci., Vol. NS-47, pp. 580-586, 2000.

[123] A. Jakšić, G. Ristić, M. Pejović, Rebound effect in power VDMOSFETs due to latent interface-trap generation, Electr. Lett., Vol. 31, pp. 1198-1199, 1995.

[124] A.B. Jakšić, M.M. Pejović, G.S. Ristić, Properties of latent interface-trap buildup in irradiated metal-oxide-semiconductor transistors determined by switching bias isothermal annealing experiments, Appl. Phys. Lett, Vol. 77, pp. 4220-4222, 2000.

[125] V. Danchenko, E.G. Stassinopouls, P.H. Fang, S.S. Brashears, Activation energies of thermal annealing of radiation-induced damage in n-and p-channels of CMOS integrated circuits, IEEE Trans. Nucl. Sci., Vol. NS-27, pp. 1658-1664, 1980.

[126] S.M. Aleksić, A. B. Jakšić, M.M. Pejović, Repeating of positive and negative high electruic stres and corresponding thermal post-stress annaling of the n-channel power VDMOSFETs, Solid State Electrn., Vol. 52, pp. 1197-1201, 2008.

[127] M.S. Park, I. Na, C.R. Wie, A comparision of ionizing radiation and high field stress effects in N-channel power vertical double-diffusen metal-oxide-semiconductor field-effect trnsistors, J. Appl. Phys., Vol. 97, 014503 (6pp) 2005.

[128] M.S. Park, C.R. Wie, Study of radiation effects in γ-ray irradiation power VDMOSFET by DCIV technique, IEEE Tran. Nucl. Sci., Vol. NS-48, pp. 2285-2293, 2001.

[129] A.K. Ward, N. Blower, L. Adams, J. Doutreleau, A. Holmes-Siedle, M. Pignol, J.J. Berneron, M. Mehlen, The meteosat-p2 radiation effects experiment, Proc. 40.th Congerss of the Inter. Astronautical Federation, Malaga, Spain, pp. 151-159, 1989.

[130] L. Adams, E.J. Daly, R. Harboe-Sorensen, A.G. Holms-Siedle, A.K. Ward, R.A. Bull, Measurements of SEU and total dose in geostationary orbit under normal and flore conditions, IEEE Trans. Nucl. Sci., Vol. NS-38, pp. 1686-1692, 1991.

[131] D.J. Glastone, L.M. Chin, A.G. Holes-Siedle, MOSFET radiation detectors used as patient radiation dose monitors during radiotherapy, 33rd Ann. Mtg. Am. Assoc. of Pyisics in Medicine, San Francisko, 1991.

[132] J.S. Leffler, S.R. Lindgrern, A.G. Holmes-Siedle, Applications of RADFET dosimeters to equipment radiation qualification and monitoring, Tran. of the American Nucl. Sosiety, Vol. 60, pp. 535-536, 1989.

[133] L.S. August, Desing criteria for high-dose MOS dosimeter for use in space, IEEE Trans. Nucl. Sci., Vol. NS-31, pp. 801-803, 1984.

[134] L.S. August, R.R. Curcle, Adventages of using a PMOS FET dosimeter in high-dose radiation effects testing, IEEE Trans. Nucl. Sci., Vol. NS-31, pp. 1113-1115, 1984.

[135] A. Jakšić, G. Ristić, M. Pejović, A. Mohammadzadeh, C. Sudre, W. Lane, Gamma-ray irradiation and post-irradiation reponse of high dose range RADFETs, IEEE Tran. Nucl. Sci., Vol. NS-49, pp. 1356-1363, 2002.

[136] G. Ristić, S. Golubović, M. Pejović, PMOS dosimeter with two-layer state oxide appeared at zero and negative bias, Electr. Lett., Vol. 30, pp. 295-296, 1994.

[137] G. Ristić, S. Golubović, M. Pejović, P-channel metal-oxide semiconductor dosimeter fading dependencies on gate bias and oxide thickness, Appl. Phys. Lett., Vol. 66, pp. 88-89, 1995.

[138] G. Ristić, A. Jakšić, M. Pejović, PMOS dosimetric transistors with two-layer oxide, Sensors and Actuators Vol. A63, pp. 123-134, 1997.

[139] G. Ristić, S. Golubović, M. Pejović, pMOS transistors for dosimetric application, Electr. Lett., Vol. 29, pp. 1644-1646, 1993.

[140] Z. Savić, B. Radjenović, M. Pejović, N. Stojadinović, The contribution of border traps to the threshold voltage shift in pMOS dosimetric transistors, IEEE Trans. Nucl. Sci., Vol. NS-42, pp. 1445-1454, 1995.

[141] G. Ristić, S. Golubović, M. Pejović, Sensitivity and fading of pMOS dosimeters with thick gate oxide, Sensors and Actuators, Vol. A51, pp. 153-158, 1996.

[142] G.S. Ristić, Thermal and UV annealing of irradiated pMOS dosimetic transistors, J. Phys. D: Appl. Phys., Vol. 42, 135101, (12pp), 2009.

[143] M.M. Pejović, M.M. Pejović and A.B. Jakšić, Radiation-sensitive field effect trasnsistors to gamma-ray irradiation, Nucl. Tecnol. and Radiat. Protection, Vol. 26, pp. 25-31, 2011.

[144] M.M. Pejović, S.M. Pejović, E.Č. Dolićanin, Đ. Lazarević, Gamma-ray irradiation and post-irradiation at room temperature response of pMOS dosimeters with thick gate oxides, Nucl. Tecnol. and Radiat. Protection, Vol. 26, 2011 (accepted to publication).

[145] M.M. Pejović, M.M. Pejović, A.B. Jakšić, Contribution of fixed traps to sensitivity of pMOS dosimeters during gamma ray irradiation and annealing at room temperature, Sensors and Actuators A: Physical, vol. 174, pp. 85-90, 2012.

# 8

# Total Dose and Dose Rate Effects on Some Current Semiconducting Devices

Nicolas T. Fourches

*Atomic Energy Commission, CEN Saclay, DSM/IRFU/SEDI, Gif sur Yvette,*
*France*

## 1. Introduction

This chapter will overview some of the aspects of total ionizing dose and dose rate effects on semiconducting devices especially those used in high energy and radiation physics. First, material aspects and interaction of Ionizing Radiation with Matter will be reviewed with emphasis on defect creation and carrier generation. Radiation induced defects are detrimental to device operation but electron-hole pair generation by impinging particles is the basis of all semiconductor detectors. In the second stage, problems related to transistor devices will be discussed with particular emphasis on gate oxide issues and on Silicon On Insulator technologies, Radiation-hardened by design. Although some of the phenomenological physics and chemistry related to ionizing radiation and its effects on Metal Oxide Semiconductors Structures were reviewed by Oldham (Oldham, 1989) and earlier authors (Ma & Dressendorfer, 1989), there is no finalized view of defect physics and chemistry in the device oxides, mainly because the related problems are very complex. However, most of the experiments regarding ionizing irradiation effects on electronics were made both at room temperature or above and at a relatively high dose rate (Messenger & Ash, 1986). For some present day practical applications in space or in high-energy physics low dose rates effects are important and are investigated to usefully complement single event effect studies. Effects at lower temperatures have been scarcely investigated (Saks et al., 1984) leading to a rather piecemeal knowledge of transport of photo-generated carriers and similarly for the activation energies of deep defects centers. As most important devices are silicon based, ionizing radiation effects studies of bulk silicon have been made in an early stage (Willardson, 1959; Sonder & Templeton, 1960; Cahn, 1959; and the Purdue Group). This is of special importance for silicon or other semiconductor detectors. Many good studies and reviews were made on semiconductors detectors mainly in the framework of Large Hadron Collider experiments (Wunstorf, 1997; Leroy, 2007). In the last part of this chapter, progress made on new pixel detectors such as Complementary Metal Oxide Semiconductor sensors will be reviewed because they constitute a very active field of research and development.

## 2. Interaction of Ionizing radiation with a semiconductor material

### 2.1 Energy deposition and electron-hole pair generation

The interaction of radiation and both non-ionizing and ionizing will be the starting point of this discussion. The main issue is related to the damage generated in the materials both by the non-ionizing and ionizing irradiation. These can be classified as permanent effects as they have a long lasting influence in a semiconductor (a few minutes to many years). Other effects should be regarded as transient effects, and have many consequences and applications. Ionization generates electron-hole pairs, which in a semiconductor are charge carriers and may be used for device operation, such as particle detectors. Generation of carriers can occur in a semiconductor with photons of energy of the order of the bandgap. At higher energies, electrons from inner shells are excited and result in a high number of photo-generated carriers. Similar processes occur with other charged particles, the total number of generated electron-hole pairs being given by a simple expression:

$$N = Ed/Eg \qquad (1)$$

where Ed is the total energy deposited and Eg is the direct bandgap.

In practice, most of these pairs recombine, either directly or indirectly. An electric field applied to the device may dissociate these pairs and lead for a semiconductor detector to a current flow that can induce a signal according to the Ramo-Shockley theorem. Silicon, germanium, diamond counters, and other compound semiconductor detectors operate on this principle and have limited bulk sensitivity to Total Ionizing Dose (TID), but are most of the time very sensitive to Non-Ionizing-Effects.

The interaction of photons in the keV energy region with silicon dioxide results in the generation of electron-hole pairs that dissociate in the presence of the electric field applied to the gate of a MOS structure. In the case of charged particles with sufficient energy similar to protons, alphas, electrons, pions or so, the interaction with the oxide also generates electron-holes pairs according to ionization models originating in the Bethe-Bloch theory (Bethe, 1930,1934).

The Bethe-Bloch formula is valid at relatively large energies whereas in the lower range of energies, LSS (Lindhard,Scharff,Schiott, 1963) and Ziegler models should be taken into consideration, hence :

$$\frac{-dE}{dx} = Kz^2 \frac{Z}{A} \frac{1}{\beta^2} \left[ \frac{1}{2} \ln \frac{2m_e c^2 \beta^2 \gamma^2 T_{max}}{I^2} - \beta^2 - \frac{\delta(\beta\gamma)}{2} \right] \qquad (2)$$

dE/dx is the linear energy transfer, $T_{max}$ is the maximum energy transmitted to a free electron in a collision, $\delta(\beta\gamma)$ is the density effect correction to ionization energy loss, the mean excitation energy whose value are derived experimentally, related to oscillator strength, the other parameters being having their normal physical meaning. Ziegler published a review (Ziegler, 1988) of the detailed stopping laws derived from the original Bethe-Bloch formula. Except at very low energies where LSS models hold with a dominant contribution of nuclear collisions, most of the energy deposited results in ionization. This leads to applications in radiation detection but also to many detrimental effects in semiconducting devices exposed to harsh ionizing environments. These effects, together with the efforts devoted to overcome them, have been studied for many decades (since the early sixties), due to the advent of the satellite era. Thus, except for some recent applications in high-energy physics, most studies were motivated by space or military purposes. Therefore, whatever particle is responsible for the energy deposition, the results are expressed in terms of Total Ionizing Dose (TID) with units most often derived from the Gy(Si) (1J/Kg) or the rad(Si). A diagram summarizing the effects of radiation on materials is shown below. Both Ionizing radiation, charged particle effects and neutral particle (Non Ionizing) effects are taken into consideration (Fig. 1). The results are mainly lattice defects that have an impact on the properties of the material.

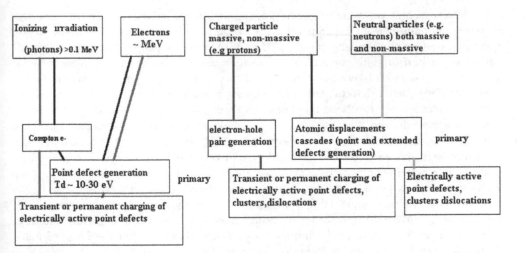

Fig. 1. Diagram summarizing the effects of ionizing radiation, charged particle and neutral particles on matter with no annealing considered.

## 2.2 Creation of defects by ionizing radiation

The point that should be discussed is whether or not ionizing irradiation directly creates point defects in the material. At high energy (~1 MeV) , the impinging particles will create energetic electrons or nuclear recoils that should result in atomic displacements (Fig. 2). This is valid either for uncharged or charged particles (neutron and protons, respectively). Defect creation has been reported even for neutrinos (Brüssler et al., 1989). Depending on the recoil energy single displacements or a displacement cascade will occur leading to the appearance of point or extended defects respectively

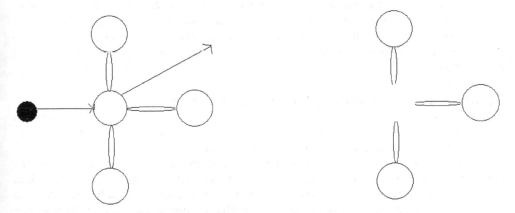

Fig. 2. Sketch of an interaction of a particle (charged or neutral, black on the schematic) with an atom in a plane lattice. A vacancy is created. Depending on the energy of the primary-knock-on the interstitial is ejected close to the vacancy and induces a Frenkel pair or result in a displacement cascade

Point defects such as vacancies or interstitials are not always stable in semiconductors such as germanium and silicon. Experiments with electron irradiation show a different picture when made either at room temperature (300 K) or at very low temperature (Mooney et al., 1983).The comparison with non-ionizing irradiation (neutron) shows a very close behavior (Fourches et al.,1991), because extended defect generation has a weak influence (Fourches,1995). Direct defect creation by ionizing particles such as electrons was widely studied because of fundamental considerations, but purely ionizing effects in semiconductors due to photons and defect creation due to photons has not been investigated so extensively (Sonder and Templeton,1960). For purely ionizing radiation such as gamma rays, this effect is clearly limited even in very sensitive HPGe detectors used for photons detection. In contrast, these materials are prone to degradation under non-ionizing irradiation, even when a very low fluence is considered (a few $10^9$ /cm$^2$) (Fourches et al.,1991a, 1991b). These non-ionizing irradiation effects are often expressed in terms of non-ionizing energy loss (NIEL) in order to reduce the damage effects to a global parameter independent of the nature of the impinging particles. These are still the basis of many studies devoted to device and process development. Recent studies made in silicon for high energy developments or CMOS device improvements have resulted in some new data on the defect introduction rate of $^{60}$Co photon irradiated silicon. According to Pintile et al. (Pintille et al. 2009), the introduction rate for room temperature induced interstitial related defect in silicon is of the order of $10^3$ cm$^{-3}$/rad. This means that to obtain defect concentrations of the order of $10^{10}$ cm$^{-3}$, the TID should reach $10^7$ rad (10 Mrad (Si)) which is a very important TID. This has two consequences. First, this means that the integrated photon flux required to degrade very defect sensitive detectors such as HPGe detectors is of the $10^{16}/10^{17}$ cm$^{-2}$ which is a high value compared to other particle effect introduction rates (Fourches et al. 1991). Direct defect creation by photons in the MeV range should be considered as a negligible effect with respect to defect production by massive particles. Even in silica, widely used in MOS devices, the defects directly induced by irradiation are in negligible concentrations compared to precursors (E' centers, positively charged particles) as long as ordinary oxides are considered. This is the reason oxygen vacancies in all the studies of ionizing radiation effects in MOS oxides focus on defect charging and charge carrier migration and annealing, properties which are mostly observed through electrical measurements.

## 2.3 Defect transformation and annealing: Device implication

Most devices using semiconductors operate at room temperature with the notable exception of cryogenic detectors such as HPGe. The concentration of defects in materials such as silicon, germanium and silicon dioxide depends significantly on the annealing processes of elementary defects created during irradiation. Most of the defects investigated have been point defects as the models used until now were simpler than for many atom defects such as dislocations and vacancy clusters (voids). Moreover, ionizing radiation studies have focused on electrical measurements either of macroscopic physical quantities or on the investigation of thermal relaxation behaviors, leading to the estimation of activation energies. The picture of the irradiated material is such that only a fraction of the defects introduced by irradiation remains in a stable form at room temperature. This has been established in a variety of semiconducting materials both elemental and compound. Most stable defects both in germanium and silicon are divacancy or vacancy related. Some doubt still exists in germanium because of the lack of symmetry sensitive experimental results such as uniaxial stress DLTS (Deep Level Transient Spectroscopy) or ESR (Electron Spin Resonance). Interstitial related defects were detected in later studies (Song et al. 1989) in silicon and interstitials related defects in germanium (Carvalho et al., 2007) were not clearly identified.

## 3. Devices questions and issues

### 3.1 Ionizing irradiation issues

### 3.1.1 Semiconductor detectors (new detector issues)

In the last twenty years, many groups undertook empirical investigations in order to determine the behavior of semiconductor detectors damaged by ionizing and charged particles. Most of these studies focused on detector characteristics directly related to operational performance. The huge leakage current increase observed after irradiation has a great impact on the overall system power management and furthermore the most important characteristic affected is the charge collection efficiency. The charge collection efficiency depends on the concentration of deep traps localized in the fiducial (i.e. the sensitive) volume of the detector (sensor). They act as recombination or trapping centers. In the full depletion mode often used in semiconductor or semi-insulating detectors the defects act as carrier generation centers. The carriers generated in the volume are responsible for the bulk leakage current of the detectors; surface leakage current related to ionization induced defects at the interface dominates in silicon detectors/sensors passivated by silicon dioxide. This is strong evidence for these effects in CMOS (Monolithic Active Pixel Sensors) sensors recently characterized (Fourches et al., 2009) which sensitivity to ionizing (gamma ray) radiation was compared to the sensitivity to neutron irradiation (Fig. 3). Clearly, the CMOS sensors are very sensitive to bulk damage, this being only dependent on the material characteristics and the bias scheme, these devices not being fully depleted. Ionizing irradiation effect (TID) can be strongly reduced by process and layout improvements (Ratti et al.2006, 2010).For other pixel detectors, in order to match future SuperLHC requirements, tolerance to TID and bulk damage can be enhanced by using structures that reduce the charge collection length (i.e; the average length taken by one carrier to be collected by the electrodes), using TSV (Through Silicon Via). Defect engineering has been fruitful in reducing the sensitivity to structural defects of many silicon detectors through impurity density modulation (ROSE Collaboration). Fewer results were published concerning high-purity germanium detectors for which early efforts were significant. Published results seems to show that annealing is not complete, although radiation induced point defects anneal out around 200°C (Mooney et al.,1983; Fourches et al.,1991) , clusters and other extended defects remain in the bulk of the crystal and only disappear at higher temperatures (Kuitunen et al.,2008). Other materials are used for semiconductor detectors, but have not attracted so much interest in terms of radiation tolerance. Diamond is the case in point but recent investigations have also been scarce (Velthuis, 2008). Their potential high radiation hardness is the result of higher atomic displacement energies, than for instance in silicon, and also the result of a higher band gap which leads in fewer electron-hole pairs generated. Applications are sought for LHC upgrades. III-V materials were also investigated (Bourgoin et al. 2001). In present times there is no definitive choice of either material or detector technology for future application in HEP (High Energy Physics), with respect to radiation hardness issues.

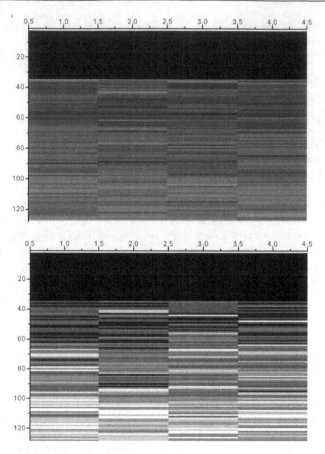

Fig. 3. comparison between ionizing (upper) and neutron (lower) irradiated pixels arrays. Note the uniform effects of ionizing (gamma) irradiation (after Fourches et al., 2008).

### 3.1.2 Microelectronic devices issues

The use of microelectronic devices, circuits, and processes in harsh ionizing irradiation environments is common for many decades now. Efforts were pursued for many years to obtain processes less sensitive to TID, Dose Rate Effects, and Single Event Effects (SEE) such as Single Event Upset, Latch-up, both destructive and non-destructive, and Single Event Gate Rupture (SEGR). A lot of review papers and books exist on that subject but this paragraph will emphasize only on key technological advancements and on the TID effects that are thought to be the dominant problem in accelerator-based experiments. SOS technologies (Silicon On Sapphire) and subsequent SOI (Silicon On Insulator) technologies were first developed to overcome SEE (Single Event Effects) with great success but some extra progress in purely bulk processes made high radiation hardness possible. Down-scaling the elementary device dimensions along with the Moore's law have decreased the sensitivity to TID provided special layout design precautions are taken. Most of the problems related to ionizing irradiation sensitivity are linked to MOS devices although bipolar and JFETs can be affected by TID (Fourches et al., 1998; Citterio et al. 1995; Dentan et al.1993,1996). Former MOS devices required gate special gate oxides, which were engineered in order to limit the total bulk positive charge induced by ionizing irradiation. Careful oxidation of the silicon has allowed a reduction of Pb (Pending Bonds) centers at the $Si/SiO_2$ interface and theretofore to a reduction of the interface charge. This results in a lower Threshold Voltage Shift (Fig. 4). It is also sometimes necessary to use closed shape MOS structures to limit leakage current when bulk processes are used.

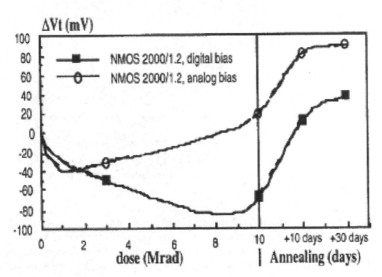

Fig. 4. Threshold voltage shift versus Total Ionizing Dose of a typical DMILL SOI NMOS device with different biasing schemes applied (out of M.Dentan et al.,1996). Digital bias corresponds to a nil source-drain voltage.

The R&D efforts made by many teams around the world have resulted in a very high radiation tolerance for n-channel SOI devices as well as p-channel ones. This has been established during the development of the DMILL technology.

## 3.2 MOS device hardening issues

The radiation hardness of MOS structures increases with the reduction of the oxide thickness, this being first observed in the 1970s and confirmed later (Saks et al., 1984). Simple models derived from classic carrier tunneling from a defect level and first used by Manzini and Modelli (Manzini & Modelli, 1983) in the reliability context show that the sensitivity to TID is negligible up to a few Mrads for oxide thicknesses less that 5 nm, which are common in modern sub-micron MOS low voltage processes.

Using the analytical expression published by the original Manzini and Modelli paper one can compute the threshold voltage shift of a MOS device versus oxide thickness for a given TID received assuming realistic e-h pair generation energies and total hole trapping with no electron trapping in the whole volume of the gate oxide. Fig. 5 summarizes the results. A commercially available computer code was used to obtain numerical results using a trapping depth of the order of 0.18 eV.

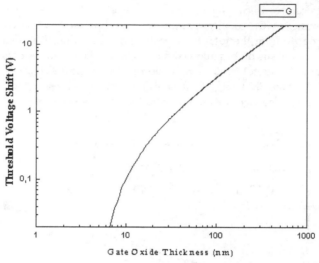

Fig. 5. Absolute threshold voltage shift versus gate Oxide Thickness (in nm) for a 100 krad
Total Ionizing Dose received assuming the conditions introduced in the text.

### 3.3 Total dose and dose rate effects; Electronic devices
### 3.3.1 Ionizing effects at moderate cryogenic temperatures
Many applications require the operation of nano-microelectronic devices below room
temperature. In the nineties, me and some of my colleagues embarked on the possibility of
making a SOI process suitable for 90 K operation. n-channel and p-channel MOSFETs had

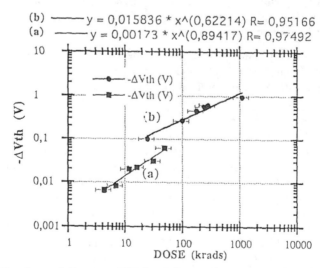

Fig. 6. Threshold voltage shift versus TID for different dose rates (a) low dose rate, (b) high
dose rate. (b) high dose rate (50 krads/hour), (a) low dose rate 0.018 krads/hour both made
at ~ 80 K on n-channel devices.( a + 5 V bias was maintained on the gate).

very good characteristics at these temperatures, making these devices usable for amplifier and detector signal readout applications (Fourches et al., 1997a,1997b). The questions arising were related to the TID radiation hardness at these temperatures. A thorough study was initiated on the effects of ionizing dose, dose rate effects, annealing and geometric effects. Fig. 6 and 7 demonstrate clearly the existence of favourable dose rate effects in the sense that a reduced dose rate results in a lower Vth shift at equivalent total ionizing dose.

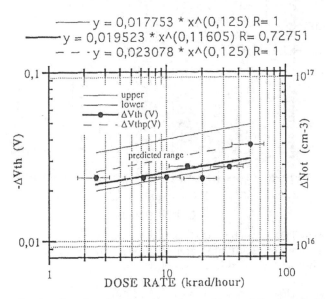

Fig. 7. Double scale plot showing the dose rate dependence of the threshold voltage shift in the range 1 krad/hour to 100 krad/hour. The net positive trapped oxide charge is indicated on the left scale. The shift follows a $D^{\prime\alpha}$ law. The law can be derived from CTRW (Continuous Time Random Walk) analytical approaches.

The TID response also strongly depends on the electric field E in the gate oxide. Fig. 8 clearly exhibits the difference in the slopes of the threshold voltage versus the Total Ionizing Dose (gamma rays) with electric Bias applied to the gate. At 0 V the dissociation of the electron-hole pairs is incomplete and a low charge generation yield results. When a moderately strong bias is applied (5V) the charge dissociation is almost complete and the slope is at its highest value. This effect is strongly reduced when the electric field is increased. Field assisted tunneling of trapped holes towards the silicon can explain the phenomenon in addition of field assisted hopping transport. The activation energy for dispersive transport is lowered when the electric field E is increased. $(E_a(E)=E_a-kE)$. This results in a faster hole hopping frequency, between random sites.

Transport of holes within a temperature range extending from liquid nitrogen to room temperature, shows a temperature and field dependence, the same being true for the conduction of electrons in the oxide, these carriers being much more mobile than holes. The holes are trapped at low temperature, where they exhibit a very low mobility. They can be trapped either at E centers (oxygen vacancies) or self-trapped as polarons, which have a very low mobility particularly below room temperature. At higher temperatures, the mobility is enhanced and they can anneal out rapidly. Experiments, carried out in the 1990s devices on DMILL thick film SOI process showed the thermal anneal of trapped holes in the gate oxide (Fig. 9).

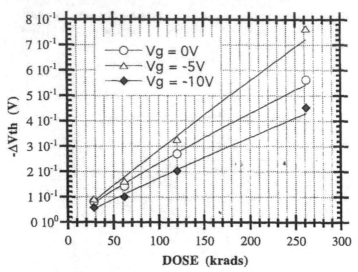

Fig. 8. Vth shift versus TID for p-channel devices with increasing voltage bias on the gate. The measurements were carried out at moderate cryogenic temperature. Note the importance of the shift with oxide thicknesses of 20 nm.

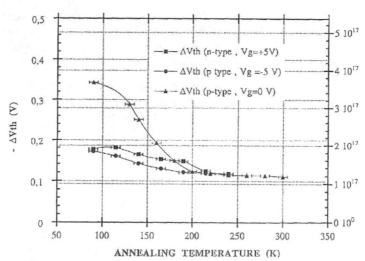

Fig. 9. Isochronal annealing of n-channel and p-channel transistors; (30 minutes for each temperature step); The Vth shift of the n and p channel devices with bias on was identical. The original Vth shift of the p- channel device with 0 V bias on was different due to a 300 krad TID compared to the 80 krad TID received by the former two devices.

The activation energy deduced from the annealing diagram is approximately 0.12 eV, this is close to the detrapping energy of the self-trapped holes and is very small compared with the silica bandgap. As this diagram shows most of the charge (at least 50 %) has annealed up to room temperature.

Early studies of hole transport in thick silica has confirmed the validity of the CTRW (Continuous Time Random Walk) theory (Sheer & Montroll, 1975) for hole-polaron transport in different conditions of field and temperature (McLean, Boesch, Jr. & Garrity in Ma & Dressendorfer,1989; McLean, Boesch, Oldham Ma & Dressendorfer,1989 ) . Analytical models were proposed a long time ago and it was reasonable to extend them to take into consideration the effects of dose rate, under continuous irradiation. Using signal theory and considering the physical processes as being causal a convolution of annealing effects and irradiation effects can be the basis of a simple analytical model. One obtains a dose and dose rate dependence of the threshold voltage shift of the form:

$$\Delta Vth \approx KoxD^{1-\alpha}D'^{\alpha} \tag{3}$$

Kox is a parameter that depends on the oxide characteristics, D the Total Ionizing Dose and D′ the dose rate. See (Fourches, 1997b ) for details.

A power law dependence on the dose and dose rate is obtained, with a value for derived from Fig. 7 of approximately 0.125, this value is lower than the values deduced from older experiments made on thicker oxides (Ma & Dressendorfer and corresponding chapter herein). Additionally it seems clear that the annealing process cannot be simply interpreted by a first order kinetic, as it would be the case for single deep traps. A saturation would occur at a very low TID level (Fourches, 2000), which is not observed in the measurements with an electric field applied. Further investigations show that saturation is clear at a much higher TID, for which threshold voltage shifts exceed 1 V. A numerical treatment was introduced (Fourches, 2000) taking into account the presence of deep hole and electron traps together with CTRW transport, which fits with the saturation observed on some devices exposed to very high TID. The existence of a strong annealing between 80-90 K and room temperature has consequences on the behavior of devices. Cryogenic operation of CMOS SOI electronics had been considered for Liquid Argon Calorimeters, but due to some of the shortcomings of the processes of the 1990's this could not be implemented.

### 3.3.2 Consequences on hardening issues at room temperature

These modeling issues have had some direct consequences on the hardening techniques to be used especially at low temperatures but also for the normal operation of MOS devices. First, it is clear that the Manzini-Modelli-Saks model is a key to these prediction issues. A good fit can be found between measurements results at low temperature and mathematical predictions leaving aside the annealing effects that result in the power law dependence. For instance a 20 nm oxide irradiated at high dose rate, leads to a Vth shift of 0.3 V at 100 krads according to our measurements. This is very close to the value deduced from Fig. 5. This experimental verification of the Manzini-Modelli model could only be made on modern hardened oxides, which exhibit a low active defect density, compared to early oxides, on which Saks et al. made the first experiments. On this basis on can deduce a roadmap for hardening issues to TID. Modern processes either bulk or SOI have thin gate oxides (at the expense of supply voltage range) and this improves radiation hardness. Leakage currents due to interface and surface effects have been somehow reduced over the years by process improvements and layout techniques.

Other data obtained at 90 K show that the sensitivity to TID of the SOI process studied reduces when the dimensions of the device decrease. This is verified by plotting the prefactor Kox of expression (3) versus the gate length of each device. The exponent of the power law found less sensitive to the gate length. Fig. 10 and 11 show the results. The fact that downscaling the device dimensions has very favourable effects on tolerance to TID had a lot of implications for future processes both bulk and SOI. In retrospect it seems clear that today nanoscale or deep submicron technologies are intrisically TID hardened on the basis of some of the scaling laws discussed here and verified experimentally in the 1990's papers (Fourches, 1997) for Silicon On Insulator processes.

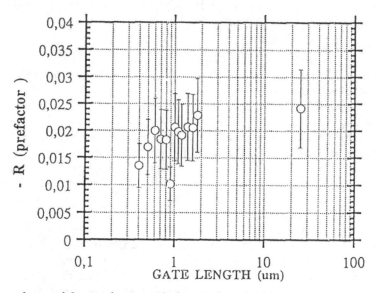

Fig. 10. Dependence of the pre-factor, with the gate length of a n-channel transistor, each transistor having received the same total ionizing dose for the same dose rates. It is clear that low gate length result in a reduced effect of radiation. The measurements were made at 90 K.

The interpretation of this effect at least at low temperature is relatively simple and was already introduced in (Fourches, 1997). Charge tunneling and simply migration is faster when the dimension of the device are lowered, resulting in an accelerated annealing of the positive charge. This should have favorable consequences even for room temperature operation of bulk device, because this simply concerns directly the properties of the gate oxide.

This is one of the reasons the TID response of MOS devices is a today an issue of now lower importance for most applications, except perhaps for the upgrades of the LHC, for Future Linear Colliders and for nuclear power plant applications. Most of the research focuses on Transient Effects, destructive or not, due to charged particles in available bulk processes. The SOI processes are immune to these transient effects by design, because of the device/device insulation systematically implemented.

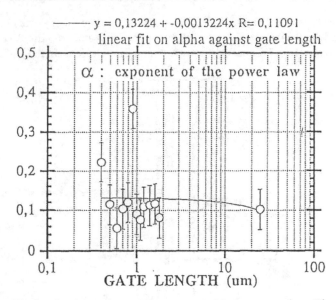

Fig. 11. Similar plot for the dependence of exponent with the power law. There is no clear exponent dependence on the gate length.

### 3.3.3 TID and dose rate effects on other devices

TID effects were reported and sustained efforts were made reduce them on bipolar transistors. DMILL technology was developed with a vertical npn bipolar transistor in order to obtain a radiation hard device, by a reduction of the passivation $Si/SiO_2$ interface. Despite this TID effects still appear with enhanced effects on npn devices due to the higher base current due to minority electron carriers, the oxide being positively charged. In addition to this dose rate effects were reported in 1991 on polysilicon bipolar transistors, and later enhanced low dose rate effects were confirmed (Pease, 2004). Studies on JFETs and MESFETs have also shown the existence of ionizing dose effects, extensive studies were not very developed (Citterio et al., 1995; Fourches, 1998), mainly because the main constraints with large and deep channel devices such as JFETs (Junction Field Effect Transistors) is their limited tolerance to displacement damage.

## 4. Irradiation: Detectors

### 4.1 Total dose and dose rate effects; Silicon detectors

In this paragraph emphasis will be on new semiconductor detectors such as CMOS sensors that are now good candidates for high radiation hardness, high spatial resolution inner pixel detectors for Vertexing purposes.

Sensitivity to Total Ionizing Dose for silicon detectors is clearly dependent on surface effects, the silicon surface being oxidized either naturally or for passivation purposes. Depending on the thickness of the oxide the sensitivity of the detector to leakage currents will be enhanced or limited. Recent studies have focused on CMOS Sensors (Fig.12) and CCDs. Sensitivity to bulk ionizing damage is limited and has not been thoroughly investigated except for charged particles (Hopkinson, 2000). The striking result is the limited Fixed

Pattern Noise increase due to ionizing irradiation, which is much less pronounced than it is when induced by displacement damage. Total Ionizing Dose induced displacements is often limited to the creation of Frenkel pairs and other point defects in the bulk that are less

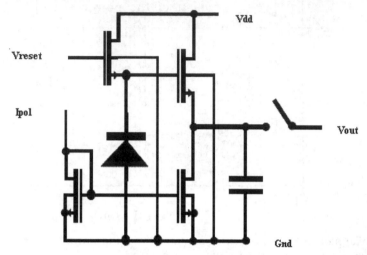

Fig. 12. Schematic structure of the CMOS sensor proposed for charged particle detection.

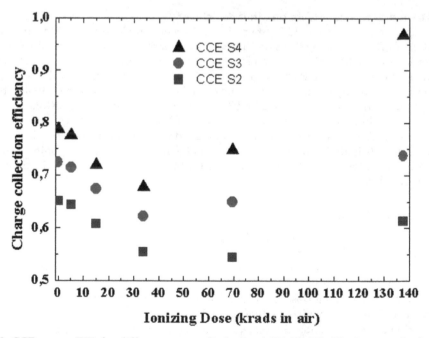

Fig. 13. CCE versus TID for different sensor diode sizes: (S4, S3, S2). The increase in the CCE along with the TID could find its origin in a high surface local electric field, which results in carrier multiplication.

effective in reducing the free drift length of the carriers than displacement cascades. For a detailed simulation study of deep defects effects on CMOS sensors see Fourches, 2009. This induces a reduction of the Charge Collection Efficiency (CCE) which characterizes the effectiveness of the sensor/detector to collect all the generated carriers. Ionizing Dose effects can reduce the charge collection efficiency through surface/interface carrier recombination. Fig. 13 seems to confirm this effect. The CMOS sensors degrade rapidly with TID. In this case the layout of the sensing diode was not optimized so high recombination and leakage current can occur even at small TID.

This Figure (Fig.13) shows that the degradation in charge collection efficiency are very fast with TID for these un-optimized pixel devices. This is due to the high surface leakage current that limits the current signal. CCE increase due to carrier multiplication has been recently reported and was the basis of possible radiation tolerant CMOS particle detectors (E. Villela et al., 2010).

The last figure (Fig.14) clearly shows that the spread of pedestals values is greater in neutron irradiated CMOS sensors than for ionizing irradiated ones. Moreover, the temporal noise increases very slowly with pedestal value, in contrast with the neutron irradiated pixels, for which low pedestal pixels correspond to low temporal noise pixels. These relatively new results obtained on very small pixels (a few tens of cubic microns) indicate that the temporal noise is very high after 137 krads and this originates in the shot noise induced by the interface leakage current. The pedestal value shift is probably due to the threshold voltage shift of the follower n-channel MOS transistors. This threshold voltage shift is much lower in the case of neutron-irradiated pixels as displacement damage has a limited impact on the threshold voltage of MOS transistors.

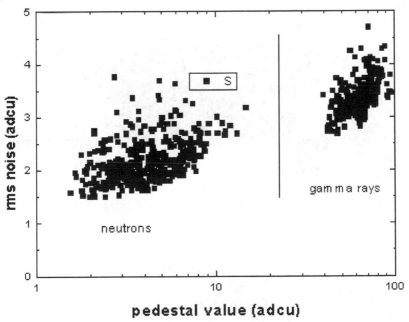

Fig. 14. Root Mean Square value of the temporal noise plotted versus the corresponding Pedestal (offset) values for pixels irradiated with neutrons (left) and pixels irradiated at a dose of 137 krads (Si)(right).

On other processes where radiation tolerance layout techniques were implemented, show a better response to ionizing irradiation. Up to now no long term dose rate effects were reported, nor studied (Fourches, 2008), on these novel CMOS detectors. But on the basis of other measurements the TID tolerance of CMOS sensors is higher than that of Charged Coupled Devices, making CMOS sensors more suitable for future ILC experiments. Radiation hardness up to several Mrads was reported on CMOS sensors using a Deep N-Well structure with charge amplifier readout (L. Ratti et al., 2010).

Other comments should follow concerning CMOS sensors and their counterparts CCDs and DEPFETs. In these devices transport is simply not due to the drift of charged carriers, carrier diffusion is also a very important component and has dramatic consequences on the overall migration length of carriers and subsequently on the signal build-up. Simulation with a software package may help obtaining predictive results but this was limited so far to displacement damage (Fourches, 2009).

## 4.2 Total dose and dose rate effects: Other detectors

Other materials have been proposed and used for particle detection and has been investigated, in terms of both ionizing irradiation-induced defects and displacement induced defects. Up to now, diamond has a good potential for charge particle detection and radiation tolerance, mainly because of the large amount of energy required to create a

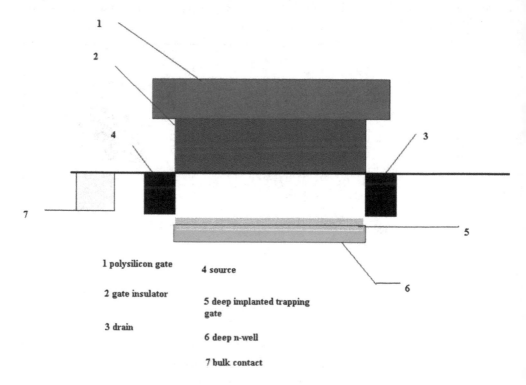

1 polysilicon gate        4 source

2 gate insulator          5 deep implanted trapping
                          gate

3 drain                   6 deep n-well

                          7 bulk contact

Fig. 15. Schematic structure of the deep trapping gate MOS detector structure.

vacancy-interstitial pair, and was proposed for LHC applications. Compound and alloy detectors have been not so successful because the density of residual defects remains high. This limits the CCE from the start, and affects the charge transport trough the fiducial zone. In spite of this, there have been recent studies of thick GaAs pixel detectors for medical applications (Bourgoin, 2001). Future proposals for high radiation hardness, both for ionizing radiation and displacement damage were recently presented (Fourches, 2010). The main idea is to reduce the fiducial value and to introduce a deep trapping gate below the channel of a MOS transistor which would act either as a detector (electron-hole pairs would be created along a charged particle track) or / and a volatile memory (Fig.15). As the trapped positive charge modulates the channel current of the MOS transistor the readout is simplified and can be implemented by a circuitry similar or to the ones used for CMOS sensors. As the deep trapping gates are made of a high density of structural defects, these result additionally in a high tolerance to displacement damage, confirmed by device simulation (Fourches, 2010).

## 5. Conclusion

Future developments about the TID effects are manifold. There is still some interest about the TID hardness at very high doses for high energy physics applications (SuperLHC). New nanoscale technologies, because of their reduced dimensions and the presumable existence of scaling laws will ease the development of new radiation hard ICs. On the detector side, similar advances are expected for reduced dimensions pixels, mainly for displacement damage but also for ionizing effects. Geiger mode Si detectors have been investigated by some researchers and are also promising (Ratti. L et al., 2006). Dose rate effects have been somehow investigated on silicon detectors for medical X ray imaging and much remains to be done to improve these sensors.

## 6. Acknowledgment

The author acknowledges the past contribution of the DMILL development team and the microelectronic group of the DAPNIA, together with the contributions of the LETI, DAM, IN2P3 and CERN. The author is grateful to the librarians for the help in bibliographic search. The author acknowledges the contributions of the CERI (Orléans), CERN, and the other members of the former R&D 29 collaboration. The author thanks R. Walker (EDELWEISS collaboration, Saclay) for proof checking part of the text.

## 7. References

Bethe H., Manchester, W. Heitler, Bristol, On The Stopping Power of Fast Particles and on the Creation of Positive Electrons, 83-112, Communicated by PAM Dirac , February 1934, *Proc. Roy. Soc.*.

Bethe H., Zur Theorie des Durchgangs Schneller Korpuskularstrahlen durch Materie, *Ann. Physik*, Vol. 5, p. 325 (1930) , Vol 397, (1930) p. 325-400

Bourgoin J.C. et al. Potential of thick GaAs epitaxial layers for pixel detectors, *Nuclear Instruments and Methods in Physics Research* A 458 (2001) 344-347,

Brüssler M., Metzner H., Sielemann R., Neutrino-Recoil Experiments in Germanium, *Mat. Sci. For. Trans. Tech. Publ. (Switzerland, )*Vols. 38-41, (1989) pp. 1205-1210

Cahn J., Irradiation Damage in Germanium and Silicon due to Electrons and Gamma Rays, *J. Appl. Phys.*, Vol. 30, N° 8, 1310-1316, August 1959

Carvalho A. et al. Self Interstitials in Germanium, *Phys. Rev. Lett.*, 99, (2007), 175502-1, 175502-4

Citterio M., Rescia S. & Radeka V., "Radiation Effects at Cryogenic Temperatures in Si-JFET, GaAs MESFET and MOSFET Devices", *IEEE Transactions On Nuclear Science*, Vol. 42, N° 6, December 1995 , p 2266-2270

Dentan, M. et al., Study of a CMOS-JFET Bipolar radiation hard analog technology suitable for high energy physics electronics, *IEEE Transactions On Nuclear Science*, Vol. 40, No. 6, 1993, 1555

Dentan M. et al , DMILL a mixed radiation hard BICMOS technology for high energy physics electronics, IEEE Transactions On Nuclear Science, Vol. 43, N° 3, June 1996, pp 1763-1767

Fourches N., et al. Design and Test of Elementary Digital Circuits Based on Monolithic SOI JFET's, *IEEE Transactions On Nuclear Science*, Vol. 45 , N° 1, February 1998, 41-49

Fourches N., Huck A. & Walter G., *IEEE Transactions on Nuclear Science* , Vol 38 , N° 6 , part 2, (1991) 1728

Fourches Nicolas T., "A Novel CMOS detector based on a deep trapping gate", Nuclear Science Symposium Record (Knoxville Tennessee), Oct. 30 2010-Nov. 6,2010, p655-658,http://dx.doi.org/10.1109/NSSMIC.2010.5873840

Fourches N. & Walter G., Bourgoin J.C., Neutron induced defects in high purity germanium, *J. Appl. Phys.* 69 (4) (1991) 2033

Fourches N., High defect density regions in neutron irradiated high purity germanium: characteristics and formation mechanisms , *J. Appl. Phys.* 77 (8) (1995) 3684

Fourches N., et al. Thick film SOI technology: characteristics of devices and performance of circuits for high energy physics at cryogenic temperatures; effects of ionizing radiation", *Nuclear Instruments and Methods in Physics Research*, A401 (1997) 229-237

Fourches N.T. et al. Radiation Induced Effects In A Monolithic Active Pixel Sensor: The Mimosa8 Chip, CEA/DSM/IRFU, CEN Saclay, 91191 GIF/YVETTE, France, (2008) http://arxiv.org/ftp/arxiv/papers/0805/0805.3934.pdf

Fourches N.T., Device Simulation of Monolithic Active Pixel Sensors: Radiation Damage Effects, *IEEE Transactions On Nuclear Science*, Vol. 56, Issue 6, (2009) 3743-3751

Fourches Nicolas T., "Charging in gate oxide under irradiation: A numerical approach" *Journal of Applied Physics* , Vol 88 , Number 9, (2000) 5410

Fourches Nicolas T., Charge buildup in metal oxide semiconductor structures at low temperature: influence of dose and dose rate, *Phys. Rev. B*, Vol 55, No 12, 1997, 7641

Hopkinson, G. R., Radiation effects in a CMOS active pixel sensor, *IEEE Transactions On Nuclear Science* ,Vol. 47, Issue 6, (2000), 2480-2484

Kuitunen K. et al., Divacancy clustering in neutron-irradiated and annealed $n$-type germanium, Phys. Rev. B 78, 033202-1, 033202-4 (2008)

Leroy, C & Rancoica Pier-Giorgio, Particle interaction and displacement damage in silicon devices operated in radiation environment, *Rep. Prog. Phys.* 70 (2007) 493-625

Lindhard J., Scharff M. and Schiott H. u, Mat. Fys. Medd. Dan. Vid. Selsk., 33, No. 14 (1963).

Ma T.P. & Dressendorfer P.V., Ionizing Radiation in MOS devices and circuits, (Wiley Interscience, New York, 1989)

Manzini S. & Modelli A., Tunneling discharge of Trapped Hole in Silicon Dioxide, Insulating Films On Semiconductors, J.F. Verweij and D.R.Wolters (editors), 1983, 112

Messenger George C. & Ash Milton S., "The Effects of Radiation on Electronic Systems", (Van Nostrand Reinhold, New York, 1986)

Mooney et al., Annealing of electron-induced defects in n-type germanium, *Phys.Rev.* B28, 6, (1983) 3272-3377

Oldham T.R, Ionizing Radiation Effects in MOS Oxides, (World Scientific, Singapore, 1999)

Pease R.L. et al., "Characterization of Enhanced Low Dose Rate Sensitivity (ELDRS), Effects Using Gated Lateral PNP Transistor Structure" , *IEEE Transactions On Nuclear Science*, Vol. 51, N°.6, December 2004, pp 3773-3780

Pintillie Iona et al. Radiation-induced point- and cluster-related defects with strong impact on damage properties of silicon detectors, *Nuclear Instruments and Methods in Physics* Research A 611 (2009) 52-68

Ratti L. et al., CMOS MAPS with fully integrated, hybrid-pixel-like analog front-end electronics, Stanford Linear Accelerator Center, Stanford, California, April 3, 2006

Ratti Lodovico et al., Front End Performance and Charge Collection Properties of Heavily Irradiated DNW MAPS, *IEEE Transactions On Nuclear Science*, Vol. 57, N°4, August 2010, p1781-1789

Saks N.S. Ancona M.G. & Modolo J.A et al., Radiation Effects On MOS capacitors with very thin oxides at 80 K, *IEEE Transactions On Nuclear Science*, Vol. NS-31, No. 6, 1984, 1249

Sher J.H. & Montroll E. W., Anomalous transit-time dispersion in amorphous solids, *Phys. Rev. B 12*, 2455-2477 (1975).

Sonder E. & L.C. Templeton L.C., Gamma irradiation of Silicon: I. Levels in n-Type Material containing oxygen, *J. Appl. Phys.*, Vol 31, No 7, 1960, 1279-1286

Song L. W., Zhan X. D., Benson B. W., & Watkins G. D., Bistable Defect in Silicon: The Interstitial-Carbon-Substitutional-Carbon Pair, *Physical, Review Lettters*, Vol 60, N° 5, February 1998, 1

Velthuis J.J. et al., "Radiation Hard diamond pixel detectors", *Nuclear Instruments and Methods in Physics Research A*, 591, (2008) 221-223

Vilella E. et al. "Readout Electronics for Low Dark Count Geiger Mode Avalanche Photodiodes Fabricated in Conventional HV-CMOS Technologies for Future Linear Colliders", Proceedings of TWEPP 2010, Aachen 20-24 September 2010

Willardson R. K. Transport Properties in Silicon & Gallium Arsenide, *J. Appl. Phys.*, Vol 30, N° 6, August 1959, 1158-1165

Wunstorf R. Radiation Hardness of Silicon Detectors: Current Status, *IEEE Transactions On Nuclear Science*, Vol. 44, N°3, June 1997, 806-814

Ziegler J.F., Stopping of Energetic Light Ions In Elemental Matter, *J. Appl. Phys., Appl.Phys. Rev.* Vol.85, N°3, 1February 1999, 1249-1272

# 9

# New Developments in the Field of Radiochemical Ageing of Aromatic Polymers

Emmanuel Richaud, Ludmila Audouin,
Xavier Colin, Bruno Fayolle and Jacques Verdu
*Arts et Metiers ParisTech, CNRS, PIMM,*
*France*

## 1. Introduction

Polymers having an aromatic backbone polymers have a high mechanical strength and a high modulus. Their aromaticity increases their resistance for use in relatively severe conditions especially in aerospace and nuclear industry for which lifetime prediction is a key issue. For example, a challenge for nuclear plants is to extend lifetime from the initially planned 40 years duration to 50 or 60 years, which makes necessary to determine lifetime by a non-empirical method. Since polymers mechanical failure originates from chain scission or crosslinking of the backbone, the ideal method of lifetime prediction would first involve the elaboration of a kinetic model for chain scission and crosslinking. Then, the changes of molecular mass would be related to the changes of mechanical properties using the available laws of polymers physics. Lifetime would be then determined using a pertinent lifetime criterion. A noticeable difficulty comes here from the fact that oxidation, which plays a key major role in chain scission, is diffusion controlled and thus heterogeneously distributed in sample thickness. It is crucial, indeed, to determine experimentally and to predict this depth distribution of chain scission and crosslinking because it will play a key role on fracture properties.

This chapter will be henceforward devoted to the effect of aromaticity on radiostability, the effect on temperature on the chain scission/crosslinking competition, the diffusion limited oxidation (which will be illustrated by the effect of dose rate, atmosphere and sample thickness), then some concluding remarks on oxidative stability of aromatic polymers and the possible link with the absence of macromolecular mobility below $T_g$. We will start by some basics of radiochemistry which are necessary for the good understanding of this paper, and especially the quantitative treatment for crosslinking and chain scission.

## 2. Basics of radiochemical degradation

Let us first consider the reaction:

$$A + h\nu \rightarrow B$$

The radiochemical yield is defined as the number of B molecules that are generated per absorbed joule:

$$G' = \frac{n}{E'} = \frac{N}{N_{Av}} \cdot \frac{1}{E \times (1.6 \times 10^{-17})} = 10^{-7}.G \tag{1}$$

where:
- $G'$ is the radiochemical yield expressed in mol $J^{-1}$,
- $n$ is the number of moles of B which is formed by the radiochemical reaction,
- $E'$ is the amount of absorbed energy in J,
- $E$ is the amount of absorbed energy in 100 eV,
- $N$ is the number of B molecules,
- $N_{Av}$ is the Avogadro's number,
- $G$ is the yield in molecules per 100 eV absorbed.

Since the absorbed dose (denoted by $\delta$ and expressed in Gy) is defined as the amount of received energy (in J) per kilogram of polymer, the following equation can be derived:

$$\frac{n}{m} = 10^{-7}.G.\delta \tag{2}$$

So that:

$$r = \frac{dc}{dt} = 10^{-7}.G.I \tag{3}$$

$r$ being the rate of a radiochemical event (in mol $l^{-1}$ $s^{-1}$), I the dose rate in Gy $s^{-1}$ and c the concentration of reacted or generated species (in mol $kg^{-1}$ even if one often considers that it is the same than in mol $l^{-1}$).

## 2. Mathematical treatment for chain scission and crosslinking

Irradiation can provoke both chain scission and crosslinking, the relative proportion of these phenomena depending on many factors which will be presented in the following. The radical coupling may lead to trifunctionnal or tetrafunctionnal crosslinking noduli called respectively Y- and H-crosslinkings (Fig. 1):

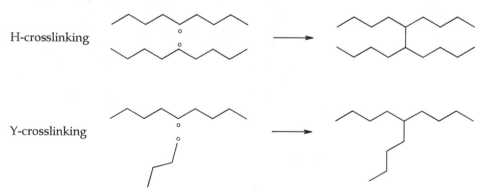

Fig. 1. Reaction of H- and Y- crosslinking.

The radiochemical yields for chain scissions G(s) and crosslinking G(x) in thick samples can be tentatively assessed using

- The Charlesby-Pinner's (Charlesby, 1960) from an analysis of the residual soluble fraction after beginning of sol-gel transition,
- Saito's equations (Saito, 1958) describing the average molar mass changes for soluble polymers before the sol-gel transition.

The corresponding mathematical treatment of these theories differs for Y- and H-crosslinking:

1. In the case of H- crosslinking mode:

$$\frac{1}{\overline{M}_N} - \frac{1}{\overline{M}_{N0}} = s - x = 10^{-7}.[G_H(s) - G_H(x)].\delta \tag{4}$$

$$\frac{1}{\overline{M}_W} - \frac{1}{\overline{M}_{W0}} = \frac{s}{2} - 2x = 10^{-7}.\left[\frac{G_H(s)}{2} - 2G_H(x)\right].\delta \tag{5}$$

$$w_S + w_S^{1/2} = \frac{G_H(s)}{2G_H(x)} + \frac{10^7}{\overline{M}_{W0}.G_H(x).\delta} \tag{6}$$

2. In the case of Y- crosslinking mode:

$$\frac{1}{\overline{M}_N} - \frac{1}{\overline{M}_{N0}} = s - x = 10^{-7}.[G_Y(s) - G_Y(x)].\delta \tag{7}$$

$$\frac{1}{\overline{M}_W} - \frac{1}{\overline{M}_{W0}} = \frac{s}{2} - x = 10^{-7}.\left[\frac{G_Y(s)}{2} - G_Y(x)\right].\delta \tag{8}$$

$$1 + 3w_S^{1/2} = \frac{2G_Y(s)}{G_Y(x)} + \frac{1.93 \times 10^7}{\overline{M}_{N0}.G_Y(x).\delta} \tag{9}$$

where:

- $w_S = 1 - w_I$ is the soluble fraction,
- $G_H(s)$, $G_H(x)$, $G_Y(s)$, and $G_Y(x)$ are the radiochemical yields expressed for 100 eV respectively for chain scissions and crosslinking for an H- and a Y- crosslinking mechanism,
- $\delta$ is the dose (Gy)
- $M_{N0}$ and $M_{W0}$ are respectively the initial number and weight average molar mass (kg mol[-1]).

## 3. Effect of aromaticity on radiostability

The degradation of aromatic polymers was studied by many authors:

1. By monitoring mechanical properties (Sasuga et al., 1985, see Table 1):

These results give a first indication on the relative radiostability of aromatic polymers. Let us first mention that these lethal doses are considerably greater than those compiled by Wilski for polymer having an aliphatic backbone (Wilski, 1987). It can be attributed to the well-known stabilizing effect of aromatic groups which was first observed on methyl methacrylate-styrene copolymers (Alexander & Charlesby, 1954, Kellman et al., 1990,

Thominette et al., 1991). However, the results of these studies cannot be directly used in kinetic models for aromatic polymers.

| Polymer | $\delta_{50\%}$ | $\delta_{20\%}$ |
|---|---|---|
| Kapton 500 | 10 | 90 |
| Upilex | 35 | 60 |
| Ultem | 1.5 | 4 |
| A-Films | 15 | 25 |
| A-Paper | 20 | 30 |
| PEEK non-cryst | 20 | 50 |
| PEEK cryst | 8 | 30 |
| U-polymer | 0.5 | 2 |
| Udel-Polysulphone | 0.75 | 1 |
| PES | 0.5 | 0.75 |
| Noryl (modified PPO) | 0.75 | 1.5 |

Table 1. Dose for reducing strain at break to 50% and 20% of initial value for various aromatic polymers submitted to $5.10^3$ Gy s$^{-1}$ in air (sample thickness ~ 0.1-0.2 mm) (Sasuga et al., 1985).

2. By radical appearance measured by Electron Spin Resonance spectroscopy (Heiland et al., 1996, see Fig. 2):

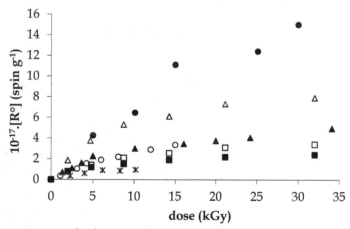

Fig. 2. Concentration in radicals versus dose measured in PPO (●), Kevlar (△), PSU (▲), PEPO (○), Ultem (□), Kapton (■), PEEK (✳) in aromatic polymers under 1 - 5 kGy h$^{-1}$ dose rate at 77 K (Heiland et al., 1996).

Data presented in Fig. 2 permit the yield in radical build-up to be calculated according to the simplified scheme for radiochemical degradation in the absence of oxygen:

| | | |
|---|---|---|
| Initiation: | Polymer (PH) + hv $\rightarrow$ P° | $r_i$ |
| Propagation: | P° + PH $\rightarrow$ PH + P° | $k_P$ |
| $\beta$-scission: | P° $\rightarrow$ P° + s | $k_R$ |
| Termination by coupling: | P° + P° $\rightarrow$ P–P + x | $k_C$ |
| Termination by disproportionnation: | P° + P° $\rightarrow$ F + PH | $k_D$ |

Where F is a double bond.
The rate of radical formation is:

$$\frac{d[P°]}{dt} = r_i - 2k_t[P°]^2 \tag{10}$$

with:

$$k_t = k_C + k_D \tag{11}$$

Fig. 2 indicates the existence of an asymptote characterizing a steady-state at which the concentration in radicals is:

$$[P°]_\infty = \sqrt{\frac{r_i}{2k_t}} \tag{12}$$

Eq. 10 gives:

$$\frac{d[P°]}{\left(1 + \dfrac{[P°]}{[P°]_\infty}\right) \cdot \left(1 - \dfrac{[P°]}{[P°]_\infty}\right)} = \frac{r_i}{[P°]_\infty} \cdot dt \tag{13}$$

Which is integrated in:

$$\ln\left(\frac{1 + \dfrac{[P°]}{[P°]_\infty}}{1 - \dfrac{[P°]}{[P°]_\infty}}\right) = \frac{2r_i}{[P°]_\infty} \cdot t \tag{14}$$

At t ~ 0, [P°] ~ 0 so that:

$$\left(1 + \frac{[P°]}{[P°]_\infty}\right)^2 = 1 + \frac{2r_i}{[P°]_\infty} \cdot t \tag{15}$$

thus:

$$G(P°) = G_i = \left(\frac{d[P°]}{dt}\right)_{t\to0} \tag{16}$$

The corresponding yields for radical build-up are given in Table 2.
Let us first mention that Table 1 results correspond to degradation under air, meanwhile Table 2 results correspond to degradation in inert environment, the possible influence of oxygen being discussed in the following. However, it can be checked that the lethal dose (or dose to reach an arbitrarily chosen threshold for a mechanical property) varies oppositely with the estimated G(P°) values.

| Polymer | G(P°) |
|---------|-------|
| PPO | 1.35 |
| Kevlar® | 1.25 |
| PSU | 0.75 |
| Bis-A-PEPO | 0.55 |
| Ultem | 0.50 |
| Kapton® | 0.40 |
| PEEK | 0.25 |

Table 2. Yields for radical build-up for aromatic polymers irradiated under 1 - 5 kGy h$^{-1}$ dose rate at 77 K.

- By gas yield emission:

Concerning the yield for total gas emission, it can be checked that there is no great difference between gamma, proton and electron-beam irradiation (Hegazy et al., 1992b, Hill & Hopewell, 1996) and that these values are considerably lower than those measured for aliphatic polymers such as PE or PP (Schnabel, 1978). An example is worth to be mentioned (Schnabel, 1978): aliphatic polysulfones have a gas emission yield equal to 39 (poly butene-1-sulfone) or 71 (poly hexene-1-sulfone) meanwhile PES or PSU gas emission yields are lower than 1, illustrating here the protective effect by aromatic rings. However, the relative proportion of each emitted gas can vary due to some differences in temperature rising under irradiation. As it will be seen below, this difference due to irradiation nature has a lower influence on the radio induced degradation of polymer than temperature, nature of environment atmosphere and oxygen gradient in sample thickness.

| I (kGy h$^{-1}$) | Gaz | Kapton | Upilex-R | Upilex-S | PEEK-c | PEEK-a | PES | U-PS | U-Polymer |
|---|---|---|---|---|---|---|---|---|---|
| | Total | 24.0 | 22.0 | 91.0 | 39.0 | 54.0 | 46.0 | 150.0 | 480.0 |
| | H$_2$ | 3.2 | 0.4 | 7.5 | - | 14.0 | 7.1 | 39.0 | 72.0 |
| | N$_2$ | 5.1 | 9.7 | 14.0 | 6.4 | - | - | - | - |
| 10 | CO | 5.4 | 2.4 | 1.4 | 12.0 | 6.0 | 16.0 | 19.0 | 220.0 |
| | CO$_2$ | 8.1 | 4.8 | 15.0 | 4.3 | 24.0 | 19.0 | 25.0 | 180.0 |
| | CH$_4$ | 1.0 | 0.1 | 0.3 | 0.2 | 0.3 | 0.3 | 16.0 | 12.0 |
| | SO$_2$ | - | - | - | - | - | 12.0 | 13.0 | - |
| | Total | 25.0 | 17.0 | 18.0 | 31.0 | 39.0 | 69.0 | 210.0 | 460.0 |
| | H$_2$ | 4.9 | 1.3 | 2.4 | 10.0 | 12.0 | 19.0 | 62.0 | 80.0 |
| | N$_2$ | 0.2 | 0.1 | 2.8 | - | - | - | - | - |
| 6120 | CO | 3.4 | 2.2 | 2.4 | 5.1 | 5.5 | 8.3 | 32.0 | 220.0 |
| | CO$_2$ | 10.0 | 3.9 | 8.8 | 9.5 | 16.0 | 16.0 | 16.0 | 120.0 |
| | CH$_4$ | 0.8 | - | 0.3 | 0.2 | 0.2 | 0.4 | 13.0 | 32.0 |
| | SO$_2$ | - | - | - | - | - | 23.0 | 43.0 | - |

Table 3. 103.G(gas) for irradiation at 30°C 100 µm films vacuum (Hegazy et al., 1992a)

3.    Let us turn to the radiation-induced crosslinking observed in PEEK and PSU.
Fig. 3 depicts the changes in thermal behavior which are induced by radio-ageing for a thick
PEEK sheath under air (Richaud et al., 2010a). As it will be seen later, the observed behavior
is comparable to the one obtained for an irradiation under vacuum. Virgin and degraded
samples were characterized by DSC heating-cooling-heating cycle.
-    In the case of virgin sample, melting endotherms for the first and the second heating
     ramps are very comparable (peak temperature and enthalpy). The crystallization from
     molten state is characterized by a sharp exotherm at c.a. 300°C.
-    In the case of 30.7 MGy irradiated sample, melting endotherm for the second heating
     ramp and crystallization exotherm are shifted to the lower temperatures with a lower
     transition enthalpy.

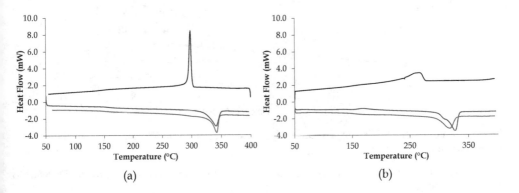

(a)                                                        (b)

Fig. 3. Heating-cooling-heating cycle by DSC for non-irradiated (a) and 30.7 MGy irradiated
PEEK sample.

During irradiation of semi-crystalline polymers, chain scission and crosslinking occur only
in amorphous phase, the crystalline phase undergoes only small changes so that melting
characteristics remain almost unchanged at the first heating scan. After a first melting
however, the sample is homogenized, and the changes of molar mass, branching ratio or
crosslink density resulting from irradiation affect melting and crystallization. Irradiation
effects become then observable by DSC. Here, the decrease of crystallization temperature
and enthalpy is explained by the occurrence of radiation-induced crosslinking (Sasuga,
1991). This latter disfavors crystallization because it lowers the transport rate of chain
segments from the melt to growing crystals. As a consequence, the degree of crystallinity
and the lamellae thickness are decreased that explains the observed decrease of melting
point and melting enthalpy.
Comparable data were obtained by Hegazy (Hegazy et al., 1992c). $T_g$ values from this study
are reported (Fig. 4):
It is well known that for a linear polymer, $T_g$ is an increasing function of molar mass.
According to Fox and Flory (Fox & Flory, 1950):

$$T_g = T_{g0} - \frac{k_{FF}}{M_N}$$   (17)

Where:
- $T_{g0}$ is the $T_g$ of a virtual infinite polymer (K).
- $M_N$ is the number average molar mass (kg mol⁻¹).
- $k_{FF}$ is a constant characteristic of the chain chemical structure (K kg mol⁻¹).

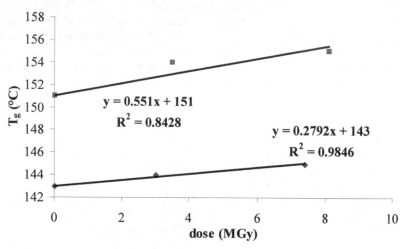

Fig. 4. $T_g$ changes for PEEK amorphous (♦) and semi-crystalline (■) samples irradiated under vacuum.

For a network, $T_g$ is an increasing function of crosslink density. According to Di Marzio (DiMarzio, 1964):

$$T_g = \frac{T_{gl}}{1 - k_{DM}.F.v_A} \qquad (18)$$

Where:
- $T_{gl}$ is the $T_g$ of a virtual linear polymer containing all the structural units of polymer except crosslinks (K).
- F is a flex parameter linked to chain stiffness (kg mol⁻¹).
- $v_A$ is the concentration in elastically active chains, which is calculated from crosslinks concentration (mol kg⁻¹).

Under vacuum, chain scission is generally negligible. Above gel point, it seems to us that $T_g$ changes are given by combining Fox-Flory relationship and Saito's equation (assuming first that s << x):

$$\frac{dT_g}{d\delta} = \frac{dT_g}{dx} \cdot \frac{dx}{d\delta} = 10^{-7}.G(x).k_{FF} \qquad (19)$$

One obtains:
- $G(x) \sim 0.012$ for amorphous PEEK,
- $G(x) \sim 0.023$ for semi-crystalline one.

It seemed to us interesting to compare radio-induced gelation for PEEK, PSU and PES (Fig. 5).

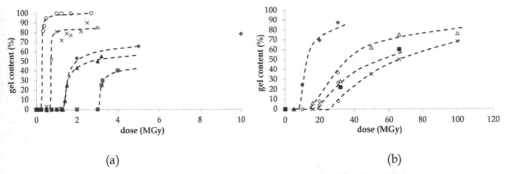

(a)                  (b)

Fig. 5. Gel formation for PSU at 210°C (○), 180°C (×), 100°C (▲), 60°C (◆), room temperature (■) from Murakami & Kudo, 2007, Richaud et al., 2011), and PEEK (■: 24 kGy h⁻¹, ◆,△,×: 60 MGy h⁻¹ for 3 different initial crystalline microstructures) samples irradiated at several temperatures (Vaughan & Stevens, 1995 , Richaud, 2010a).

From the above given mechanistic scheme where chain scission results from radical rearrangement and crosslinking results from radical coupling scheme, the following stationary rate expressions can be derived:
- Rate of chain scission:

$$\frac{ds}{dt} = \frac{k_R}{2} \cdot \left( \frac{r_i}{k_C + k_D} \right)^{1/2} \tag{20}$$

- Rate of crosslinking:

$$\frac{dx}{dt} = \frac{r_i}{2} \frac{k_C}{k_C + k_D} \tag{21}$$

Assuming that disproporationation is negligible compared to coupling for polymers under study, it comes:

$$\frac{dx}{dt} = \frac{r_i}{2} \tag{22}$$

So that : $G_i = 2G(x)$ in the absence of oxygen.

It seems that this relationship disagrees with experimental results in the case of aromatic polymers which will be discussed next.

As a conclusion of this short review of experimental results, it is clear that aromatic polymers belong the family of relatively radiostable polymers because primary radiochemical events, presumably radiolytic chain scissions of C-H bonds, have a radiochemical yield less than 0.5, i.e. 5 to 50 times lower than for aliphatic polymers. Aromatic groups are actually able to dissipate a great part of the absorbed energy into reversible processes (fluorescence, phosphorescence...). Data militate in favour of the following ranking:

A comparison of the rate of gel content increase (Fig. 4) for samples irradiated at various temperatures and dose rates shows that gelation is more sudden in the case of PSU and more progressive in PEEK which means that radiation generated radicals would be more reactive for PSU than PEEK.

Data also show unambiguously:

- The protective effect of aromaticity: Kapton®, which is the most aromatic polymer, presents a very limited degradation, or even no significant degradation (Kang et al., 2008, Richaud et al, 2010b).
- The role of $SO_2$ moiety: experimental results for radiochemical yields values (Horie & Schnabel, 1984) shows that aliphatic polysulfones are among the less stable polymers with G(s) values on the order of 10. High yields of $SO_2$ emission are also observed indicating thus the probable existence of the weak (Li & Huang, 1999, Molnár et al., 2005) carbon-sulfone bond cleavage. It seems reasonable to suppose that such events occur also in PSU even though the yield is considerably lower owing to the well-known protective effect of aromatic nuclei illustrated for instance in studies on isobutylene-styrene.
- The role of isopopylidene group: gelation is undoubtedly due to the coupling of alkyl radicals. The difference in gelation rate would be explained by the nature of radicals: in PEEK, only aryl radicals can react meanwhile both aryl and primary methyl radical can react in PSU.

Theoretically, it seems that yields in gas emission G(gas) is half of yield radicals G(P°) as illustrated by proposal of degradation mechanisms below:

1.  In isopropylidene containing materials:

Then radicals react by coupling to give a gaseous molecule:

$$H° + H° \rightarrow H_2$$
$$CH_3° + H° \rightarrow CH_4$$
$$CH_3° + CH_3° \rightarrow C_2H_6$$

2.  In sulfone containing polymer:

Analogous mechanisms could lead to carbon monoxide formation from PEEK. Data for Kapton®, PEEK and PSU (Table 1 and Table 2) seem also to show that:

$$G(P°) >> 2G(x) \text{ and } 2G(gas)$$

PEEK and PSU values gathered in this work are compared with those for polyethylene (Khelidj 2006), and those compiled for cyclohexane and benzene (Ferry, 2008) in Table 4:

| thickness | T (K) | dose rate | atmosphere | $\delta_{50\%}$ (MGy) | $\delta_{gel}$ (MGy) | G (gas) |
|-----------|-------|-----------|------------|-------------------|------------------|---------|
|           | 298   |           |            | > 0.3             |                  |         |
| 100 μm    | 423   | 0.1 kGy s$^{-1}$ | He   | > 0.15            |                  |         |
|           | 463   |           |            | 0.1               |                  |         |
|           | 503   |           |            | 0.1               |                  |         |
|           | 298   |           |            |                   | 3                | 0.025   |
| 50 μm     | 373   | 1.7 Gy s$^{-1}$ | vacuum |                 | 1.25             | 0.07    |
|           | 453   |           |            |                   | 0.5              | 0.17    |
|           | 483   |           |            |                   | < 0.25           | 2       |

Table 4. Yields for radicals, gas emission, crosslinking and double bond formation (corresponding to disproportionnation process). *: G(H$_2$) is expected to be close to G(gas), **: G(F) is expected to be negligible in aromatic polymers.

These results call for the following comments:
1.  The effect of aromaticity on radiostability is confirmed.
2.  The ratio $G(P°)/G(gas)$ is surprisingly lower than 1 for cyclohexane, is fairly close to 2 for benzene and PE, but increases for PSU and PEEK.

A possible explanation is the existence of very stable radicals that would be detected by ESR but would contribute neither to gas emission nor to crosslinking (see later) as for example:

Let us also note that the absence of contribution to gas emission seems easily explainable because the probability for generating a short and volatile segments from such macroradicals is very low.
3.  The ratio $G(P°)/(G(x)+G(F))$ is also clearly greater than 2 for PSU and particularly for PEEK meanwhile it is close to 1 in PE and in cyclohexane. The comparison with benzene suggests that this behaviour is due to the macromolecular structure of radicals preventing them to react by coupling to give crosslinking.

In fact, even if a negative concavity is observed in Fig. 1 and indicates the existence of a termination process even at 77 K, the relatively low value of $G(x)$ indicates that segmental mobility would be reduced below $T_g$, which will be discussed in a first section dedicated to temperature effect and another one dealing with oxidizability.

## 4. Effect of temperature

Data in Table 5 confirm that lethal dose (here. the necessary dose to half the initial elongation at break) and gel dose vary oppositely with temperature meanwhile gas emission yield would increase.

| thickness | T (K) | dose rate | atmosphere | $\delta_{50}\%$ (MGy) | $\delta_{gel}$ (MGy) | G(gas) |
|-----------|-------|-----------|------------|----------------------|----------------------|--------|
| 100 μm | 298 |  |  | > 0.3 | - | - |
| 100 μm | 423 | 0.1 kGy s$^{-1}$ | He | > 0.15 | - | - |
| 100 μm | 463 |  |  | 0.1 | - | - |
| 100 μm | 503 |  |  | 0.1 | - | - |
| 50 μm | 298 |  |  | - | 3 | 0.025 |
| 50 μm | 373 | 1.7 Gy s$^{-1}$ | vacuum | - | 1.25 | 0.07 |
| 50 μm | 453 |  |  | - | 0.5 | 0.17 |
| 50 μm | 483 |  |  | - | < 0.25 | 2 |

Table 5. Effect of temperature on lethal dose for PES (Sasuga & Hagiwara, 1987 and Murakami & Kudo, 2007).

As previously suggested, the only knowledge of changes in macroscopic (engineering) properties does not permit to ascribe the right cause of failure. Yields in chain scission and crosslinking permit to elucidate the degradation mechanism.

Let us first comment the $T_g$ changes in PES (Li et al., 2006) presented in Fig. 6:

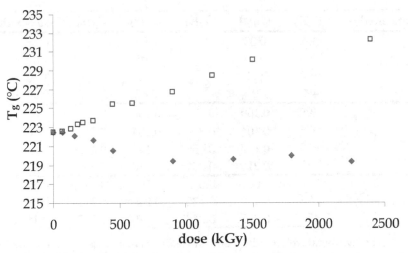

Fig. 6. Change in $T_g$ for PES submitted to electron beam irradiation at room temperature (◆) or 230°C (□) (Li et al., 2006).

Before sol-gel transition, $T_g$ changes are given by combining Fox-Flory law (Fox & Flory, 1950) with Saito's equation (Saito, 1958) irrespectively of H- or Y-crosslinking mode:

$$T_g - T_{g0} = -k_{FF}.\left(\frac{1}{M_N} - \frac{1}{M_{N0}}\right) = -10^{-7}.k_{FF}.[G(s) - G(x)].\delta \qquad (23)$$

It turns into:
- for a pure crosslinking mechanism:

$$\frac{dT_g}{d\delta} = 10^{-7}.k_{FF}.G(x) \qquad (24)$$

- for a pure chain scission mechanism:

$$\frac{dT_g}{d\delta} = -10^{-7}.k_{FF}.G(s) \qquad (25)$$

Results presented in Fig. 5 indicate that crosslinking predominates above $T_g$, whereas chain scission predominates below $T_g$. Comparable exploitation can be done from other published results (Murakami & Kudo, 2007, Brown & O'Donnell, 1979):

All data converge towards the fact that $G(x)$ increases with temperature. Despite some scattering, it also suggests that $G(s)/G(x)$ decreases with temperature. According to Brown and O'Donnell (Brown & O'Donnell, 1979), it falls to 0 above $T_g$.

The results in Table 6 can be explained as follows: irradiation creates macroradicals. At low temperature, they react by an unimolecular process which generates a chain scission (possibly accompanied by gas emission). At high temperature, macromolecular mobility is sufficient to permit macroradicals to react by coupling to give a crosslinking. According to Zhen (Zhen, 2001) most polymers can only crosslink above their melting point.

| dose rate | method | T (K) | $G_H(s)$ | $G_Y(s)$ | $G_H(x)$ | $G_Y(x)$ | G(s)/G(x) | thickness |
|---|---|---|---|---|---|---|---|---|
| 8 Gy s$^{-1}$ | Sol gel | 308 | 0.03 | | 0.04 | | 0.75 | 3 mm |
| | | 353 | 0.05 | | 0.05 | | 1.00 | 3 mm |
| | | 398 | 0.3 | | 0.2 | | 1.50 | 3 mm |
| | | 493 | 0 | | 0.67 | | 0.00 | 3 mm |
| 1.7 Gy s$^{-1}$ | GPC | 298 | 0.100 | 0.373 | 0.100 | 0.301 | 1.00 | 50 μm |
| | | 373 | 0.104 | 0.318 | 0.104 | 0.313 | 1.00 | 50 μm |
| | | 423 | 0.147 | 0.455 | 0.147 | 0.442 | 1.00 | 50 μm |
| | | 453 | 0.216 | 0.601 | 0.216 | 0.647 | 1.00 | 50 μm |
| 1.7 Gy s$^{-1}$ | Sol gel | 298 | 0.067 | 0.065 | 0.104 | 0.147 | 0.64 | 50 μm |
| | | 373 | 0.170 | 0.138 | 0.245 | 0.312 | 0.69 | 50 μm |
| | | 453 | 0.075 | 0.268 | 0.170 | 0.575 | 0.44 | 50 μm |

Table 6. Effect of temperature on radiochemical yields in chain scissions and crosslinking (envisaging the possibility of H- or Y- crosslinking mode) (Murakami & Kudo, 2007, Brown & O'Donnell, 1979).

Let us now discuss of the possibility of Y- and H-crosslinking.
-    At 150°C, Hill (Hill et al., 1998) unambiguously showed that PSU crosslinks in Y-mode. A comparable analysis was performed by Li (Li et al., 2006). A proposal of mechanism is shown below:

-    Kudo (Murakami & Kudo, 2007) proposed a Y-mechanism also. However, by reexaminating his results using the above equations, his conclusion is questioned.
-    Richaud (Richaud et al., 2011) proposed a H- crosslinking mechanism at 60°C, basing on the observation that if isopropopylidene groups are generated, the →C–CH$_2$° group is considerably more reactive than aryl one and should react by coupling to give →C–CH$_2$–CH$_2$–C← crosslinking bridges corresponding more to a H- crosslinking mechanism.

As a conclusion, it seems that elevating the temperature promotes the crosslinking. However, the nature of this latter and presumably the role of temperature remain unclear.

## 5. Effect of atmosphere and sample thickness

Results in Fig. 7 (Sasuga & Hagiwara, 1987) clearly show that oxygen accelerates the degradation of aromatic polymers.

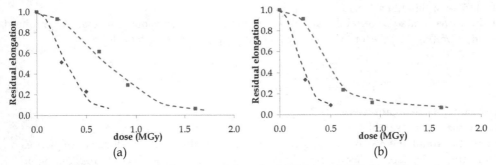

(a)                                        (b)

Fig. 7. Residual elongation as a function of dose for PSU (a) and PES (b) submitted to g-rays under air (■) or 0.7 MPa oxygen (◆).

The nature of the responsible process is suggested in a study on the degradation of PET films monitored by $T_g$ measurements (Burillo et al., 2007): a strong $T_g$ decrease is observed in air at the beginning and is certainly due to chain scission whereas a moderate $T_g$ decrease is observed for irradiation under vacuum. As proposed by Sasuga (Sasuga, 1988), polymer radio-degraded under anaerobic conditions would undergo mainly crosslinking meanwhile they would undergo chain scissions when they are degraded under aerobic (oxidative) conditions.

The effect of atmosphere nature and the effect of thickness have the same origin linked to oxygen diffusion: thick samples present a diffusion limited oxidation (DLO), i.e. that their surface undergoes an oxidative degradation leading to chain scission meanwhile bulk undergoes an anaerobic radio-ageing generating radicals reacting only by coupling.

This effect was quantified by comparing the degradation of thin and thick samples by means of GPC (for molar mass assessment and subsequently chain scission and crosslinking yields assessment) and $T_g$ (Richaud et al., 2011). $T_g$ decreases for both 2 mm and 200 μm thick samples, but more significantly for thin films (Fig. 8) which is not surprising since $T_g$ decreases with chain scissions and increases with crosslinking.

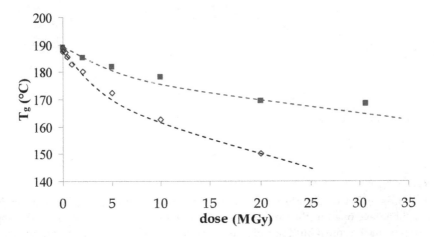

Fig. 8. $T_g$ changes with dose for thin films (◆) or thick samples (■).

1.   Case of thin films

Thin films undergo certainly a "pure" random chain scissions process. A graphical estimation (Fig. 5) gave for the thin samples (using $k_{FF} \sim 220$ K kg mol$^{-1}$ for PSU):

$$\frac{dT_g}{d\delta} \sim \frac{T_g(0 \text{ MGy}) - T_g(5.0 \text{ MGy})}{5.0 \times 10^6} = 3.2 \times 10^{-6} \text{ K Gy}^{-1}$$

from which one obtains $G(s) \sim 0.15$

2.   Case of thick samples

Here, the simultaneous crosslinking compensates partially chain scissions consequences on $T_g$ changes. For ideal networks having no dangling chains, the effect of crosslinking on $T_g$ obeys the DiMarzio's law (DiMarzio,1964 – equation (18)) with:

-   $T_{gl} \sim 473$ K
-   $F_{PSU} = 0.1105$ kg mol$^{-1}$
-   $k_{DM}$ is the DiMarzio's constant: $k_{DM}$ is close to 1
-   $\nu$ is the elastically active chains concentration. For ideal tetrafunctional networks:

$$\nu = 2.x \tag{26}$$

x being the concentration of crosslinks events. Supposing that DiMarzio's equation can be applied for non-ideal networks, Eq. 23 turns into the Fox-Loshaek (Fox & Loshaek, 1955) equation for low conversion of the crosslinking process:

$$T_g = T_{gl} + k_{FL}.x \tag{27}$$

with the Fox-Loshaek constant $k_{FL}$ being equal to:

$$k_{FL} = 2.T_{gl}.F.k_{DM} \tag{28}$$

so that:                                     $k_{FL} \sim 104$ K mol kg$^{-1}$

Finally, the global changes of $T_g$ would be given by:

$$\frac{dT_g}{d\delta} = \left(\frac{\partial T_g}{\partial s}\right).\frac{ds}{d\delta} + \left(\frac{\partial T_g}{\partial x}\right).\frac{dx}{d\delta} \tag{29}$$

so that:

$$\frac{dT_g}{d\delta} = 10^{-7}\left(k_{FL}.G(x) - k_{FF}.G(s)\right)$$

Using $G(s) \sim 0.15$, one can simulate the experimental $T_g$ decrease:

$$\frac{dT_g}{d\delta} = 15.6 \times 10^{-7} \text{ K Gy}^{-1}$$

with $G(x) \sim 0.15$.

In other words, effect of sample thickness is explained by the effect of oxygen diffusion (which will be developed in the next section): irradiation generates radicals. In the absence of oxygen (i.e. for the core a bulk sample), these later react together by coupling which gives a crosslinking phenomenon. When oxygen is present (which is the case for thin samples or the superficial layer of a bulk sample), it orientates the degradation towards an oxidative mechanism with the formation of peroxy radicals POO°, and POOH of which decomposition induces chain scissions.

## 6. Effect of dose rate

This effect is illustrated by some results by Tavlet (Tavlet, 1997) presented in Fig. 9:

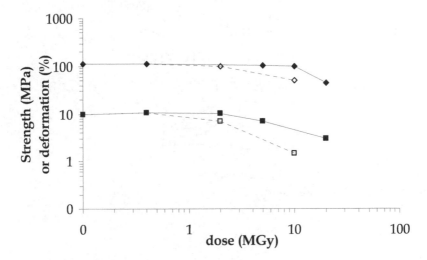

Fig. 9. Radiation degradation in an epoxy resin on strain (■,□), stress (◆,◇). Open symbols correspond to long term (low dose rate) ageing (Tavlet, 1997).

There are two effects to be distinguished:
- An increase in temperature involved by enhancing the dose rate,
- A physical and chemical effect linked to Diffusion Limited Oxidation, which will be described below.

From a simplified theory of diffusion controlled oxidation (Audouin et al, 1994), a rough estimation of oxygen consumption rate $r_{OX}$ in films or superficial layers of thick samples can be done:

$$r_{OX} = \frac{D_{O2}.[O_2]_S}{TOL^2}$$ (30)

where:
- TOL is the thickness of oxidized layer (m),
- $D_{O2}$ is the oxygen diffusion coefficient in polymer amorphous phase (m$^2$ s$^{-1}$),
- $[O_2]_S$ is the equilibrium concentration under atmospheric pressure given by Henry's law:

$$[O_2] = s_{O2} \times P_{O2}$$ (31)

$s_{O2}$ being the solubility coefficient of oxygen in polymer amorphous phase.

$r_{OX}$ can be expressed from a simplified mechanistic scheme for radio-thermal oxidation:

$$PH + h\nu \rightarrow P^\circ + \tfrac{1}{2}\, H_2 \qquad\qquad\qquad\qquad\qquad r_i$$
$$P^\circ + O_2 \rightarrow P^\circ \qquad\qquad\qquad\qquad\qquad\qquad\quad k_2$$
$$POO^\circ + PH \rightarrow POOH + P^\circ \qquad\qquad\qquad\quad k_3$$
$$POO^\circ + POO^\circ \rightarrow \text{inact. prod.} + O_2 \qquad\qquad k_6$$

$k_i$ being the rate constants (expressed in l mol$^{-1}$ s$^{-1}$).

The oxidation rate for oxygen consumption can be expressed as:

$$r_{OX} = -\frac{d[O_2]}{dt} = -\,k_2.[P^\circ][O_2] + k_6.[POO^\circ]^2 \tag{32}$$

Using the classical steady-state assumption:

$$k_2.[P^\circ][O_2] = r_i + k_3.[POO^\circ].[PH] \tag{33}$$

$$r_i = 2k_6.[POO^\circ]^2 \tag{34}$$

it comes:

$$r_{OX} = \frac{r_i}{2} + k_3 \cdot \sqrt{\frac{r_i}{2k_6}} \cdot [PH] \tag{35}$$

knowing that: $\qquad\qquad\qquad\qquad r_{OX} = 10^{-7}.G(P^\circ).I$

I denoting the dose rate. Consequently, if:

-    I is very high

$$\Rightarrow \quad r_{OX} \approx \frac{r_i}{2} \tag{36}$$

$$\Rightarrow \quad r_{OX} \propto I \tag{37}$$

$$\Rightarrow \quad TOL \propto I^{-1/2}$$

-    I is moderate

$$\Rightarrow \quad r_{OX} \approx k_3 \cdot \sqrt{\frac{r_i}{2k_6}} \cdot [PH] \tag{38}$$

$$\Rightarrow \quad r_{OX} \propto I^{1/2} \tag{39}$$

$$\Rightarrow \quad TOL \propto I^{-1/4} \tag{40}$$

It is thus demonstrated that radiothermal oxidation is characterized by a skin-core structure: superficial layers are oxidized meanwhile bulk not. Results by Tavlet (Tavlet, 1997) are thus explained as follows: for irradiation performed under a relatively elevated dose rate, only a very thin polymer layer is degraded and time for drop of elongation at break is longer than for irradiation performed under a moderate dose rate. This reasoning is confirmed by the comparison of a bulk sample (4 mm) for which no dose rate effect is observed undoubtedly because the thickness of degraded superficial layer is negligible compared to the total thickness of bulk material. Examples of Oxidized layers measured for radiochemical degradation of aromatic polymers (Richaud et al., 2010a) are presented in Fig. 10 and confirm the existence of a Diffusion Limited Oxidation.

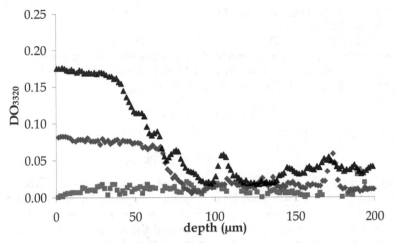

Fig. 10. Oxidation profiles in PEEK sheaths submitted to 10 MGy (■), 20 MGy (♦), 30.7 MGy (▲) γ-rays exposure (Richaud et al., 2010a).

## 7. On the radio-induced oxidizability of aromatic polymers

It seemed to us interesting to compare the $k_3^2/k_6$ ratio estimated from radiodegradation of aromatic polymers with polypropylene or polyethylene ones. Let us recall that $k_3^2/k_6$ describes the oxidizability of a polymer irrespectively of the oxidation mode (i.e. photo, thermo- or radio-induced). $k_3^2/k_6$ can be calculated from Eq. (32) using:
- [PH] = ρ/M, ρ being the polymer density and M the molar mass of repetitive unit,
- $r_i = 10^{-7}.G(P°).I$ from Heiland's results (Table 2),
- $r_{OX}$ from TOL measured for example in FTIR mapping mode (Eq (30)).

One can use with a good approximation the values of oxygen permeation of BisPhenol A Polycarbonate values for in PEEK and PSU (Van Krevelen & Te Nijenhuis, 2009) at 60°C :
- $D_{O2} = 8.1 \times 10^{-8}$ m$^2$ s$^{-1}$
- $s_{O2} = 4 \times 10^{-8}$ mol l$^{-1}$ Pa$^{-1}$

For samples exposed under 1.5 bar oxygen, TOL were found close to:
- 100 μm in PSU (Richaud et al., 2011),
- 70 μm in PEEK (Fig. 10).

|  | TOL | $r_{OX}$ | $G_i$ | $r_i$ | [PH] | BDE | $k_3^2/k_6$ | $k_3$ | $k_6$ |
|---|---|---|---|---|---|---|---|---|---|
|  | (m) | (mol l$^{-1}$ s$^{-1}$) | (mol/100 eV) | (mol l$^{-1}$ s$^{-1}$) | (mol l$^{-1}$) | (kJ mol$^{-1}$) | (l mol$^{-1}$ s$^{-1}$) | (l mol$^{-1}$ s$^{-1}$) | (l mol$^{-1}$ s$^{-1}$) |
| PEEK | 1.00E-04 | 5.2E-06 | 0.25 | 1.7E-07 | 12.9 | 460 | 1.8E-06 | 1.0E-04 | 5.6E-03 |
| PSU | 7.00E-05 | 1.1E-05 | 0.75 | 5.0E-07 | 5.6 | 410 | 1.3E-05 | 9.5E-03 | 6.8E+00 |

Table 7. Kinetic parameters of PEEK and PSU estimated from DLO theory.

so that one can calculate (Table 6):
- $k_3^2/k_6 = 1.3 \times 10^{-5}$ l mol$^{-1}$ s$^{-1}$            for PSU
- $k_3^2/k_6 = 1.8 \times 10^{-6}$ l mol$^{-1}$ s$^{-1}$            for PEEK

These orders of magnitude are surprisingly on comparable or higher than values reported in PP (Richaud et al., 2006) and PE (Khelidj, 2006). $k_3$ can be conveniently estimated from the paper by Korcek (Korcek et al., 1972) using the relations:

$\log k_3^{s-POO°} = 16.4 - 0.2 \times BDE(C-H)$    at 30°C

$\log k_3^{\text{t-POO}^\circ} = 15.4 - 0.2 \times \text{BDE(C-H)}$   at 30°C

$E_3^{\text{s-POO}^\circ} = 0.55.(\text{BDE(C-H)} - 65)$

$E_3^{\text{t-POO}^\circ} = 0.55.(\text{BDE(C-H)} - 62.5)$

s-POO° and t-POO° representing respectively secondary and tertiary peroxy radicals. Given the difference in Bond Dissociation Energies for C-H (denoted by BDE(C-H)) in benzene (Davico et al., 1995) or in -CH₃ (Korcek et al., 1972), it is clear that:

$$k_3(\text{PEEK}) < k_3(\text{PSU}) < k_3(\text{PE}) < k_3(\text{PP})$$

Thus, it suggests that (Table 7):

$$k_6(\text{PEEK, PSU}) \ll k_6(\text{PE, PP})$$

From a chemical point of view, the reactivity of aryl or primary peroxy radicals is not low enough to justify such values (Table 6). A possible explanation is that the termination between two POO° radicals is a process involving two rare species. The mobility of segments favouring the coupling could thus control the kinetics of the reaction. The low order of magnitude below $T_g$ could be hence justified. Aromatic polymers could be interesting substrate for studying the Waite's theory:

$$-\frac{dC}{dt} = 4\pi N_A r_0 D \left(1 + \frac{r_0}{\sqrt{\pi D t}}\right) C^2 \tag{41}$$

where:
- $r_0$ is the cage radius, which corresponds to the maximal distance permitting the reaction: It would be on the order of $r_0 \sim 1\text{-}10$ nm (Emanuel & Buchachenko, 1990),
- D is the diffusion coefficient of radical species. On the order of $D \sim 10^{-22} - 10^{-20}$ m² s⁻¹ (Emanuel & Buchachenko, 1990),
- $N_A$ is Avogadro's number.

According to this theory, kinetic constant depends on the diffusivity of a radical and in other words of macromolecular mobility.

## 8. Conclusions

A comparison of yields for gas emission, gel dose or critical dose for reaching an unacceptable level of mechanical properties shows that aromatic polymers radiochemical behaviour can be first explained by their aromatic character which is doubly favourable:
- Yield for radical formation is very low compared with aliphatic polymers, because aromatic groups are able to dissipate a great part of the absorbed energy into reversible processes (fluorescence, phosphorescence...).
- In the case of radio-oxidation, the propagation rate constant of hydrogen abstraction by peroxy radicals ($k_3$) is a decreasing function of the C-H bond dissociation energy, that takes a high value for aromatic hydrogen.

These positive effects are partially counterbalanced by the relatively low termination rate linked to the low segmental mobility in glassy state. It is noteworthy that P° + P° termination (which leads to crosslink when oxygen is lacking) is more efficient than expected because P° radicals can diffuse independently of segmental mobility by the valence migration process ($P_1^\circ + P_2H \rightarrow P_1H + P_2^\circ$).

This chapter also illustrates the effect of dose rate, thickness and oxygen concentration in polymers layers, which are linked by the theory of Diffusion Limited Oxidation: chain scissions predominate in aerobic degradation (i.e. thin sample, and low dose rate) and crosslinking in case of anaerobic degradation (thick sample, high dose rate). The effect of temperature can be considered as not partially understood.

# 9. References

Alexander A. & Charlesby A. (1954). Energy Transfer in Macromolecules Exposed to Ionizing Radiations. *Nature*, Vol.173, No.4404, (1954), pp. 578-579, ISSN 0028-0836

Audouin, L.; Langlois, V.; Verdu, J., de Bruijn, J.C.M. (1994). Role of oxygen diffusion in polymer ageing: kinetic and mechanical aspects. *Journal of Materials Science*, Vol.29, No.3, (January 1994), pp. 569-583, ISSN 0022-2461

Brown, J.R. & O'Donnell, J.H. (1979). Effect of gamma radiation on two aromatic polysulfones II. A comparison of irradiation at various temperatures in air-vacuum environments. *Journal of Applied Polymer Science*, Vol.23, No.9, (May 1979), pp. 2763-1775, ISSN 0021-8995

Burillo, G.; Tenorio, L.; Bucio, E.; Adem, E. & Lopez, G.P. (2007). Electron beam irradiation effects on poly(ethylene terephthalate). *Radiation Physics and Chemistry*, Vol.76, No.11-12, (November-December 2007), pp. 1728-1731, ISSN 0969-806X

Charlesby, A. (1960) *Atomic Radiation and Polymer*, Pergamon Press, Oxford, UK

Davico, G.E.; Bierbaum, V.M.; DePuy, C.H.; Ellison, G.B. & Squires, R.R. (1995). The C-H Bond Energy of Benzene. *Journal of American Chemical Society*, 1995, Vol.117, No.9, March 1995, pp. 2590–2599, ISSN 0002-7863

DiMarzio, E.A. (1964). On the second-order transition of a rubber. *Journal of Research of the National Bureau of Standards-A. Physics and Chemistry*, Vol .68A, No.6, 1964, pp 611-617, ISSN 0091-0635

Emanuel, N.M. & Buchachenko A.L. (1987). *Chemical Physics of Polymer Degradation and Stabilization (New Concepts in Polymer Science)*, VNU Science Press, ISBN 978-9067640923, Utrecht, The Netherlands

Ferry, M. (2008). *Comportement des verres cyclohexane/benzène et des copolymères éthylène/styrène sous rayonnements ionisants: Transferts d'énergie et d'espèces entre les groupements aliphatiques et aromatiques*, Thèse. Université de Caen

Fox, T.G. & Flory, P.J. (1950). Second-Order Transition Temperatures and Related Properties of Polystyrene. I. Influence of Molecular Weight. *Journal of Applied Physics*, Vol.21, No.6, (1950), pp. 581-591, ISSN 0021-8979

Fox, T.G. & Loshaek S. (1955). Influence of molecular weight and degree of crosslinking on the specific volume and glass temperature of polymers. *Journal of Polymer Science*, Vol.15, No.80 , (February 1955), pp. 371-390, ISSN 1099-0518

Hegazy, E.S.A.; Sasuga, T.; Nishii, M. & Seguchi, T. (1992a). Irradiation effects on aromatic polymers: 1. Gas evolution by gamma irradiation. *Polymer*, Vol.33, No.14, (1992), pp. 2897-2903, ISSN 0032-3861

Hegazy, E.SA.; Sasuga, T.; Nishii, M. & Seguchi, T. (1992b). Irradiation effects on aromatic polymers: 2. Gas evolution during electron-beam irradiation. *Polymer*, Vol.33, No.14, (1992), pp. 2904-2910, ISSN 0032-3861

Hegazy, E.S.A.; Sasuga, T. & Seguchi, T. (1992c). Irradiation effects on aromatic polymers: 3. Changes in thermal properties by gamma irradiation. *Polymer*, Vol.33, No.14, (1992), pp. 2911-2914, ISSN 0032-3861

Heiland, K.; Hill, D.J.T.; Hopewell, J.L.; Lewis, D.A., O'Donnell, J.H., Pomery, P.J. & Whittaker A.K. (1996). Measurement of Radical Yields To Assess Radiation Resistance in Engineering Thermoplastics, In: Polymer Durability, R.L. Clough, N.C. Billingham, K.T. Gillen, (Ed.), 637-649, American Chemical Society, ISBN 9780841231344, Washington DC, USA

Hill, D.J.T. & Hopewell, J.L. (1996). Effects of 3 MeV proton irradiation on the mechanical properties of polyimide films. *Radiation Physics and Chemistry*, Vol.48, No.5, (November 1996), pp. 533-537, ISSN 0969-806X

Hill, D.J.T.; Lewis, D.A.; O'Donnell, J.H. & Whittaker, A.K. (1998). The crosslinking mechanism in gamma irradiation of polyarylsulfone: evidence for Y-links. *Polymers for Advanced Technologies*, Vol.9, No.1, (January 1998), pp. 45–51, ISSN 1042-7147

Horie, K. & Wolfram Schnabel, W. (1984). On the kinetics of polymer degradation in solution: Part XI — Radiolysis of poly(olefin sulfones). *Polymer Degradation and Stability*, Vol.8, No.3, (1984), pp. 145-159, ISSN 0141-3910

Kang, P.H; Jeon, Y.K.; Jeun, J.P.; Shin, J.W. & Nho, Y.C. (2008). Effect of electron beam irradiation on polyimide film. *Journal of Industrial and Engineering Chemistry*, Vol.14, No.5, (September 2008), pp. 672-675, ISSN 1226-086X

Kellman, R.; Hill, D.T.J.; Hunter, D.S.; O'Donnell, J.H. & Pomery, P.J. (1991). Gamma Radiolysis of Styrene-co-Methyl Acrylate Copolymers. An Electron Spin Resonance Study, In: *Radiation Effects on Polymers*, R.L. Clough, S.W. Shalaby (Ed.), 119-134, American Chemical Society, ISBN 9780841221659, Washington DC, USA

Khelidj, N. (2006). *Vieillissement d'isolants de câbles en polyéthylène en ambiance nucléaire*, Thèse ENSAM de Paris

Korcek, S.; Chenier, J.H.B.; Howard, J.A. & Ingold, K.U. (1972). Absolute Rate Constants for Hydrocarbon Autoxidation. XXI. Activation Energies for Propagation and the Correlation of Propagation Rate Constants with Carbon–Hydrogen Bond Strengths. *Canadian Journal of Chemistry*, (1972), Vol.50, No.14, pp. 2285-2297, ISSN 0008-4042

Li, X.G. & Huang, M.R. (1999). Thermal degradation of bisphenol A polysulfone by high-resolution thermogravimetry. *Reactive and Functional Polymers*, Vol.42, No.1, (September 1999), pp. 59-64, ISSN 1381-5148

Li, J.; Oshima, A.; Miura, T. & Washio, M. (2006). Preparation of the crosslinked polyethersulfone films by high-temperature electron-beam irradiation. *Polymer Degradation and Stability*, Vol.91, No.12, (December 2006), pp. 2867-2873, ISSN 0141-3910

Molnár, G.; Botvay, A.; Pöppl, L.; Torkos, K.; Borossay, J.; Máthé, Á. & Török, T. (2005). Thermal degradation of chemically modified polysulfones. *Polymer Degradation and Stability*, Vol.89, No.3, (September 2005), pp. 410-417, ISSN 0141-3910

Murakami, K. & Kudo, H. (2007). Gamma-rays irradiation effects on polysulfone at high temperature. *Nuclear Instruments and Methods in Physics Research Section B: Beam*

*Interactions with Materials and Atoms*, Vol.265, No.1, (December 2007), pp. 125-129, ISSN 0168-583X

Richaud, E. ; Farcas, F. ; Bartoloméo, P. ; Fayolle, B. ; Audouin, L. & Verdu, J. (2006). Effect of oxygen pressure on the oxidation kinetics of unstabilised polypropylene. *Polymer Degradation and Stability*, Vol.91, No.2, (February 2006), pp. 398-405, ISSN 0141-3910

Richaud, E.; Ferreira, P.; Audouin, L.; Colin, X.; Verdu, J. & Monchy-Leroy, C. (2010a). Radiochemical ageing of poly(ether ether ketone). *European Polymer Journal*, Vol.46, No.4, (April 2010), pp. 731-743, ISSN 0014-3057

Richaud, E.; Audouin, L.; Colin, X.; Monchy-Leroy, C. & Verdu, J. 2010. Radiochemical Ageing of Aromatic Polymers PEEK, PSU and Kapton®. *AIP Conference Proceedings*, Vol.1255, (June 2010), pp. 10-12, ISSN 1551-7616

Richaud, E.; Colin, X.; Monchy-Leroy, C.; Audouin, L. & Verdu, J. (2011). Diffusion-controlled radiochemical oxidation of bisphenol A polysulfone. *Polymer International*, Vol.60, No.3, (March 2011), pp. 371–381, ISSN 0959-8103

Saito, O. (1958). On the Effect of High Energy Radiation to Polymers I. Cross-linking and Degradation. *Journal of the Physical Society of Japan*, Vol.13, No.2, (February 1958), pp. 198-206, ISSN 0031-9015

Sasuga, T.; Hayakawa, N.; Yoshida, K. & Hagiwara, M. (1985). Degradation in tensile properties of aromatic polymers by electron beam irradiation. *Polymer*, Vol.26, No.7, (July 1985), pp.1039-1045, ISSN 0032-3861

Sasuga, T. & Hagiwara, M. (1987). Radiation deterioration of several aromatic polymers under oxidative conditions. *Polymer*, Vol.28, No.11, (October 1987), pp. 1915-1921, ISSN 0032-3861

Sasuga, T. (1988). Oxidative irradiation effects on several aromatic polyimides. *Polymer*, Vol.29, No.9, (September 1988), pp. 1562-1568, ISSN 0032-3861

Sasuga, T. (1991). Electron irradiation effects on dynamic viscoelastic properties and crystallization behaviour of aromatic polyimides. *Polymer*, Vol.32, No.9, (1991), pp. 1539-1544, ISSN 0032-3861

Schnabel, W. (1978). Degradation by High Energy Irradiation, In: *Aspects of Degradation and Stabilization of Polymers*, H.H.G. Jellinek (Ed.), 149-190, Elsevier, ISBN13 978-0444415639, Oxford and New York

Tavlet, M. (1995). Aging of organic materials around high-energy particle accelerators. *Nuclear Instruments and Methods in Physics Research Section B: Beam Interactions with Materials and Atoms*, Vol.131, No.1-4, (August 1997), pp. 239-244, 0168-583X

Thominette, F.; Metzger, G.; Dalle, B. & Verdu, J. (1991). Radiochemical ageing of poly(vinyl chloride) plasticized by didecylphthalate. *European Polymer Journal*, Vol.27, No.1, (1991), pp. 55-59, ISSN 0014-3057

Van Krevelen, D.W. & Te Nijenhuis, K. (2009). *Properties of polymers, their correlation with chemical structure; their numerical estimation and prediction from additive group contributions*, 4th Edition, Elsevier, ISBN 978-0-08-054819-7, Amsterdam, The Netherlands

Vaughan, A.S. & Stevens, G.C. (1995). On crystallization, morphology and radiation effects in poly(ether ether ketone). *Polymer*, Vol.36, No.8, (1995), pp. 1531-1540, ISSN 0032-3861

Wilski, H. (1987). The radiation induced degradation of polymers. *Radiation Physics and Chemistry*, Vol.29, No.1, (1987), pp. 1-14, ISSN 0969-806X

Zhen, S.J. (2001). The effect of chain flexibility and chain mobility on radiation crosslinking of polymers. *Radiation Physics and Chemistry*, Vol.60, No.4-5, (2001), pp. 445-451, ISSN 0969-806X

# Permissions

The contributors of this book come from diverse backgrounds, making this book a truly international effort. This book will bring forth new frontiers with its revolutionizing research information and detailed analysis of the nascent developments around the world.

We would like to thank Dr. Mitsuru Nenoi, for lending his expertise to make the book truly unique. He has played a crucial role in the development of this book. Without his invaluable contribution this book wouldn't have been possible. He has made vital efforts to compile up to date information on the varied aspects of this subject to make this book a valuable addition to the collection of many professionals and students.

This book was conceptualized with the vision of imparting up-to-date information and advanced data in this field. To ensure the same, a matchless editorial board was set up. Every individual on the board went through rigorous rounds of assessment to prove their worth. After which they invested a large part of their time researching and compiling the most relevant data for our readers. Conferences and sessions were held from time to time between the editorial board and the contributing authors to present the data in the most comprehensible form. The editorial team has worked tirelessly to provide valuable and valid information to help people across the globe.

Every chapter published in this book has been scrutinized by our experts. Their significance has been extensively debated. The topics covered herein carry significant findings which will fuel the growth of the discipline. They may even be implemented as practical applications or may be referred to as a beginning point for another development. Chapters in this book were first published by InTech; hereby published with permission under the Creative Commons Attribution License or equivalent.

The editorial board has been involved in producing this book since its inception. They have spent rigorous hours researching and exploring the diverse topics which have resulted in the successful publishing of this book. They have passed on their knowledge of decades through this book. To expedite this challenging task, the publisher supported the team at every step. A small team of assistant editors was also appointed to further simplify the editing procedure and attain best results for the readers.

Our editorial team has been hand-picked from every corner of the world. Their multi-ethnicity adds dynamic inputs to the discussions which result in innovative outcomes. These outcomes are then further discussed with the researchers and contributors who give their valuable feedback and opinion regarding the same. The feedback is then collaborated with the researches and they are edited in a comprehensive manner to aid the understanding of the subject.

Apart from the editorial board, the designing team has also invested a significant amount of their time in understanding the subject and creating the most relevant covers. They scrutinized every image to scout for the most suitable representation of the subject and create an appropriate cover for the book.

The publishing team has been involved in this book since its early stages. They were actively engaged in every process, be it collecting the data, connecting with the contributors or procuring relevant information. The team has been an ardent support to the editorial, designing and production team. Their endless efforts to recruit the best for this project, has resulted in the accomplishment of this book. They are a veteran in the field of academics and their pool of knowledge is as vast as their experience in printing. Their expertise and guidance has proved useful at every step. Their uncompromising quality standards have made this book an exceptional effort. Their encouragement from time to time has been an inspiration for everyone.

The publisher and the editorial board hope that this book will prove to be a valuable piece of knowledge for researchers, students, practitioners and scholars across the globe.

# List of Contributors

**Mark D. Hammig**
University of Michigan, USA

**Amany A. El-Kheshen**
National Research Centre, Glass Research Department, Egypt
Chemistry Department, Faculty of Science, Taif University, KSA

**Hans Riesen and Zhiqiang Liu**
The University of New South Wales, Australia

**Ahmed M. Maghraby**
National Inst. of Standards (NIS) – Radiation Dosimetry Department, Giza, Egypt

**K. Farah and A. Mejri, F. Hosni**
National Center for Science and Nuclear Technology, Sidi-Thabet, Tunisia

**K. Farah**
ISTLS, 12, University of Sousse, Tunisia

**A. H. Hamzaoui**
National Centre for Research in Materials Science, Borj Cedria, Hammam-Lif, Tunisia

**B. Boizot**
Laboratory of Irradiated Solids, UMR 7642 CEA-CNRS - Polytechnic School, Route de Saclay, Palaiseau, France

**Christopher J. Mertens**
NASA Langley Research Center, Hampton, Virginia, USA

**Brian T. Kress**
Dartmouth College, Hanover, New Hampshire, USA

**Michael Wiltberger**
High Altitude Observatory, National Center for Atmospheric Research, Boulder, Colorado, USA

**W. Kent Tobiska**
Space Environment Technologies, Pacific Palisades, California, USA

**Barbara Grajewski**
National Institute for Occupational Safety and Health, Cincinnati, Ohio, USA

**Xiaojing Xu**
Science Systems and Applications, Inc., USA

**Momčilo Pejović and Milica Pejović**
University of Niš, Faculty of Electronic Engineering, Niš, Serbia

**Predrag Osmokrović and Koviljka Stanković**
University of Belgrade, Faculty of Electrical Engineering, Belgrade, Serbia

**Nicolas T. Fourches**
Atomic Energy Commission, CEN Saclay, DSM/IRFU/SEDI, Gif sur Yvette, France

**Emmanuel Richaud, Ludmila Audouin, Xavier Colin, Bruno Fayolle and Jacques Verdu**
Arts et Metiers ParisTech, CNRS, PIMM, France

Printed in the USA
CPSIA information can be obtained
at www.ICGtesting.com
JSHW011500221024
72173JS00005B/1155

9 781632 381439